Springer Series in Statistics

Perspectives in Statistics

More information about this series at http://www.springer.com/series/1383

Johann Pfanzagl

Mathematical Statistics

Essays on History and Methodology

Johann Pfanzagl
Mathematical Institute
University of Cologne
Cologne
Germany

ISSN 0172-7397 ISSN 2197-568X (electronic)
Springer Series in Statistics
ISSN 2522-0411 ISSN 2522-042X (electronic)
Perspectives in Statistics
ISBN 978-3-642-31083-6 ISBN 978-3-642-31084-3 (eBook)
DOI 10.1007/978-3-642-31084-3

Library of Congress Control Number: 2017948627

Printed on acid-free paper

This Springer imprint is published by Springer Nature
The registered company is Springer-Verlag GmbH Germany
The registered company address is: Heidelberger Platz 3, 14197 Berlin, Germany

I would like to thank two people who were crucial to shaping my life:
My grammar school teacher who convinced my mother that she should send me to secondary school; and a judge at the military court in Linz who sided with my aversion against serving in the German army and was, therefore, ready to look upon my desertion as just a field trip.

Preface

This is not a textbook on mathematical statistics; it is just a collection of essays on problems that were of topical interest in the middle of the twentieth century. The choice of this period is far from arbitrary: It marked the onset of mathematical statistics.

The main purpose of these essays is not to list who did what and when, but to present the background for the concepts and theorems which evolved during this time. Some readers will find that sometimes there are too many details in the description of the status quo. Yet, only a detailed description of the status quo provides the necessary information for the evaluation of new results.

Wolfgang Wefelmeyer has carefully read the manuscript and suggested a large number of corrections and improvements.

A fair discussion of an achievement would be impossible if it was "politically incorrect" to say that even highly respected scholars made mistakes, that they presented insufficient proofs, that they had missed certain opportunities, had misunderstood the relevance of certain results, and had overlooked priorities. I presume that the reader of this book is familiar with the stature of various scholars. Hence there seems to be no need to bolster every critical remark using the mantra "... but he is an honourable man".

Native speakers will find many parts of the text lacking the elegance of idiomatic English. I apologize for this, but it seems to be a necessity to sacrifice beauty of language in the interest of having a common language for science.

Acknowledgements. I thank Barbara Wehmeyer for typing several versions of a sometimes not very legible manuscript. I also thank a reviewer for a number of perceptive comments.

Cologne, Germany
July 2017

Johann Pfanzagl

Contents

Chapter 1
Introduction

In the years between 1940 and 1970, say, mathematical statistics has undergone a dramatic change. Not only has the amount of publications increased by a factor of 10 (starting from about 1.500 pages a year in 1940); it is even more the level of mathematical sophistication which has changed. Comparing a volume of the *Annals of Statistics* of 1970 with a volume of 1940 (then called *Annals of Mathematical Statistics*) one would not believe it to be the same journal.

The following essays try to portray the historical developments during this period for selected areas of mathematical statistics. Our emphasis is on conceptualal issues; this justifies the restriction to basic models mostly based on independent and identically distributed observations. To describe what has happened fifty years ago is, however, not the main purpose. What is still of interest today is to see how the emergence of refined mathematical techniques influenced the subject of mathematical statistics. Whether these refined techniques were fully understood and their power fully exploited by all contemporary statisticians is an interesting second aspect.

Around 1940, mathematical statistics was mainly restricted to parametric families, and the highlights were theorems about unbiased estimators and families with monotone likelihood ratios. As a new generation of statisticians, familiar with the techniques of measure theory, entered the stage, it soon became clear that compelling intuitive ideas like "sufficiency" or "asymptotic optimality" were not so easy to translate into constructs which are accessible to a mathematical treatment. It turned out that the solution of seemingly meaningful problems can be hampered by difficulties of purely mathematical nature (such as the existence of regular conditional probabilities), difficulties which are in no inherent relationship with the problem itself, and which became —even so—the favourite subjects of certain statisticians.

The following essays are restricted to subjects which show some inherent relationship: Descriptive statistics, sufficiency, estimation, asymptotics. The omission of important topics like "Bayesian theory", "decision theory", "robustness" takes the restricted competence of the author into account. Before going into the details, we try to set forth the basic ideas of our approach.

J. Pfanzagl, *Mathematical Statistics*, Springer Series in Statistics,
DOI 10.1007/978-3-642-31084-3_1

A Methodological Manifesto

Assertions and assumptions on statistical procedures should be of operational significance, in other words: Their relations to reality should be expressible in terms of probabilities and, perhaps, costs. This requires converting inexact, intuitive concepts (like independence, concentration, information) into concepts which are operationally meaningful. The invention of such constructs is limited by what is feasible from the mathematical point of view. Insisting on operational significance excludes the sole reliance on "principles" (like "maximum likelihood") which rest upon authority or metaphysics.

Assertions based on a particular loss function are operationally significant if the loss function represents reality. Results based on the quadratic loss function will, therefore, be operationally significant only under special circumstances. That they are easy to prove, or nice looking, does not give them any meaning.

Assertions on posterior distributions are operationally significant if the prior distribution admits an interpretation in terms of reality, be it as a "physical probability" based on past experience (as, for instance, in acceptance sampling) or a description of the state of mind (hopefully based on some empirical evidence). Prior distributions justified by formal arguments might result in nice posterior distributions. This by itself does not give them any meaning, though.

Converting an intuitive concept into a meaningful mathematical construct may turn out to be difficult. An example of a successful conversion is the concept of a "sufficient statistic, containing all information in the sample". It is made precise by the idea that for any function which could be computed from the sample there exists a (randomized) function of the sufficient statistic which has the same distribution, for every probability measure in the underlying family. Like it or not: For mathematical reasons, this interpretation is confined to functions taking their values in a complete separable metric space. (See Sect. 2.2 for details.)

An example of a condition without operational significance is the requirement that the estimator for the sample size n should be an element of a "convergent" sequence.

"Robustness" is an example of a convincing idea which is difficult to cast in mathematical terms: Since models are never exact, the performance of statistical procedures should be insensitive to small departures from the model. Yet, what are small departures from the model, and what are small changes in the performance? Consider e.g. a real-valued functional κ of a distribution P in some family $\mathfrak{P}$ and a sequence of estimators $\kappa^{(n)}$, $n \in \mathbb{N}$, based on n i.i.d. observations, with joint distribution P^n. The suggestion of Hampel (1971, p. 1890) to define "qualitative robustness" of $\kappa^{(n)}$, $n \in \mathbb{N}$, by equicontinuity (with respect to n) of the map $P \to P^n \circ \kappa^{(n)}$ proved inadequate for grasping the phenomenon of "robustness" in its complexity. Hampel's concept of qualitative robustness was supplemented by more detailed concepts of departures from the model (like ε-contamination), additional measures of robustness (like breakdown point and gross error sensitivity), and, finally, principles for finding a balance between robustness and efficiency. Although this network of concepts and principles did not result in a coherent theory, it attracted the attention of competent scholars. For informations about the present state of art see Rieder (1994) and Jurecková and Sen (1996).

Here is an example of a totally misleading intuitive idea. Assume that $(x_1, \dots, x_n)$ is a realization from some normal distribution $N(\mu, 1)^n$ with mean 0 and variance 1. If the sample mean $\overline{x}_n$ equals 10, it seems much more likely that μ is in the interval (8, 12) than in the interval (157, 161), say. Evident as this appears, the mathematician will discover soon the impossibility of expressing this in a mathematically correct way without introducing a prior probability of μ. Starting in R.A. Fisher (1930) took numerous occasions to present his idea of a "fiducial probability". That means: Given a sample from P_ϑ^n, $\vartheta \in \Theta \subset \mathbb{R}$, it is possible to compute a distribution of the unknown parameter ϑ. If Fisher had tried to describe such a distribution in mathematical terms, he would have run into insuperable difficulties. Let Φ denote the standard normal distribution function. For samples $(x_1, \dots, x_n)$ from $N(\mu, 1)^n$ with $\mu \in \mathbb{R}$ unknown, the probability of $\mu < \overline{x}_n + n^{-1/2}\Phi^{-1}(\alpha)$ is equal to α. Applying this relation for $\alpha = \Phi(n^{1/2}(t - \overline{x}_n))$ one arrives at the conclusion that $\mu < t$ holds with probability $\Phi(n^{1/2}(t - \overline{x}_n))$, or that $t \to \Phi(n^{1/2}(t - \overline{x}_n))$ is the (fiducial) distribution function of μ. This sounds plausible. If one tries to write this down in formulas (in particular: replacing the word "probability" by $N(\mu, 1)^n$), the absurdity of this argument becomes patent. The fallacy in Fisher's reasoning was elaborated by Neyman[1] (1941, Sects. 4 and 5); see also Lehmann (1995). Neither Neyman nor any other critic was able to convince Fisher of his mistake. Even in Fisher (1959), when mathematical rigor was standard already, Fisher says (p. 56):

> The treatment in this book ... does ... rely on a property inherent in the semantics of the word "probability".

In this connection, it might be opportune to remember Fisher as a notorious inventor of concepts lacking operational significance. "Fiducial probability" is just one example of Fisher's propensity to unsupported concepts. "Likelihood" and "amount of information" are other examples. Fisher (1935, p. 40) says:

> A mathematical quantity of a different kind [i.e., different from probability] which I have termed *mathematical likelihood* appears to take its place as a measure of rational belief ...

Referring to Fisher's idea of the "amount of information" in the discussion of Fisher's paper (1935), Bowley (p. 57) says:

> I must confess to dislike the method of nomenclature that leads to such a phrase as [see Fisher, p. 46] "there are 1 and no less units of information to be extracted from the data, if we equate the information extracted to the variance of our estimate ...". The measurement on this basis of the amount of knowledge seems to me to have the same dangers as treating the correlation coefficient or its square as the amount of covariation. In both cases a definite meaning is attached to the maximum called unity, and to the minimum called zero. In neither case does an intermediate value correspond to ... anything otherwise definable."

---

[1] Curiously, the very first sentence in Neyman's paper is: "The theory of confidence intervals was started by the author in 1930." However, forerunners are Laplace (1812) and Poisson (1837). If there had been any doubts about the interpretation of "covering probability", they were settled by Wilson (1927).

Fisher's abuse of language was carried on by some of his adherents, so for instance by C.R. Rao, who gives the following definition (1962, p. 77) for i.i.d. observations $(x_1, \ldots, x_n)$ with density $p(\cdot, \vartheta)$:

> A statistic is said to be efficient if its asymptotic correlation with the derivative of log likelihood is unity. The efficiency of any statistic may be measured by ϱ^2, where ϱ is the asymptotic correlation with $[n^{-1/2} \sum_{\nu=1}^{n} \partial_\vartheta \log p(x_\nu, \vartheta)]$.

One cannot but sympathize with Lindley when he says, in the discussion of C.R. Rao's paper (p. 68):

> Professor Rao follows in the footsteps of Fisher in basing his thesis on intuitive considerations of estimation that people like myself who lack such penetrating intuition, cannot aspire to.

This and other attempts at defining first and second order efficiency will be discussed in Sects. 5.9 and 5.18.

The Increasing Role of Mathematics

Prior to the middle of the 20th century, the collection of statistically relevant results was moderate: The theory of sufficiency on the onset, the Neyman–Pearson Lemma (which Wolfowitz, 1969, p. 747, calls "pretty trivial to prove and not difficult to discover") establishing the existence of "uniformly most powerful" tests for certain parametric families, abortive attempts at proving consistency and asymptotic efficiency of maximum likelihood estimators, together with some scattered nonparametric procedures (like rank tests). The use of more advanced mathematical tools made it possible to fully develop the theory of sufficiency, together with its application to the theory of unbiased estimators, to clarify the role of monotone likelihood-ratios for the existence of uniformly most powerful tests and optimal median unbiased estimators, etc.

Assertions are turned into theorems by specifying the regularity conditions under which they are true. Until the middle of the 20th century, it was common to assume that all operations which are carried out are legitimate (e.g. that derivatives can be taken unde the integral, that suprema are attained, that solutions of the likelihood equations can be chosen measurable etc.). To clarify the conditions under which such operations are legitimate is a matter of course for a theory which considers itself as a part of mathematics, but the fruit of this clearing work did not bring basically new findings, apart from, maybe, some counterexamples (e.g. that maximum likelihood sequences are not necessarily consistent).

The problematic point: Most theorems are based on conditions which cannot be supported by empirical evidence. That the probability measure belongs to a certain parametric family can be supported by a model of the data-generating process in exceptional cases only (such as the radioactive decay, for instance). For nonparametric families $\mathfrak{P}$, moment conditions like boundedness of moments of some order or smoothness conditions on the densities, say a Lipschitz-condition for some derivative of the density, are chosen in regard to mathematical convenience. Progress usually consists in presenting proofs under weaker conditions, the validity of which remains open, though.

Refinements of the mathematical techniques did not only expose gaps in the preceding literature. They also turned up technical assumptions the relevance of which is hard to understand from the intuitive point of view, for instance the conditions for the existence of a regular conditional probability.

For scholars with a firm background in mathematics it was natural to consider statistical theory as a part of mathematics. From this point of view it is legitimate to generalize theorems with a clear statistical interpretation to a more abstract framework. This is not without danger, though. Isolating the abstract core of an argument might kill the original idea, thus causing a handicap for generalizations in other directions. Assertions valid in the abstract framework might be without an interpretation in terms of the original problem and, therefore, without operational significance. Realizing the shape which Hájek's Convolution Theorem took in the publications of Le Cam (see e.g. 1979), the statistician cannot avoid thinking of Lucretius:

> When a thing changes its nature, at that moment comes the death of what it was before.

The following example from test theory illustrates the usefulness of refined mathematical techniques, and their limitations.

Example. Given a family $\mathfrak{P}$ of probability measures which is interpreted as a "hypothesis", and a probability measure P_0 outside $\mathfrak{P}$, the "alternative", the question arises whether a most powerful test exists, i.e., whether there is a critical function φ_0 which maximizes the expectation $\int \varphi d P_0$ in the class Φ_α of all critical functions φ fulfilling $\int \varphi d P \leq \alpha$ for $P \in \mathfrak{P}$. The mathematical problem is to show that for any sequence of critical functions $\varphi_n : X \to [0, 1]$, $n \in \mathbb{N}$, there exists a subsequence $\mathbb{N}_0$ and a critical function $\varphi_0 : X \to [0, 1]$ such that $(\int \varphi_n d P)_{n \in \mathbb{N}_0} \to \int \varphi_0 d P$ for $P \in \mathfrak{P}$ and $P = P_0$. This so-called "weak compactness theorem" occurs already in Banach (1932). It was proved independently by Lehmann (1959, p. 354, Theorem 3) under the assumption that the underlying σ-field $\mathscr{A}$ is countably generated and $\mathfrak{P}|\mathscr{A}$ is dominated. The same result is obtained by Nölle and Plachky (1967, p. 182, Satz) for arbitrary $\mathscr{A}$ (by considering the sub-σ-field generated by a sequence φ_n, $n \in \mathbb{N}$, with $\int \varphi_n d P_0$, $n \in \mathbb{N}$, approaching the supremum), and by Landers and Rogge (1972, p. 339, Theorem) for arbitrary $\mathscr{A}$ and without domination of $\mathfrak{P}$, using that Φ_α is convex, and that, therefore, the P_0-weak closure of Φ_α coincides with the P_0-strong closure. Even mathematically advanced textbooks like Schmetterer (1974, p. 14, Theorem XI), Witting (1985, p. 207, Korollar 2.15) or Lehmann (1986, p. 576, Theorem 3) withhold this result from the reader, so that their existence theorems for most powerful (or most stringent) tests are confined to dominated hypotheses. □

Whether the supremum is attained or not is irrelevant from the practical point of view; in any case, it can be approximated as closely as one likes. Moreover, the results mentioned above assert the existence of a critical function for which the supremum is attained, without giving any advice how such an optimal critical function can be obtained. After all, the real problems lie somewhere else: In a real testing problem, there is not one, but a whole class of alternatives and the question is: For which kind of models does there exist a critical function φ_0 in Φ_α such that $\int \varphi_0 d P_0 = \sup\{\int \varphi d P_0 : \varphi \in \Phi_\alpha\}$, simultaneously for every P_0 in the class of alternatives. For i.i.d. products in a one-parameter family, say $\{P_\vartheta^n : \vartheta \in \Theta\}, \Theta \subset \mathbb{R}$, this is—under weak regularity conditions—the case if and only if the family is exponential, and the hypothesis $\vartheta \le \vartheta_0$ is tested against the alternatives $\vartheta > \vartheta_0$.

These results extend to certain exponential families with more than one parameter as far as similar tests are concerned, say tests of the hypothesis $\vartheta = \vartheta_0, \eta \in H$ against alternatives (ϑ, η) with $\vartheta > \vartheta_0$.

The situation is different if the tests are not required to be similar. As an example, we mention a result due to Lehmann and Stein (1948), referring to the family $\{N(\mu, \sigma^2)^n : \mu \in \mathbb{R}, \sigma^2 > 0\}$. There exists a test of level α for the hypothesis $\mu \in \mathbb{R}$, $\sigma^2 \le \sigma_0^2$ which is most powerful against all alternatives $\mu \in \mathbb{R}, \sigma^2 > \sigma_0^2$. As against that, the test of the hypothesis $\mu \in \mathbb{R}, \sigma^2 \ge \sigma_0^2$ which is most powerful against an alternative (μ_1, σ_1^2) with $\sigma_1^2 < \sigma_0^2$ depends on μ_1 and is, therefore, useless for practical purposes.

The Role of "Leading Personalities"

Robert K. Merton's hypothesis (1973, p. 356) "all scientific discoveries are in principle multiples" also finds support in the realm of mathematical statistics. The standard example from statistical theory: the Cramér-Rao inequality with (at least) five discoverers. In accordance with "Stigler's law of eponymy" it is not named after Aitken and Silverstone (1942), Fréchet (1943) or Darmois (1945), but after Cramér (1946) and Rao (1945). Another example of this kind are "Cramér's lemmas" dating from 1946 which are due to Slutsky (1925).

The presence of such multiple discoveries is of methodological interest; it shows that an idea occurs if there is the time for it. This leads to the question: Are there any unique ideas in the period under consideration? If we fancy away the contributions of one of the outstanding scholars: Would the present picture of mathematical statistics be any different?

This might be true for R.A. Fisher's concept of "sufficiency". Remarkably, this invention is due to a scholar with extraordinary intuition, but limited mathematical background. He might have had trouble with his concept of sufficiency if he had been familiar with Kolmogorov's concept of conditional probabilities.

It appears difficult to find any more examples. Even a concept tied to a particular name like "Efron's bootstrap" was bound to come up some time or other. (See, Stigler (1999), Chap. 7, for the prehistory of the bootstrap.) The situation is different with Wald's decision theory, developed by Blackwell and Le Cam to a theory of experiments. One may doubt whether the basic ideas would have developed to respectable theories without the efforts of two outstanding scholars.

There are outstanding mathematicians who showed some interest in statistics, like Fréchet, Linnik, Kolmogorov, Cramér, Doob, Halmos,[2] ... In spite of important contributions to measure theory and probability theory, however, they contributed

[2]Having been undecided between probability and statistics, Halmos made up his mind as early as 1937: "I'll take probability, and to hell with Fisher" (see Halmos 1985, p. 65). See, however, his fundamental contribution to the concept of "sufficiency" in Halmos and Savage (1949).

nothing to statistics which is unique in the sense that "nobody else could have done it".

Universal Theories

Statisticians with a strong preference for abstract results might be disappointed by the fact that attempts at developing a universal theory which "solves" all statistical problems—like Wald's "Decision Theory", or Le Cam's "Theory of Experiments"—are not accepted by statisticians interested in problems of practical relevance.

Since the problems dealt with in statistical theory are so diverse, no experienced statistician would seriously consider the possibility of placing them all in the Procrustean bed of a coherent theory. It required the courage of a mathematician with limited experience in statistics to build a theory on three principles, each of which is unreasonable by itself:

(i) Given a decision space $(D, \mathscr{D})$, the consequence drawn on the basis of a sample x from a distribution P can be evaluated by a loss function $a \rightarrow \ell(a, P)$.

(ii) The performance of a randomized decision function $D : X \times \mathscr{D} \rightarrow [0, 1]$ at P is evaluated by the expected loss, or risk, $R(D, P) = \iint \ell(a, P) D(x, da) P(dx)$.

(iii) The overall performance of a statistical procedure on a family $\mathfrak{P}$ of probability measures is evaluated by the maximal expected loss $\sup_{P \in \mathfrak{P}} R(D, P)$.

Wald's conception of "decision making" as the prototype of a statistician's usual activity was obviously inspired by hypotheses testing (more precisely by acceptance sampling). This impression is confirmed by the content of the lectures Wald gave in (1941a) at Notre Dame University: About 25 pages are on testing and confidence intervals, and 4 on estimation. Even in the final shape Wald gave to his decision theory in (1950), his endeavours to include estimation in this framework seem to be inadequate. Around (1941b), Wald's methodological background on estimation theory was confined to some papers by Fisher; in the finished version of his decision theory of Wald (1950) he added just one more paper: Pitman (1939). Motivated by Pitman's paper (p. 401) he suggests to evaluate an estimator $\hat{\vartheta}$ by the loss function $1_{\{|\hat{\vartheta} - \vartheta| > t\}}$. This ignores other aspects which might be relevant for the evaluation of estimators, such as unbiasedness, or measures of concentration other than the probability of $|\hat{\vartheta} - \vartheta| \leq t$. Even now, textbooks which introduce the conceptual framework of decision theory (like Heyer 1982, pp. 16–24 or Witting 1985, pp. 1–17) make no use of it as soon as it comes to estimation theory.

Among the "principles" constituting the conceptual framework of decision theory, the "minimax-principle" is the most unreasonable one. With his endeavours to justify a mathematically fruitful idea from the methodological point of view, Wald seems to be ill at ease. In (1939), Wald's approach to the evaluation of statistical procedures is Bayesian. In (1943) he changes his position. Since an "a priori distribution ... is usually unknown ... it seems of interest to consider a decision function which minimizes the maximum risk" (p. 267), and similarly in ((Wald, 1950), p. 27) "it is perhaps not unreasonable for the experimenter to behave as if Nature wanted to maximize the risk". As Jimmy Savage never stopped telling, Wald's attitude towards the minimax principle was just "let's try whether something reasonable comes out of it", a fact confirmed by Wolfowitz (1952, p. 8).

The minimax principle was refuted by philosophers on philosophical grounds (Carnap 1952, Sect. 25, pp. 81–90) and criticized by statisticians on account of statistical and arguments: Hodges and Lehmann (1950, pp. 190/1) determine the minimax estimator of the parameter p in the Binomial distribution under quadratic loss. It turns out that the risk of the minimax estimator is larger than the risk of the usual mean-unbiased estimator k/n, except for a small interval about $p = 1/2$ (which shrinks to $\{1/2\}$ as n tends to infinity). For more on the history of the minimax principle see L.D. Brown (1964). We conclude our remarks on "Decision Theory" with an important voice, Fisher (1959, p. 101):

> The idea that this responsibility [i.e., the interpretation of observations] can be delegated to a giant computer programmed with Decision Functions belongs to the phantasy of circles rather remote from scientific research.

Some Basic Concepts

The basic problem in statistical theory is to draw conclusions from a realization of a random variable to probability measure. In particular: Given a functional $\kappa : \mathfrak{P} \to Y$, which function $\hat{\kappa} : X \to Y$ should be used as an estimator of $\kappa(P)$? Given a distance function, say $D : Y \times Y \to [0, \infty)$, we judge the quality of $\hat{\kappa}$ as an estimator of $\kappa(P)$ by the distribution $P \circ (x \to D(\kappa(P), \hat{\kappa}(x)))$ of the distance between $\kappa(P)$ and $\hat{\kappa}(x)$. An estimator $\hat{\kappa}_0$ is better than an estimator $\hat{\kappa}_1$ if, for every $P \in \mathfrak{P}$, the distribution $P \circ D(\kappa(P), \hat{\kappa}_0)$ is more concentrated about zero than the distribution $P \circ D(\kappa(P), \hat{\kappa}_1)$.

If $\hat{\kappa}_0$ is better than $\hat{\kappa}_1$ in this respect, it is tempting to expect that, given a realization x from an unknown $P \in \mathfrak{P}$, the unknown value $\kappa(P)$ of the functional will be closer to $\hat{\kappa}_0(x)$ than to $\hat{\kappa}_1(x)$. How can this be made operational? The appropriate answer is: By a *confidence procedure*, i.e. a map assigning to $x \in X$ a set $K(x)$ such that $P\{x : \kappa(P) \in K(x)\}$ is close to one for every $P \in \mathfrak{P}$. The minimal requirement for a confidence procedure is that the *confidence coefficient* $\inf_{P \in \mathfrak{P}} P\{x : \kappa(P) \in K(x)\}$ is greater than zero.

Given a distance function D, a confidence procedure for κ can be defined by $K(x) = \{y \in Y : D(y, \hat{\kappa}(x) \le t)\}$ for some fixed $t > 0$. The relation $\kappa(P) \in K(x)$ is then equivalent to $D(\kappa(P), \hat{\kappa}(x)) \le t$. Hence the covering probability of $\kappa(P)$ by $K(x)$ is the same as the probability that $D(\kappa(P), \hat{\kappa}(x)) \le t$. The properties of this confidence procedure are determined by the properties of the distance D. For example, $K(x)$ is convex for every x if $y \to D(y, z)$ is convex for every z. Similarly, $K(x)$ is symmetric about $\hat{\kappa}(x)$ for every x if $y \to D(y, z)$ is symmetric about z for every z. These are properties that hold in particular for the Euclidean distance $D(y, z) = \|y - z\|$.

References

Aitken, A. C., & Silverstone, H. (1942). On the estimation of statistical parameters. *Proceedings of the Royal Society of Edinburgh Section A*, *61*, 186–194.
Banach, S. (1932). Théorie des opérations linéaires. Monografie Matematyczne 1, Subwencji Funduszu Kultury Narodowej, Warszawa.
Brown, L. D. (1964). Sufficient statistics in the case of independent random variables. *The Annals of Mathematical Statistics*, *35*, 1456–1474.
Carnap, R. (1952). *The continuum of inductive methods*. Chicago: University Chicago Press.
Cramér, H. (1946). *Mathematical methods of statistics*. Princeton: Princeton University Press.
Darmois, G. (1945). Sur les limites de la dispersion de certaines estimations. *Revue de l'Institut International de Statistique*, *13*, 9–15.
Fisher, R. A. (1930). Inverse probability. *Mathematical Proceedings of the Cambridge Philosophical Society*, *26*, 528–535.
Fisher, R. A. (1935). The logic of inductive inference (with discussion). *Journal of the Royal Statistical Society*, *98*, 39–82.
Fisher, R. A. (1959). *Statistical methods and scientific inference* (2nd ed.). Edinburgh: Oliver and Boyd.
Fréchet, M. (1943). Sur l'extension de certaines évaluations statistiques au cas de petits échantillons. *Revue de l'Institut International de Statistique*, *11*, 182–205.
Halmos, P. R. (1985). *I want to be a mathematician: An automathography*. Berlin: Springer.
Halmos, P. R., & Savage, L. J. (1949). Application of the Radon-Nikodym theorem to the theory of sufficient statistics. *The Annals of Mathematical Statistics*, *20*, 225–241.
Hampel, F. R. (1971). A general qualitative definition of robustness. *The Annals of Mathematical Statistics*, *42*, 1887–1896.
Heyer, H. (1982). *Theory of statistical experiments, Springer series in statistics*. Berlin: Springer.
Hodges, J. L, Jr., & Lehmann, E. L. (1950). Some problems in minimax point estimation. *The Annals of Mathematical Statistics*, *21*, 182–197.
Jurecková, J., & Sen, P. K. (1996). *Robust statistical procedures: Asymptotics and interrelations*. New York: Wiley.
Landers, D., & Rogge, L. (1972). Existence of most powerful tests for undominated hypotheses. *Z. Wahrscheinlichkeitstheorie verw. Gebiete*, *24*, 339–340.
Laplace, P. S. (1812). *Théorie analytique des probabilités*. Paris: Courcier.
Le Cam, L. (1979). On a theorem of J. Hájek. In J. Jurečkova (Ed.), Contributions to statistics: J. Hájek memorial volume (pp. 119–137). Prague: Akademia.
Lehmann, E. L. (1959). *Testing statistical hypotheses*. New York: Wiley.
Lehmann, E. L. (1986). *Testing statistical hypotheses* (2nd ed.), Wiley series in probability and mathematical statistics: Probability and mathematical statistics. New York: Wiley.
Lehmann, E. L. (1995). Neyman's statistical philosophy. *Probability and Mathematical Statistics-PWN*, *15*, 29–36.
Lehmann, E. L., & Stein, C. (1948). Most powerful tests of composite hypotheses. I. Normal distributions. *The Annals of Mathematical Statistics*, *19*, 495–516.
Merton, R. K. (1973). *The sociology of science: Theoretical and empirical investigations*. Chicago: University of Chicago press.
Neyman, J. (1941). Fiducial argument and the theory of confidence intervals. *Biometrika*, *32*, 128–150.
Nölle, G., & Plachky, D. (1967). Zur schwachen Folgenkompaktheit von Testfunktionen. *Z. Wahrscheinlichkeitstheorie verw. Gebiete*, *8*, 182–184.
Poisson, S.-D. (1837). *Récherches sur la probabilité des jugements*. Paris: Bachelier.
Pitman, E. J. G. (1939). Tests of hypotheses concerning location and scale parameters. *Biometrika*, *31*(1/2), 200–215.
Rao, C. R. (1945). Information and accuracy attainable in the estimation of statistical parameters. *Bulletin of Calcutta Mathematical Society*, *37*, 81–91.

Rao, C. R. (1962). Apparent anomalies and irregularities in maximum likelihood estimation (with discussion). *Sankhyā: The Indian Journal of Statistics: Series A*, *24*, 73–102.
Rieder, H. (1994). *Robust asymptotic statistics*. Berlin: Springer.
Schmetterer, L. (1974). *Introduction to mathematical statistics, Translation of the 2nd* (German ed.). Berlin: Springer.
Slutsky, E. (1925). Über stochastische Asymptoten und Grenzwerte. *Metron*, *5*(3), 3–89.
Stigler, S. M. (1999). *Statistics on the table: The history of statistical concepts and methods*. Cambridge: Harvard University Press.
Wald, A. (1939). Contributions to the theory of statistical estimation and testing hypotheses. *The Annals of Mathematical Statistics*, *10*, 299–326.
Wald, A. (1941a). On the Principles of Statistical Inference. Four Lectures Delivered at the University of Notre Dame, February 1941. Published as Notre Dame Mathematical Lectures 1, 1952.
Wald, A. (1941b). Asymptotically most powerful tests of statistical hypotheses. *The Annals of Mathematical Statistics*, *12*, 1–19.
Wald, A. (1943). Tests of statistical hypotheses concerning several parameters when the number of observations is large. *Transactions of the American Mathematical society*, *54*, 426–482.
Wald, A. (1950). *Statistical decision functions*. New York: Wiley.
Wilson, E. B. (1927). Probable inference, the law of succession, and statistical inference. *Journal of the American Statistical Association*, *22*, 209–212.
Witting, H. (1985). Mathematische statistik I: parametrische verfahren bei festem stichprobenumfang. Teubner
Wolfowitz, J. (1952). Abraham Wald, 1902–1950. *The Annals of Mathematical Statistics*, *23*, 1–14.
Wolfowitz, J. (1969). Reflections on the future of mathematical statistics. In R. C. Bose, et al. (Eds.), *Essays in probability and statistics* (pp. 739–750). Chapel Hill: University of North Carolina Press.

Chapter 2
Sufficiency

2.1 The Intuitive Idea

Fisher developed his concept of "sufficiency" in the context of a parametric family of probability measures $\{P_\vartheta : \vartheta \in \Theta\}$ with $\Theta \subset \mathbb{R}^k$ and an i.i.d. sample of size n. To explain his idea of the "sufficiency" of a statistic S, Fisher alternately uses wordings like

> (i) The whole information to be obtained from the sample $(x_1, \ldots, x_n)$ is already included in $S(x_1, \ldots, x_n)$. (Fisher 1920, p. 769; Fisher 1922, p. 316.)
> (ii) The whole information to be obtained from [another] statistic T is already included in S. (Fisher 1920, p. 768; Fisher 1922, pp. 316/1.)
> (iii) For a given value $S(x_1, \ldots, x_n)$, the distribution of T is independent of ϑ. (Fisher 1920, p. 768; Fisher 1922, pp. 316/7.)

Comparisons between estimators based on the asymptotic variance have a long history, dating back to Laplace and Gauss. To express in mathematical terms the idea that "S contains all information included in T", one needs the joint distribution of S and T. In this way, Laplace (1818) came across the example of two estimators S and T such that no linear combination of S and T has a smaller asymptotic variance than S itself (Stigler 1973). This could be considered as a restricted version of Fisher's idea that "S contains all information included in T".

How did Fisher arrive at his idea of "sufficiency"? Unaware (as often) of what others had done before him, Fisher considered the estimators

$$S(x_1, \ldots, x_n) = \left(n^{-1} \sum_{\nu=1}^{n} (x_\nu - \overline{x}_n)^2\right)^{1/2}$$

and

$$T(x_1, \ldots, x_n) = \sqrt{\pi/2}\, n^{-1} \sum_{\nu=1}^{n} |x_\nu - \overline{x}_n|$$

J. Pfanzagl, *Mathematical Statistics*, Springer Series in Statistics,
DOI 10.1007/978-3-642-31084-3_2

for the parameter σ in the family $\mathfrak{P} = \{N(\mu, \sigma^2)^n : \mu \in \mathbb{R}, \sigma^2 > 0\}$. With computations for the sample size $n = 4$, Fisher (1920, Sect. 7, pp. 768–770) came to the conclusion that the conditional distribution of T, given S, does not depend on σ [and not on μ].

Fisher's definition that "the whole of the information respecting σ, which the sample provides, is summed up in the value of S" (p. 769), isolated from the special context of a parametric family, an i.i.d. sample and real-valued statistics S and T, could be phrased as follows:

Definition 2.1.1 A statistic $S : X \to Y$ is sufficient for $\mathfrak{P}$ if the conditional distribution of any statistic $T : X \to Z$ under P, given S, is independent of $P \in \mathfrak{P}$.

In Fisher's publications, the probability measures are defined by their Lebesgue densities on $\mathbb{R}^m$, and the statistics are maps from $\mathbb{R}^m$ to $\mathbb{R}^p$ (usually with $p = 1$). This makes it possible to define conditional distributions in an elementary way by using Lebesgue densities. Fisher could not foresee the mathematical difficulties this idea would encounter in a more general framework. Within the restricted framework where such technical problems do not arise, Fisher's idea of the conditional distribution, of T within $\{x \in X : S(x) = y\}$ seems to have been accepted by other scholars without reserve (see e.g. Bartlett 1937, Sect. 7, pp. 275/6).

Yet, the equivalence of the statements (ii) and (iii) above is not so obvious. If one thinks of S as an estimator of a parameter ϑ (as Fisher always did), the distribution of a combined statistic $\psi(S(x), T(x))$ will differ from the distribution of $S(x)$, and whether it is better or not depends on how the quality of the estimator is evaluated. (Laplace, after all, considered $\psi(S(x), T(x)) = (1 - c)S(x) + cT(x)$ only.)

The ultimate justification for the equivalence of (ii) and (iii) came much later (see Halmos and Savage 1949, Sect. 10, pp. 239–241). If the, conditional distribution of T, given S, does not depend on P, then this conditional distribution can be used to generate a random variable which has, for every $P \in \mathfrak{P}$, the same distributions as T. Hence for any statistical procedure based on $T(x)$ there exists a (randomized) procedure depending on x through $S(x)$ only which has the same performance, for every $P \in \mathfrak{P}$. When Fisher developed his ideas about sufficiency, the mathematical tools needed to make this idea precise were not yet available. Yet, this need not have been an obstacle at this level of rigor. After all, in special examples this randomization takes on a concrete shape.

Example For $\mathfrak{P} = \{N(\mu, \sigma^2)^n : \mu \in \mathbb{R},\ \sigma > 0\}$, a sufficient statistic is

$$S(x_1, \ldots, x_n) = (\mu_n(x_1, \ldots, x_n), s_n(x_1, \ldots, x_n))$$

with

$$\mu_n(x_1, \ldots, x_n) = n^{-1} \sum_{\nu=1}^{n} x_\nu$$

and

$$s_n(x_1,\ldots,x_n) = \left(n^{-1}\sum_{\nu=1}^{n}(x_\nu - \mu_n(x_1,\ldots,x_n))^2\right)^{1/2}.$$

For $\nu = 1,\ldots,n$ let

$$\psi_\nu(u_1,\ldots,u_n) := (u_\nu - \mu_n(u_1,\ldots,u_n))/s_n(u_1,\ldots,u_n).$$

Then the distribution of

$$(\mu_n(x_1,\ldots,x_n) + \psi_\nu(u_1,\ldots,u_n)s_n(x_1,\ldots,x_n))_{\nu=1,\ldots,n}$$

under $N(\mu,\sigma^2)^n \times N(0,1)^n$ is $N(\mu,\sigma^2)^n$. (Example 2 in Kumar and Pathak 1977. See Pfanzagl 1981, for a generalization.) □

We now, discuss the, mathematical difficulties connected with the generalization of Fisher's idea to a more general framework. Let $(X,\mathscr{A})$, $(Y,\mathscr{B})$ and $(Z,\mathscr{C})$ be measurable spaces and $S : X \to Y$ and $T : X \to Z$ measurable maps. Let $P|\mathscr{A}$ be a probability measure. Speaking of the conditional distribution of T within the partition $\{x \in X : S(x) = y\}$ requires, in mathematical terms, the existence of a *regular conditional probability*, i.e. a Markov kernel $M|Y \times \mathscr{C}$ such that, for every $C \in \mathscr{C}$, $M(\cdot, C)$ is a conditional expectation of $1_{T^{-1}C}$, given S, under P, or in other words:

$$\int M(S(x),C)1_B(S(x))P(dx) = P(S^{-1}B \cap T^{-1}C) \quad \text{for every } B \in \mathscr{B}. \tag{2.1.1}$$

The question whether the idea of the conditional distribution of T, given S, can always be expressed by means of a Markov kernel was answered in the affirmative by Doob (1938, Theorem 3.1, pp. 95/8) for $T(x) = x$ under the assumption that $\mathscr{A}$ is countably generated, but his proof is valid for $(X,\mathscr{A}) = (\mathbb{R},\mathbb{B})$, with $\mathbb{B}$ the Borel σ field, only. It appears that the existence of a Markov kernel was not a trivial problem then. Even Halmos published an erroneous result (1941, p. 390, Theorem 1), based on Doob's Theorem. A counterexample by Dieudonné[1] (1948, p. 42) demonstrated that the existence of a Markov kernel can be guaranteed only under additional restrictions, such as the compact approximation of P. For the more general case of maps $T : (X,\mathscr{A}) \to (Z,\mathscr{C})$, the restrictive assumptions have to be

---

[1]It is interesting to observe how the authorship of a nontrivial example was lost as time went on. Dieudonné's example appears in various textbooks as an exercise with reference to the author (e.g. Doob 1953, p. 624; Halmos 1950, p. 210, Exercise 4 and p. 292, Reference to Sect. 48; Dudley 1989, p. 275, Problem 6 and Note Sect. 10.2, p. 298). It also appears in Ash 1972, p. 267, Problem 4, without reference to Dieudonné, and in Romano and Siegel (1986), pp. 138/9, Example 6.13, as an "Example given in Ash". In Lehmann and Casella (1998, p. 35) it was presented as an "Example due to Ash, presented by Romano and Siegel (1986)".

placed upon $P \circ T|\mathscr{C}$. The existence of a Markov kernel is guaranteed if $(Z, \mathscr{C})$ is Polish (i.e., if Z is a complete separable metric space and $\mathscr{C}$ its Borel field). More general concepts like "perfect measures" are hardly used in statistical theory.

Fisher's idea of sufficiency presumes for every $T:X \to Z$ the existence of a Markov kernel, not depending on P, which fulfills (2.1.1) for every $P \in \mathfrak{P}$. To evade the problems with the existence of a Markov kernel, one may start from the following, less restrictive definition.

Definition 2.1.2 A map $S : (X, \mathscr{A}) \to (Y, \mathscr{B})$ is *sufficient* if for every $A \in \mathscr{A}$ there is a conditional expectation of 1_A, given S, which is independent of $P \in \mathfrak{P}$. That means: For every $A \in \mathscr{A}$ there exists $\psi_A : (Y, \mathscr{B}) \to (\mathbb{R}, \mathbb{B})$ such that

$$\int \psi_A(S(x))1_B(S(x))P(dx) = P(S^{-1}B \cap A) \qquad (2.1.2)$$

for every $B \in \mathbb{B}$ and every $P \in \mathfrak{P}$.

This definition first appears in Halmos and Savage (1949, pp. 232/3). Since the connection with Fisher's concept of "sufficiency" is not so obvious, one would expect a clearcut theorem asserting that the conditional distribution of any real-valued statistic with respect to a sufficient statistic can always be represented by a Markov kernel independent of P. The absence of such a theorem is all the more surprising as the authors discuss randomization based on a conditional distribution. Their opinion (see p. 230) that "... conditional probabilities are sufficiently tractable for most practical and theoretical purposes, and the requirement that they should behave like probability measures in the strict sense ... is almost never needed" corresponds to the somewhat casual treatment of this problem in Sect. 10 of this paper.

What is missing in the paper by Halmos and Savage is a theorem saying that if S is sufficient in the sense of Definition 2.1.2, then for every $T : (X, \mathscr{A}) \to (Z, \mathscr{C})$ there exists a Markov kernel, not depending on P, such that (2.1.1) holds for every $P \in \mathfrak{P}$. This theorem occurs first in Bahadur (1954, p. 434, Theorem 5.1) for T real-valued. It is now standard for statistics T attaining their values in a Polish space. Hence the randomization procedure works for virtually all applications. Critical functions and unbiased estimators are, perhaps, the only instances where the conditional expectations with respect to a sufficient statistic do a better job than randomization.

Remark In Definition 2.1.2, the σ-field $\mathscr{B}$ occurs three times; (i) in the $\mathscr{A}$, $\mathscr{B}$-measurability of S, (ii) in the $\mathscr{B}, \mathbb{B}$-measurability of ψ_A, and (iii) in relation (2.1.2). If Y is a Euclidean space, the pertaining Borel field is a natural choice for $\mathscr{B}$. If Y is an abstract space, the presence of an unspecified σ-field $\mathscr{B}$ is irritating. This, presumably, motivated Bahadur to introduce the concept of a "sufficient transformation", which uses for $\mathscr{B}$ the σ-field induced on Y by S, i.e., $\mathscr{B}_S := \{B \subset Y : S^{-1}B \in \mathscr{A}\}$, the largest σ-field which renders $x \to S(x)$ $\mathscr{A}$-measurable.

Most authors ignore the σ-field $\mathscr{B}$ which shows up in the Definition 2.1.2 of sufficiency. (An exception is Witting 1966, who discusses this problem in footnote

1, p. 121.) Does the sufficiency of S really depend on the —more or less arbitrary— σ-field $\mathscr{B}$?

The operational significance of sufficiency depends on relation (2.2.2). Using Proposition 1.10.25 in Pfanzagl (1964, p. 60), which connects (2.1.2) with (2.2.2), it can be shown that (2.1.2) holds for *every* countably generated σ-field $\mathscr{B} \subset \mathfrak{P}(Y)$ containing $\{y\}$ for $P \circ S$-a.a. $y \in Y$ if it holds for *some* σ-field sharing these properties. Presumably this result also holds true under less restrictive regularity conditions.

The concept of a sufficient transformation was almost entirely neglected in the literature. Sato (1996) seems to have been the first author to reflect upon the relationship between a sufficient statistic and a sufficient transformation. He shows (p. 283, Lemma) that the two concepts are identical if $(X, \mathscr{A})$ and $(Y, \mathscr{B})$ are Euclidean and $\mathfrak{P}$ is dominated. This results from the special nature of the spaces $(X, \mathscr{A})$ and $(Y, \mathscr{B})$: If $f : (\mathbb{R}^m, \mathbb{B}^m) \to (\mathbb{R}^k, \mathbb{B}^k)$, then for every $B \subset \mathbb{R}^k$ with $f^{-1}B \in \mathbb{B}^m$, there exists $B_0 \in \mathbb{B}^k$ with $B_0 \subset B$ and $\lambda^m((f^{-1}B)\Delta(f^{-1}B_0)) = 0$, so that $\{B \in \mathbb{R}^k : f^{-1}B \in \mathbb{B}^m\}$ is not much larger than $f^{-1}\mathbb{B}^k$. (See also Bahadur 1955b, p. 493, Lemma 5.)

2.2 Exhaustive Statistics

Originally, the concept of "sufficiency" rests on the existence of statistics S such that for any statistic T the conditional distribution, given S, does not depend on $P \in \mathfrak{P}$. The mathematically precise version led to a Markov kernel fulfilling (2.1.1) for every $P \in \mathfrak{P}$, which, in turn, justifies the concept of sufficiency by the possibility of randomization. What if one disregards the idea of the conditional distribution of T, given S, and starts from the randomization procedure immediately? This requires for the statistic T a Markov kernel $M \mid Y \times \mathscr{C}$ such that

$$\int M(S(x), C)P(dx) = P(T^{-1}C) \quad \text{for } C \in \mathscr{C} \text{ and } P \in \mathfrak{P}. \tag{2.2.1}$$

We shall call S *exhaustive* if relation (2.2.1) holds with $T(x) = x$, i.e., if P itself can be obtained by a randomization procedure based on $S(x)$. More precisely: If there exists a Markov kernel $M|Y \times \mathscr{A}$ such that

$$\int M(S(x), A)P(dx) = P(A) \quad \text{for } A \in \mathscr{A} \text{ and } P \in \mathfrak{P}. \tag{2.2.2}$$

This corresponds to the concept of "sufficiency" introduced by Blackwell (1951) for the comparison of experiments and occurs, therefore, in the literature as Blackwell sufficiency (see e.g. Heyer 1972, Sect. 4, or Strasser 1985, p. 102, Definition 23.3).

If $(X, \mathscr{A})$ is Polish, any sufficient statistic is exhaustive: Specialize (2.1.1) for $B = Y$. Since (2.2.2) is all one needs for recovering P, it is of interest to look for statistics which are exhaustive (but not necessarily sufficient). If $\mathfrak{P}$ is dominated, there are none: If relation (2.2.2) is true (no matter where the Markov kernel comes from), the statistic S is sufficient. This follows from a result obtained by Sacksteder (1967, p. 788, Theorem 2.1) on the comparison of experiments (see also Heyer 1969, p. 39, Satz 5.2.1). Without domination this follows from Roy and Ramamoorthi (1979, p. 50, Theorem 2) under the assumption that $\mathscr{B}$ is countably generated. For dominated families, the simplest way to show that exhaustivity implies, sufficiency is to use a result of Pfanzagl (1974, p. 197, Theorem) which implies, in particular, that S is sufficient if for every $A \in \mathscr{A}$ there exists a critical function φ_A such that $\int \varphi_A(S(x))P(dx) = P(A)$ for every $P \in \mathfrak{P}$.

It is tempting to conclude from (2.2.2) that $M(\cdot, A)$ itself is a conditional expectation of 1_A, given S, i.e., that (2.1.2) holds with $\psi_A = M(\cdot, A)$. This is, however, not generally true.

Example Let $X = \{-1, 0, 1\} \times \{-1, 1\}$ be endowed with the σ-field $\mathfrak{P}(X)$ of all subsets of X. For $\vartheta \in (0, 1)$ let

$$P_\vartheta(\{\eta, \varepsilon\}) := (1 - |\eta|)(1 - \vartheta)/2 + |\eta|\vartheta/4 \quad \text{for } \eta \in \{-1, 0, 1\},\ \varepsilon \in \{-1, 1\}.$$

Let $S : X \to Y = \{0, 1\} \times \{-1, 1\}$ be defined by

$$S(\eta, \varepsilon) := \begin{cases} (0, \varepsilon), \\ (|\eta|, \eta\varepsilon), \end{cases} \quad \eta \in \begin{cases} \{0\}, \\ \{-1, 1\}. \end{cases}$$

The Markov kernel $M|Y \times \mathfrak{P}(X)$, defined by

$$M((\rho, \delta), \{\eta, \varepsilon\}) := ((1 - \rho)(1 - |\eta|) + \rho|\eta|/2)1_{\{\delta\}}(\varepsilon)$$

for $\rho \in \{0, 1\}$, $\eta \in \{-1, 0, 1\}$ and $\varepsilon, \delta \in \{-1, 1\}$, fulfills relation (2.2.2), which is to say that

$$\int M(y, \{\eta, \varepsilon\})P \circ S(dy) = (1 - |\eta|)(1 - \vartheta)/2 + |\eta|\vartheta/4 = P_\vartheta(\{\eta, \varepsilon\}),$$

without being a conditional probability, given S. Notice that S is neither minimal sufficient ($S(\eta, \varepsilon) = |\eta|$ is sufficient), nor does it fulfill (2.2.3): We have $M((1, \delta), S^{-1}\{1, \delta\}) = 1/2$. □

There is an exception: Relation (2.1.2) is automatically fulfilled if S is minimal sufficient. This follows from Sacksteder (1967, p. 793, Theorem 8.1). For a simple proof see Pfanzagl (1994, p. 16, Proposition 1.4.9).

The "minimality condition" comes in as some sort of deus ex machina. There is however an inherent reason why a Markov kernel M that fulfills (2.2.2) should be a conditional probability, given S: If the idea that $M(y, \cdot)$ describes the distribution of x on $\mathscr{A}$ within $\{x \in X : S(x) = y\}$ is taken seriously, this demands that

$$M(y, S^{-1}B) = 1_B(y) \quad \text{for } B \in \mathscr{B} \text{ and } y \in Y. \tag{2.2.3}$$

Relation (2.1.2) implies that (2.2.3) is, in fact, true for $\mathfrak{P} \circ S$-*a.a.* $y \in Y$. If $\mathscr{B}$ is countably generated, relation (2.2.3) implies

$$M(y, S^{-1}\{y\}) = 1 \quad \text{for } P \circ S\text{-a.a. } y \in Y.$$

The relevant point is the converse: (2.2.3) and (2.2.2) together imply (2.1.2), i.e., $M(\cdot, A)$ is necessarily a conditional expectation of 1_A, given S, with respect to P. (Blackwell and Dubins 1975, p. 741/2. For a detailed proof see Pfanzagl 1994, p. 60, Lemma 1.10.24 and Proposition 1.10.25.)

Of course, one would prefer to have a version of M such that (2.2.3) holds for all (rather than just for $\mathfrak{P} \circ S$-a.a.) $y \in Y$. According to Blackwell and Ryll-Nardzewski (1963, p. 223, Theorem 1) this cannot be generally achieved, even if $(X, \mathscr{A})$ is Polish.

The idea that random variables with distribution P can be obtained from random variables with distribution $P \circ S$ by means of a randomization device (not depending on P) if S is sufficient was extended to the comparison of "experiments" by Blackwell (1951, 1953). Let $\mathscr{X} = (X, \mathscr{A}, \{P_i : i \in I\})$ and $\mathscr{Y} = (Y, \mathscr{B}, \{Q_i : i \in I\})$ be two "experiments" with an arbitrary index set I. Then $\mathscr{Y}$ is called "sufficient for $\mathscr{X}$" if there exists a Markov kernel $M|Y \times \mathscr{A}$ such that

$$P_i(A) = \int M(y, A) Q_i(dy) \quad \text{for } A \in \mathscr{A} \text{ and } i \in I.$$

If $S : (X, \mathscr{A}) \to (Y, \mathscr{B})$ is exhaustive, then the experiment $(Y, \mathscr{B}, \{P_i \circ S : i \in I\})$ is sufficient for $(X, \mathscr{A}, \{P_i : i \in I\})$.

The problems that are discussed in connection with sufficient statistics (like the characterization of sufficiency) can also be discussed at this more general level, with corresponding results. The application of the more general results to the case of sufficient statistics offers no additional insights. It is just the same with the further generalization of the concept of an experiment from probability measures on a σ-field of sets to linear operators on a lattice of functions. Reader interested in such generalizations should consult books like Heyer (1982), Strasser (1985), LeCam (1986), Bomze (1990) and Torgersen (1991).

2.3 Sufficient Statistics—Sufficient σ-Fields

Sufficient sub-σ-fields have been introduced by Bahadur (1954, p. 430), following a suggestion of L.J. Savage (see p. 431):

Definition 2.3.1 $\mathscr{A}_0 \subset \mathscr{A}$ is *sufficient* for $\mathfrak{P}$ if for every $A \in \mathscr{A}$ there exists a conditional expectation of 1_A, given $\mathscr{A}_0$, which is independent of $P \in \mathfrak{P}$.

That means: For every $A \in \mathscr{A}$ there exists a function $\psi_A : (X, \mathscr{A}_0) \to (\mathbb{R}, \mathbb{B})$ such that

$$\int \psi_A(x) 1_{A_0}(x) P(dx) = P(A_0 \cap A) \quad \text{for every } A_0 \in \mathscr{A}_0.$$

Fisher's idea of sufficiency was related to statistics S that are estimators (of some parameter ϑ). As soon as it became clear that an interpretation of S as an estimator is irrelevant for the idea of "sufficiency", it was also clear that the values taken by the statistic S are irrelevant, too: Any statistic $\hat{S}$ which is one-one with S serves the same purpose. If the image-space of S is irrelevant, the same is true for the σ-field with which the image-space of S might be endowed. Consequently, Bahadur suggested the concept of a sufficient transformation as a map from X to Y, and that we base the definition of sufficiency on $\mathscr{A}_0 := \mathscr{A} \cap S^{-1}(\mathfrak{P}(Y))$.

> The evident notational simplifications which result from studying a statistic in terms of the subfield induced by it suggest the possibility of taking a sufficient subfield rather than a sufficient statistic to be the basic concept in the formal exposition. (Bahadur 1954, p. 430.)

When Bahadur wrote these lines, he was not fully aware of the troubles connected with the duplicity between "sufficient statistics" and "sufficient sub-σ-fields", still present in many textbooks. Obviously, $S:(X, \mathscr{A}) \to (Y, \mathscr{B})$ is sufficient iff $S^{-1}\mathscr{B}$ is sufficient. The problem lies with the sufficiency of sub-σ-fields that are not inducible by means of a statistic. Bahadur (1954) was still uncertain about this point, which, however, was soon clarified: According to a Lemma of Blackwell (see Lemma 1 in Bahadur and Lehmann 1955, p. 139) a subfield $\mathscr{A}_0 \subset \mathscr{A}$ cannot be induced by a statistic if $\{x\} \in \mathscr{A}_0$ for every $x \in X$. The situation is much more favourable if $\mathscr{A}$ is countably generated. Under the additional assumption that $\mathfrak{P}$ is dominated, Bahadur (1955, Lemmas 3 and 4, pp. 492–493) proved for any sub-σ-field $\mathscr{A}_0 \subset \mathscr{A}$ the existence of a statistic $f : (X, \mathscr{A}) \to (\mathbb{R}, \mathbb{B})$ such that $\mathscr{A}_0 = f^{-1}(\mathbb{B})$ $(\mathfrak{P})$. More important in connection with sufficiency is a result of Burkholder (1961, p. 1200, Theorem 7): If $\mathscr{A}$ is countably generated, then for any sufficient sub-σ-field $\mathscr{A}_0 \subset \mathscr{A}$ there exists a statistic $f:(X, \mathscr{A}) \to (\mathbb{R}, \mathbb{B})$ such that $\mathscr{A}_0 = f^{-1}(\mathbb{B})$.

Hence for $\mathscr{A}$ countably generated, the concepts of "sufficient statistic" and "sufficient sub-σ-field" are equivalent. Yet even in this case, puzzling things may occur. If a sufficient statistic $S : (X, \mathscr{A}) \to (Y, \mathscr{B})$ is the contraction of a statistic $S' : (X, \mathscr{A}) \to (Y', \mathscr{B}')$ (i.e., $S = g \circ S'$ with $g : (Y', \mathscr{B}') \to (Y, \mathscr{B})$), then S' is sufficient, too. In contrast, a sub-σ-field containing a sufficient sub-σ-field is not necessarily sufficient itself: For $(X, \mathscr{A}) = (\mathbb{R}, \mathbb{B})$, the σ-field $\mathbb{B}_0$ of all sets in $\mathbb{B}$ symmetric about 0 is sufficient for the family of all probability measures on $\mathbb{B}$ which are symmetric about 0. Burkholder (1961, pp. 1192/3, Example 1), constructs a sub-σ-field of $\mathbb{B}$ containing $\mathbb{B}_0$, which fails to be sufficient. Such a paradox is impossible if $\mathfrak{P}$ is dominated (Bahadur 1954, p. 440, Theorem 6.4).

Instead of making the theory of sufficiency more elegant, the introduction of sufficient sub-σ-fields gave rise to mathematical problems which have nothing to do with the idea of sufficiency as such. Whereas the interpretation of a sufficient statistic is clear, neither Bahadur (1954, pp. 431/2) nor any of his followers have so far been able to explain the operational significance of a sufficient sub-σ-field. Bahadur's argument (1954, pp. 431/2) is based on the partitions induced by the σ-fields $\mathscr{A}$

and $\mathscr{A}_0$, say $\pi(x) := \bigcap\{A \in \mathscr{A} : x \in A\}$ and $\pi_0(x) := \bigcap\{A \in \mathscr{A}_0 : x \in A\}$, respectively: "If $\mathscr{A}_0$ is sufficient, a statistician who knows only $\pi_0(x)$ is as well off as one who knows $\pi(x)$." That this argument works if $\mathscr{A}_0$ is the sub-σ-field induced by S, (so that $\pi_0(x) = S^{-1}\{S(x)\}$) is not a convincing argument for replacing the concept of a sufficient statistic by the concept of a sufficient sub-σ-field, and what if, in the general case, $\pi_0(x) = \pi(x) = \{x\}$? Leaving aside all mathematical aspects, we still encounter a problem with the interpretation of sufficiency. If one thinks of $S(x) \in Y$ as "containing all information contained in x" one might feel uneasy if the definition of sufficiency refers to some σ-field of subsets of Y.

2.4 The Factorization Theorem

Let S and T be two real-valued statistics with a joint density $p_\vartheta(s,t)$. According to Fisher (1922, p. 331) "the factorization of p_ϑ into factors involving (ϑ, s) and (s,t), respectively, is merely a mathematical expression of the condition of sufficiency". More precisely (Fisher 1934, p. 288): "If $p_\vartheta(s,t) = g_\vartheta(s)h(s,t)$, the conditional distribution of T, given S, will be independent of ϑ" (hence S is sufficient in Fisher's sense).

One of the important achievements of Neyman was to show that this factorization is even necessary. Starting from the definition that S is sufficient if the conditional distribution of any other statistic T is independent of P (Neyman 1935, p. 325), Neyman asserts in Teorema II, p. 326, that the factorization is necessary and sufficient for the sufficiency of S. His theorem refers to a parametric family, and to i.i.d. samples. Written in our notations this is

$$p_\vartheta(x_1,\ldots,x_n) = g_\vartheta(S(x_1,\ldots,x_n))h(x_1,\ldots,x_n).$$

Neyman was unaware that Fisher had already shown that the factorization implies sufficiency. Neyman's proof of the nontrivial necessity part of the theorem (Sect. 7, pp. 328–332) is spoiled by numerous regularity conditions (including, for instance, the differentiability of S). Obviously, Neyman was not familiar with the measure-theoretic tools developed by Kolmogorov (1933). The first mathematically "up to date" paper was that of Halmos and Savage (1949). Their Theorem 1 (p. 233) asserts that for a dominated family $\mathfrak{P}$, a necessary and sufficient condition for the sufficiency of $S : (X, \mathscr{A}) \to (Y, \mathscr{B})$ is the existence of a dominating measure, say μ_0, such that every $P \in \mathfrak{P}$ admits an $S^{-1}\mathscr{B}$-measurable version of $dP/d\mu_0$. Expressed with reference to an arbitrary dominating measure μ, and with explicit use of the sufficient statistic S, this is their Corollary 1, p. 234.

Factorization Theorem *Assume that $\mathfrak{P}$ is dominated by μ ($\mathfrak{P} \ll \mu$). Then S is sufficient if and only if there exists a nonnegative $\mathscr{A}$-measurable function $h : (X, \mathscr{A}) \to ([0,\infty), \mathbb{B} \cap [0,\infty))$ and for every $P \in \mathfrak{P}$ a nonnegative $\mathscr{B}$-measurable function*

$g_P : (Y, \mathscr{B}) \to ([0, \infty), \mathbb{B} \cap [0, \infty))$ *such that* $x \to g_P(S(x))h(x)$ *is a* μ-density of P.

An important step in the proof of this theorem is Lemma 7, p. 232, which asserts that for any dominated family $\mathfrak{P}$ there exists a countable subset $P_k \in \mathfrak{P}, k \in \mathbb{N}$, such that $P_* := \sum_{k=1}^{\infty} 2^{-k} P_k$ dominates $\mathfrak{P}$. (Notice that the clever proof of this lemma serves just one purpose: To avoid the assumption that $\mathscr{A}$ is countably generated, in which case the assertion becomes almost trivial.)

It appears that the essence of this Lemma was not fully understood in Schmetterer (1966), the first textbook offering a detailed proof of the Factorization Theorem. In Satz 7.1, p. 249, Schmetterer requires the existence of a dominating probability measure, say $\hat{P}$, equivalent to $\mathfrak{P}$ (i.e., $P(A) = 0$ for $P \in \mathfrak{P}$ implies $\hat{P}(A) = 0$). According to Satz 5.2, p. 221, a mangled version of the Halmos–Savage Lemma, such a $\hat{P}$ exists for every dominated family $\mathfrak{P}$. However, Schmetterer's Satz 5.2 omits the essential point: that $\hat{P}$ is a convex combination of a countable subset of $\mathfrak{P}$, so that sufficiency for $\mathfrak{P}$ implies sufficiency for $\mathfrak{P} \cup \{\hat{P}\}$. This slip has survived in the translated version (see Schmetterer 1974, p. 214, Theorem 7.3 and p. 208, Theorem 7.1).

It led to some confusion that Halmos and Savage included in the formulation of their Corollary 1 that h is μ-integrable. In fact, if a factorization holds with *some* nonnegative, measurable h, then a factorization also holds with a μ-integrable h: If P_* is the dominating measure from Lemma 7, then the μ-density $g_P(S(x))h(x)$ can be rewritten as $(g_P(S(x))/g_*(S(x)))h_*(x)$, with $g_*(y) := \sum_{k=1}^{\infty} 2^{-k} g_{P_k}(y)$ and $h_*(x) := g_*(S(x))h(x)$. This holds since $g_*(y) = 0$ implies $g_{P_k}(y) = 0$ for $k \in \mathbb{N}$; hence $P_*\{g_* \circ S = 0\} = 0$. Note that $\int g_{P_k}(S(x))h(x)\mu(dx) = 1$ for every $k \in \mathbb{N}$ implies $\int h_*(x)\mu(dx) = 1$. The irritation resulting from the μ-integrability condition for h was accentuated by an example of Bahadur (1954, p. 438) for which the natural factorization holds with (the non-integrable) $h(x) \equiv 1$. It survived in numerous publications, and Bahadur was praised, for removing the integrability condition in his Corollary 6.1 on p. 438. (See Lehmann and Scheffé 1950, p. 332, Theorem 6.2.; Torgersen 1991, p. 65, Remark 1; Dudley 1999, p. 193, Notes, to Sect. 5.1.) Zacks (1971) has two factorization theorems: Theorem 2.3.2, pp. 44/5, with, and Theorem 2.3.3, p. 48, without μ-integrability.

2.5 Completeness

Let $\mathfrak{F}$ be a family of measurable functions $f : (X, \mathscr{A}) \to (\mathbb{R}^k, \mathbb{B}^k)$, and $\mathfrak{P}$ a family of probability measures $P|\mathscr{A}$.

Definition 2.5.1 The family $\mathfrak{P}$ is $\mathfrak{F}$-complete if for every $f \in \mathfrak{F}$, the relation $\int f dP = 0$ for every $P \in \mathfrak{P}$ implies $f = 0$ $\mathfrak{P}$-a.e.

Versions useful in statistical theory are *bounded completeness* ($\mathfrak{F} =$ all bounded functions), *completeness* ($\mathfrak{F} =$ all $\mathfrak{P}$-integrable functions), and *2-completeness* ($\mathfrak{F} =$ all square integrable functions).

The concept of "completeness" has its origins in two statistical problems. Lehmann and Scheffé (1950) deserve credit for introducing completeness as a unifying concept.

(i) Let $\varphi_k(x_1, \ldots, x_k)$ be an unbiased estimator of some functional κ, i.e.,

$$\kappa(P) = \int \varphi_k(x_1, \ldots, x_k) P(dx_1) \ldots P(dx_k) \quad \text{for } P \in \mathfrak{P}.$$

Halmos (1946, p. 40, Theorem 5) shows that for every sample size $n \geq k$, the symmetrized version of φ_k, i.e., $\varphi_k^{(n)}(x_1, \ldots, x_n) := \sum \varphi_k(x_{i_1}, \ldots, x_{i_k})$, with the summation extending over all k-tuples $(i_1, \ldots, i_k)$, is of minimal variance among all unbiased estimators. His argument consists of two steps:

(a) For any unbiased estimator f_n, the symmetrized version $f_n^{(n)}$ is unbiased and of equal or smaller variance (see proof of his Theorem 5).

(b) If the family $\mathfrak{P}$ contains all probability measures with finite support, then there is only one symmetric unbiased estimator. This follows from Lemma 2, p. 38, which asserts—in modern terminology—that under the conditions on $\mathfrak{P}$, the order statistic is complete.

It is surprising that Halmos—more familiar with the concept of conditional expectations than practically anybody else— overlooked the fact that the symmetrized version is just the conditional expectation with respect to the order statistic $(x_{n:1}, \ldots, x_{n:n})$. The idea that taking the conditional expectation reduces the variance in general was used by C.R. Rao and by Blackwell just one year later, in 1947.

Hoel (1951, p. 301, Theorem) shows that for exponential families, unbiased estimators based on the sufficient statistic are unique (and therefore optimal, thanks to Blackwell's theorem). His proof essentially shows that an exponential family with one real parameter is complete.

(ii) Neyman (1937) pointed out that a sufficient statistic $S : (X, \mathscr{A}) \to (Y, \mathscr{B})$ can be used to construct a "similar" critical region $C \in \mathscr{A}$. In our notation this amounts to choosing C such that the conditional expectation of 1_C, given S, is constant. Since one is interested in obtaining critical regions of high rejection power, the question occurs whether every similar critical region is of this type (now called "Neyman structure"). To answer this question, Lehmann and Scheffé (1947, p. 383) introduce the concept of what is now called "bounded completeness": all similar critical regions are of Neyman structure if $\mathfrak{P} \circ S$ is boundedly complete and they give a first proof of the bounded completeness of the sufficient statistic for exponential families. Independently of the paper by Lehmann and Scheffé, Sverdrup (1953, p. 67, Theorem 1) offers a precise proof that, for exponential families, every similar test is of Neyman structure (which amounts to proving bounded completeness for exponential families).

Of course, there are instances where $\mathfrak{P} \circ S$ fails to be boundedly complete. A prominent example is the Behrens–Fisher problem. In such cases there are similar tests which are not of Neyman structure: Any bounded function $f : (Y, \mathscr{B}) \to (\mathbb{R}, \mathbb{B})$ fulfilling $\int f(S(x))P(dx) = 0$ for $P \in \mathfrak{P}$ can be transformed into a critical function, say φ, fulfilling $\int \varphi(S(x))P(dx) = 0$ for $P \in \mathfrak{P}$. How to obtain good critical regions in such a case was discussed by Wijsman (1958).

The question whether the two concepts of completeness (bounded and plain) originating from two different statistical problems are, in fact, different, was answered in the affirmative by the example of a family which is boundedly complete without being complete (Lehmann and Scheffé 1950, p. 312, Example 3.1).

An important result put forward by Lehmann and Scheffé (1950, p. 316, Theorem 3.1) asserts that a sufficient statistic S is *minimal* sufficient (Definition 2.6.1) if $\mathfrak{P} \circ S$ is boundedly complete. (The conditions for this theorem remain somewhat vague in this paper. It is certainly enough that $\mathfrak{P}$ is dominated and $(Y, \mathscr{B})$ Polish. A precise proof may be found in Heyer 1982, p. 41, Theorem 6.15.) Consequently, the search for a sufficient statistic S with $\mathfrak{P} \circ S$ boundedly complete requires the search for a minimal sufficient statistic. This will be dealt with in Sect. 2.6.

What makes the concept of (bounded) completeness manageable are the simple criteria for it to hold, especially the following:

Theorem 2.5.2 *For an exponential family $\mathfrak{P}$ with densities*

$$x \to C(P)h(x)\exp\left[\sum_{i=1}^{m} a_i(P)T_i(x)\right], \quad T_i : (X, \mathscr{A}) \to (\mathbb{R}, \mathbb{B}), \quad i = 1, \dots, m,$$

the family $\{P \circ (T_1, \dots, T_m) : P \in \mathfrak{P}\}$ is complete if the set $\{(a_1(P), \dots, a_m(P) : P \in \mathfrak{P}\}$ has a nonempty interior.

A first version of this theorem appears in Lehmann and Scheffé (1955, pp. 223/4, Theorem 7.3). In Lehmann and Scheffé (1950, pp. 313–315) the completeness for various exponential families had been proved by means of ad hoc arguments using power series expansions, Laplace and Mellin transformations.

The condition that $\{(a_1(P), \dots, a_m(P)) : P \in \mathfrak{P}\}$ has a nonempty interior is sufficient but not necessary for the completeness of $\{P \circ (T_1, \dots, T_m) : P \in \mathfrak{P}\}$. Examples of complete "curved" exponential families can be found in Messig and Strawderman (1993). For a particularly simple example see Pfanzagl (1994, p. 96, Example 2.7.3).

What matters for applications is the completeness of sufficient statistics for i.i.d. products. Since i.i.d. products of exponential families are exponential, too, this problem is solved by the theorem above.

Another important result is the [bounded] completeness of the order statistic for certain nonparametric families, referred to as "symmetric [bounded] completeness".

Theorem 2.5.3 $\{P^n : P \in \mathfrak{P}\}$ *is* symmetrically [boundedly] complete *if for every [bounded] symmetric (i.e., permutation invariant) function $f_n : (X^n, \mathscr{A}^n) \to (\mathbb{R}, \mathbb{B})$, the relation $\int f_n dP^n = 0$ for $P \in \mathfrak{P}$ implies $f_n = 0$ P^n-a.e. for every $P \in \mathfrak{P}$.*

A precise proof may be found in, Heyer (1982, p. 41, Theorem 6.15).

It is intuitively clear that $\{P^n : P \in \mathfrak{P}\}$ will be symmetrically complete if $\mathfrak{P}$ is sufficiently rich. The first results of this kind came from Halmos (1946; for $\mathfrak{P}$ consisting of all discrete distributions with finite support) and Fraser (1954; for $\mathfrak{P}$ consisting of all uniform distributions over a finite number of intervals).

Developing an idea of Lehmann (1959, p. 152, Problem 12), and Mandelbaum and Rüschendorf (1987, p. 1233) proved the following

Polarization Lemma *Assume that $\mathfrak{P}$ is weakly convex in the sense that for arbitrary $P_i \in \mathfrak{P}$, $i = 0, 1$, the convex combination $(1-\alpha)P_0 + \alpha P_1 \in \mathfrak{P}$ for some $\alpha \in (0,1)$. Then for every symmetric function $f_n : (X^n, \mathscr{A}^n) \to (\mathbb{R}, \mathbb{B})$, the relation $\int f_n dP^n = 0$ for $P \in \mathfrak{P}$ implies $\int f_n d(P_1 \times \cdots \times dP_n) = 0$ for arbitrary $P_i \in \mathfrak{P}$, $i = 1, \ldots, n$.*

(A simpler proof for convex—rather than weakly convex—families can be found in, Pfanzagl 1994, p. 20, Lemma 1.5.9.)

Since the [bounded] completeness of $\mathfrak{P}|\mathscr{A}$ implies the [bounded] completeness of $\{P_1 \times \cdots \times P_n : P_i \in \mathfrak{P}\}$, $i = 1, \ldots, n$ (see Landers and Rogge 1976, p. 139, Theorem; improving an earlier result of, Plachky 1977, concerning bounded completeness), the Polarization Lemma implies the following theorem (see Mandelbaum and Rüschendorf 1987, p. 1239, Theorem 7):

If $\mathfrak{P}$ is [boundedly] complete and weakly convex, then $\{P^n : P \in \mathfrak{P}\}$ is symmetrically [boundedly] complete for every $n \in \mathbb{N}$.

According to Mattner (1996, p. 1267, Theorem 3), this implies that $\{P^n : P \in \mathfrak{P}\}$ is symmetrically complete if $\mathfrak{P}|\mathbb{B}$ contains all P with unimodal Lebesgue densities.

In this connection we mention a result of Hoeffding (1977) concerning families that are symmetrically incomplete, but symmetrically boundedly complete, which generalizes an earlier result of Fraser (1954, p. 48, Theorem 2.1).

If $\int u\,dP = 0$ for some $u : (X, \mathscr{A}) \to (\mathbb{R}, \mathbb{B})$ and some $P|\mathscr{A}$, then

$$f_n(x_1, \ldots, x_n) := \sum_{\nu=1}^{n} u(x_\nu) h_{n-1}(x_{n\cdot\nu}) \tag{2.5.1}$$

is symmetric and fulfills $\int f_n dP^n = 0$ if $h_{n-1} : X^{n-1} \to \mathbb{R}$ is a P^{n-1}-integrable, symmetric function and $x_{n\cdot\nu} := (x_1, \ldots, x_{\nu-1}, x_{\nu+1}, \ldots, x_n)$.

Let now $\mathfrak{P}_u$ be the family of all P dominated by μ that fulfill $\int u\,dP = 0$. A result of Hoeffding (1977, p. 279, Theorem 1B) implies that any symmetric function f_n fulfilling $\int f_n dP^n = 0$ for $P \in \mathfrak{P}$ can be represented by (2.5.1). Hence (Theorem 2B, p. 280) the family $\{P^n : P \in \mathfrak{P}_u\}$ is symmetrically boundedly complete if u is unbounded.

Refined mathematical techniques have been used to obtain a great variety of complete and/or boundedly complete families (for instance Bar-Lev and Plachky 1989 or Isenbeck and Rüschendorf 1992). Referring, so to speak, to the sample size 1, these results seem to be answers waiting for questions.

2.6 Minimal Sufficiency

If sufficient statistics can be used for a reduction of the data, the aspiration to maximal reduction leads to the concept of a "minimal suff icient statistic", which is the contraction of any sufficient statistic.

Definition 2.6.1 The sufficient statistic $S_0 : (X, \mathscr{A}) \to (Y_0, \mathscr{B}_0)$ is *minimal sufficient* for $\mathfrak{P}$ if for any sufficient statistic $S : (X, \mathscr{A}) \to (Y, \mathscr{B})$ there exists a function $H : (Y, \mathscr{B}) \to (Y_0, \mathscr{B}_0)$ such that $S_0 = H \circ S$ $(\mathfrak{P})$.

The concept of a minimal sufficient statistic was introduced by Lehmann and Scheffé (1950, Sect. 6). Their interest in minimal sufficient statistics was motivated by applications to the theory of similar tests and unbiased estimators. The theory in this field becomes particularly clear if there exists a sufficient statistic, say S, such that $\mathfrak{P} \circ S$ is (boundedly) complete. Since the bounded completeness of $\mathfrak{P} \circ S$ for a sufficient statistic S implies that S is minimal sufficient, the applications of sufficiency for certain problems in testing and estimation motivated the interest of Lehmann and Scheffé (1950) in minimal sufficient statistics. Lehmann and Scheffé (1950, p. 316, Theorem 3.1) show that "S sufficient and $\mathfrak{P} \circ S$ boundedly complete" implies that S is minimal sufficient if a minimal sufficient statistic exists. Bahadur (1957, p. 217, Theorem 3) shows that "S sufficient and $\mathfrak{P} \circ S$ boundedly complete" implies that S is minimal sufficient.

In their Sect. 6, pp. 327ff, Lehmann and Scheffé suggest a workable technique for obtaining a minimal sufficient statistic. Their basic assumption is that there exists a countable subfamily $\mathfrak{P}_0 \subset \mathfrak{P}$ which is dense with respect to the sup-distance. This implies the existence of a dominating measure, say, μ, equivalent to $\mathfrak{P}$. (Recall that, conversely, the existence of such a subfamily follows for dominated families if $\mathscr{A}$ is countably generated.) To determine a minimal sufficient statistic, Lehmann and Scheffé introduce the "operation ϑ" which amounts to determining for $x_0 \in X$ the set of all x such that $q_P(x)/q_P(x_0)$ is independent of $P \in \mathfrak{P}$. (The fact that this is formulated in terms of a parametric family is of no relevance.) This means that determining a function $S|X$ such that $S(x) = S(x_0)$ implies $q_P(x) = q_P(x_0)$ for $P \in \mathfrak{P}$. According to Theorem 6.3, p. 336, the resulting statistic S is minimal sufficient if the "operation ϑ" is applied with $\mathfrak{P}$ replaced by $\mathfrak{P}_0$. The restriction to a countable subfamily is required since the densities are unique μ-a.e. only.

Presuming that a sufficient statistic $S : (X, \mathscr{A}) \to (Y, \mathscr{B})$ has been found, i.e., that there exists a μ-density of P of the form $x \to g_P(S(x))h(x)$, the basic idea of Lehmann and Scheffé can be put into the following form:

Assume that $\mathscr{A}$ is countably generated and $(Y, \mathscr{B})$ Polish. If there exists a countable dense subfamily $\mathfrak{P}_0 \subset \mathfrak{P}$ such that

$$g_P(y') = g_P(y'') \text{ for } P \in \mathfrak{P}_0 \quad \text{implies} \quad y' = y'',$$

then S is minimal sufficient.

See Pfanzagl (1994), p. 14, Theorem 1.4.4. A related result appears in Sato (1996, pp. 381/2, Theorem). Translated from parametric families to the case of a general family, Sato's result reads as follows: Let $(X, \mathscr{A})$ and $(Y, \mathscr{B})$ be Euclidean spaces, $\mathfrak{P}$ a dominated family on $\mathscr{A}$ and $\mathfrak{P}_0$ a countable dense subfamily. Assume that the densities are continuous in the sense that $d(P_n, P_0) \to 0$ implies $q_{P_n} \to q_{P_0}$ μ-a.e. If for arbitrary x', x'', the relation $S(x') = S(x'')$ implies that $q_P(x')/q_P(x'')$ is independent of $P \in \mathfrak{P}$, then S is minimal sufficient.

The duplicity between sufficient statistics and sufficient sub-σ-fields extends to the concept of "minimality".

Definition 2.6.2 A sufficient sub-σ-field $\mathscr{A}_*$ is *minimal sufficient* if $\mathscr{A}_* \subset \mathscr{A}_0$ $(\mathfrak{P})$ for any sufficient σ-field $\mathscr{A}_0$.

According to Burkholder (1961, p. 1197, Corollary 3) this property is equivalent to the following: If $\mathscr{A}_0 \subset \mathscr{A}_*$ $(\mathfrak{P})$ is sufficient, then $\mathscr{A}_0 = \mathscr{A}_*$ $(\mathfrak{P})$. Bahadur (1954, p. 439, Theorem 6.2) establishes the existence of a minimal sufficient σ-field for dominated families, starting from Theorem 1 in Halmos and Savage (1949, p. 233): The σ-field generated by $\{x \in X : g_P(x) < r\}$, $r \in (0, \infty)$, $P \in \mathfrak{P}\}$ is minimal sufficient. Bahadur (1955b, p. 495, Corollary 1) shows that for a dominated family on a Euclidean space, the σ-field induced by a minimal sufficient statistic is minimal sufficient. This result is contrasted by an example of Bahadur and Lehmann (1955, p. 140) showing that, in general, the σ-field induced by a minimal sufficient statistic might fail to be minimal sufficient, even if a minimal sufficient statistic exists. Pitcher (1957) presented a family of probability measures for which neither a minimal sufficient statistic nor a minimal sufficient σ-field exists. This still left open the question whether the existence of a minimal sufficient statistic implies the existence of a minimal sufficient σ-field or vice versa. This question was answered to the negative by two counterexamples in a short but profound paper by Landers and Rogge (1972). Specializing a result of Rogge (1972, p. 210, Theorem 4) on perfect measures one obtains that for a dominated family on a Polish space a minimal sufficient statistic induces a minimal sufficient σ-field and vice versa.

It was obviously hard for statisticians with mainly mathematical interests to accept the fact that neither minimal sufficient statistics nor minimal sufficient σ-fields exist in general. The paper by Pitcher (1965) is a good example of this discomfiture. In it he defines a property of a family of probability measures, called "compactness", which is more general than domination, and which allows the construction of a minimal sufficient statistic. Even a mathematically minded statistician is tempted to question the relevance of an artificial concept that is bare of intuitive appeal and hard to verify and leads to a result beyond any practical relevance. In spite of this, various authors followed the path first opened by Pitcher's paper. Among these, the most interesting one seems to be Hasegawa and Perlman (1974) which invalidates Pitcher's construction by means of a counterexample (pp. 1054/5, Sect. 5). The authors suggest a concept related to Pitcher's "compactness" that implies the existence of a minimal sufficient σ-field.

For families of product measures $\{P^n : P \in \mathfrak{P}\}$, the order statistic is always sufficient. Hence the relevant question is whether there is a (minimal) sufficient statistic coarser than the order statistic. For exponential families, the statistic $(x_1, \ldots, x_n) \to$

$(\sum_{\nu=1}^{n} T_1(x_\nu), \ldots, \sum_{\nu=1}^{n} T_k(x_\nu))$ is sufficient and complete, hence minimal sufficient if the set $\{(a_1(P), \ldots, a_k(P)) : P \in \mathfrak{P}\} \subset \mathbb{R}^k$ has a nonempty interior. (For *curved* exponential families, this statistic may be minimal sufficient without being complete.)

An interesting result of a general nature is provided by Mattner (2001, p. 3402, Theorem 1.5). If $\mathfrak{P}$ is a dominated convex family admitting the minimal sufficient σ-field $\mathscr{A}_0$, then the σ-field of all permutation invariant subsets of $\mathscr{A}_0^n$ is minimal sufficient for $\{P^n : P \in \mathfrak{P}\}$. (Note: Recall the analogous result mentioned in Sect. 2.1.5 on (not necessarily dominated) convex families that are [boundedly] complete.)

Concerning location and/or scale parameter families on $\mathbb{B}$, one could, roughly speaking, say that under suitable regularity conditions on the densities, the order statistic is minimal sufficient, except for particular families, like $\{N(\mu, \sigma^2)^n : \mu \in \mathbb{R}, \sigma^2 > 0\}$ or $\{\Gamma(\mu, \sigma^2)^n : \mu \in \mathbb{R}, \sigma^2 > 0\}$. A recent result in this direction that requires the use of more subtle mathematics was put forward by Mattner (2000, pp. 1122/3, Theorem 1.1) who shows under a weak regularity condition on the Lebesgue density of P that for the location and scale parameter family generated by P the order statistic is minimal sufficient for the sample size n, unless $\log p$ is a polynomial of degree less than n.

2.7 Trivially Sufficient Statistics

If the vague notion that S is sufficient if "$S(x)$ contains all information contained in x" is taken seriously, it only produces new problems. If the problem is just to regain from $S(x)$ a random variable equivalent to x: Why use the randomization device instead of using for S a bijective function from X to $\mathbb{R}$, say? If $X = \mathbb{R}^n$ and $S : \mathbb{R}^n \to \mathbb{R}$ is a bijective map, it is possible to regain from $S(x_1, \ldots, x_n)$ the original sample $(x_1, \ldots, x_n)$ itself (rather than a random variable with the same distribution as $(x_1, \ldots, x_n)$). Why do statisticians use $S(x_1, \ldots, x_n) = (\sum_{\nu=1}^{n} x_\nu, \sum_{\nu=1} 1^n x_\nu^2)$ as a sufficient statistic for the family $\{N(\mu, \sigma^2)^n : \mu \in \mathbb{R},\ \sigma > 0\}$ rather than an injective map $S : \mathbb{R}^n \to \mathbb{R}$, which is trivially sufficient for the larger family of all probability measures on $\mathbb{B}^n$, and is 1- rather than 2-dimensional? Of course, S should be "regular" in some sense, continuous at least, and it should be interpretable in some operational sense.

Denny (1964, p. 95, Theorem) proves the existence of a uniformly continuous function $S_n : \mathbb{R}^n \to (0, 1)$ with nondecreasing partial functions $x_\nu \to S_n(x_1, \ldots, x_n)$, $\nu = 1, \ldots, n$, such that S_n is injective on a subset $D_n \subset \mathbb{R}^n$ with $\lambda^n(D_n^c) = 0$. Such a function S_n is sufficient for the family of all probability measures on $\mathbb{B}^n$ with λ^n-density. This result poses a serious problem for the concept of "sufficiency". Yet it is ignored in virtually all textbooks on mathematical statistics. Romano and Siegel (1986, Sect. 7.1, pp. 158/9) even give examples for which, as they think, "no single [i.e., real-valued] continuous sufficient statistic" exists.

If attention is confined to i.i.d. products, the order statistic is always sufficient. Mattner (1999, p. 399, Theorem 2.2) constructs a uniformly continuous and strictly increasing function $T : \mathbb{R} \to (0, 1)$ such that $\sum_{\nu=1}^{n} T(x_\nu)$ is equivalent to the order statistic on $\mathbb{R}^n$, for arbitrary probability measures on $\mathbb{B}^n$ with Lebesgue density λ^n. More precisely, there exists a subset $D \subset \mathbb{R}$ with $\lambda(D^c) = 0$ such that, for every $n \in \mathbb{N}$, the following is true: For $(x_1, \ldots, x_n)$, $(y_1, \ldots, y_n) \in D^n$, the relation $\sum_{\nu=1}^{n} T(x_\nu) = \sum_{\nu=1}^{n} T(y_\nu)$ implies that $(y_1, \ldots, y_n)$ is a permutation of $(x_1, \ldots, x_n)$. Hence, for $(x_1, \ldots, x_n) \in D^n$, one can regain the original sample $(x_1, \ldots, x_n)$ up to a permutation. This implies that $\sum_{\nu=1}^{n} T(x_\nu)$ is sufficient for the family of all i.i.d. products $P^n|\mathbb{B}^n$ with $P \ll \lambda$. But note that these functions T are not exactly what one would consider a decent statistic. In spite of being strictly increasing they have derivative 0 λ-a.e.

2.8 Sufficiency and Exponentiality

Multidimensional exponential families $\mathfrak{P}$, with densities

$$x \to C(P)h(x) \exp\left[\sum_{i=1}^{m} a_i(P)T_i(x)\right], \quad T_i : (X, \mathscr{A}) \to (\mathbb{R}, \mathbb{B}), \quad i = 1, \ldots, m,$$

are the most familiar example of families admitting a sufficient statistic for every sample size:

$$(x_1, \ldots, x_n) \to \left(\sum_{\nu=1}^{n} T_1(x_\nu), \ldots, \sum_{\nu=1}^{n} T_m(x_\nu)\right) \quad \text{is sufficient for } \{P^n : P \in \mathfrak{P}\}.$$

This, of course, is not the only example: If $P_\vartheta|\mathbb{B} \cap (0, \infty)$ has Lebesgue density $\vartheta^{-1} 1_{(0,\vartheta)}$, then $(x_1, \ldots, x_n) \to x_{n:n}$ is sufficient for $\{P_\vartheta^n : \vartheta > 0\}$.

In spite of such isolated examples, the idea soon came up that under certain regularity conditions (in particular if all members of the family $\mathfrak{P}$ have the same support), a sufficient statistic for every sample size exists only if the family is exponential. This idea occurs in vague form in Fisher (1934, Sect. 2.5, pp. 293/4). Readers who are not satisfied with the outline of a proof offered by Fisher may consult Hald (1998, p. 728).

Pitman (1936, p. 569) was among the first statisticians to attempt something approaching a detailed proof. To illustrate the level of mathematical sophistication at this time we follow Pitman's argument more closely.

Let $\mathfrak{P} = \{P_\vartheta : \vartheta \in \Theta\}$, $\Theta \subset \mathbb{R}$, be a family with Lebesgue densities $p(\cdot, \vartheta)$. Following Fisher's arguments related to the idea of "maximum likelihood", Pitman starts from the assumption that $\sum_{\nu=1}^{n} \partial_\vartheta \log p(x_\nu, \vartheta)$ is a function of the sufficient statistic $S_n : \mathbb{R}^n \to \mathbb{R}$, i.e., there exists a function $\psi_\vartheta : \mathbb{R} \to (0, \infty)$ such that

$$\sum_{\nu=1}^{n} \varphi_\vartheta(x_\nu) = \psi_\vartheta(S_n(x_1, \ldots, x_n)) \qquad \text{with} \quad \varphi_\vartheta(x) := \partial_\vartheta \log p(x, \vartheta). \tag{2.8.1}$$

Assuming (tacitly) that ψ_ϑ has for some ϑ_0 a differentiable inverse, say f_0, Pitman concludes that

$$S(x_1, \ldots, x_n) = f_0\left(\sum_{\nu=1}^{n} \varphi_{\vartheta_0}(x_\nu)\right),$$

which implies, with $\Psi_\vartheta := \psi_\vartheta \circ f_0$,

$$\sum_{\nu=1}^{n} \varphi_\vartheta(x_\nu) = \Psi_\vartheta\left(\sum_{\nu=1}^{n} \varphi_{\vartheta_0}(x_\nu)\right) \quad \text{for } \vartheta \in \Theta. \tag{2.8.2}$$

Differentiating this relation with respect to x_{ν_0} he obtains

$$\partial_{x_{\nu_0}} \varphi_\vartheta(x_{\nu_0}) = \Psi'_\vartheta\left(\sum_{\nu=1}^{n} \varphi_{\vartheta_0}(x_\nu)\right) \partial_{x_{\nu_0}} \varphi_{\vartheta_0}(x_{\nu_0}).$$

Since this relation holds for every $\nu_0 = 1, \ldots, n$, it follows that $\Psi'_\vartheta(y)$, considered as a function of y, is constant, i.e., $\Psi'_\vartheta(y) = a(\vartheta)$, hence $\Psi_\vartheta(y) = a(\vartheta)y + b(\vartheta)$. From (2.8.2), written for $n = 1$,

$$\varphi_\vartheta(x) = a(\vartheta)\varphi_{\vartheta_0}(x) + b(\vartheta).$$

Since $\varphi_\vartheta(x) = \partial_\vartheta \log p(x, \vartheta)$, this implies that q_ϑ is exponential.

Apart from the many regularity conditions assumed implicitly (such as the differentiability of ψ_ϑ and the existence of an inverse), Pitman's argument has two more serious defects: The Factorization Theorem

$$\prod_{\nu=1}^{n} p(x_\nu) = g_P(S_n(x_1, \ldots, x_n))h(x_1, \ldots, x_n) \tag{2.8.3}$$

implies

$$\sum_{\nu=1}^{n} \varphi_P(x_\nu) = \psi_P(S_n(x_1, \ldots, x_n)) \tag{2.8.4}$$

with $\varphi_P(x) = \log p(x)/p_0(x)$ and $\psi_P(y) = \log g_P(y)/g_{P_0}(y)$. Hence it is not necessary to assume that $\mathfrak{P}$ is a parametric family in order to obtain a relation like (2.8.1), and to assume differentiability with respect to ϑ in order to get rid of the factor h from (2.8.3). Unjustified, too, is the assumption that relation (2.8.3), or the resulting relation (2.8.4), holds for all (rather than μ^n-a.a.) $(x_1, \ldots, x_n) \in X^n$.

There are several papers that disregard this point and try to solve the functional equation (2.8.4) under minimal regularity conditions. It suffices to consider the case $n = 2$. Let $\varphi : X \to \mathbb{R}$ and $S : X^2 \to \mathbb{R}$ be functions such that

$$\varphi(x_1) + \varphi(x_2) = \psi(S(x_1, x_2)) \quad \text{for all } x_i \in X,\ i = 1, 2. \tag{2.8.5}$$

The problem is: Given S, what can be said about pairs of functions (φ, ψ) for which (2.8.5) holds true? It is clear that this can be solved under restrictive conditions on the functions S, φ and ψ only. Considering the origins of this problem, conditions of operational significance can be placed upon φ (corresponding to conditions on the densities), and conditions on the sufficient statistic S. Conditions on the function ψ can hardly be justified from an operational point of view.

The intended result is that the solution φ is unique up to a linear transformation. More precisely: If (φ_i, ψ_i), $i = 0, 1$, are two solutions, then

$$\varphi_1(x) = \alpha\varphi_0(x) + \beta \quad \text{and} \quad \psi_1(y) = \alpha\psi_0(y) + 2\beta.$$

The consequence: If the equation

$$\varphi_P(x_1) + \varphi_P(x_2) = \psi_P(S(x_1, x_2))$$

holds for $P \in \mathfrak{P}$ with a function S not depending on P, then

$$\varphi_P(x) = \alpha(P)\varphi_{P_0}(x) + \beta(P).$$

Since $\varphi_P(x) = \log p(x)/p_0(x)$, exponentiality follows.

Under the assumption that S is continuous, it was shown that the continuous solution φ is unique upon a linear transformation (Brown 1964, p. 1461, Theorem 4.1) for X an interval, and with a nicer proof for X a region in $\mathbb{R}^k$ by Barndorff-Nielsen and Pedersen (1969, p. 198, (i)). This result was further generalized to X an arcwise connected Hausdorff space (Denny 1970, p. 404, Corollary 3.2) under the assumption that the continuous φ is not constant on some open set, and to φ which are locally bounded rather than continuous. (See Pfanzagl 1971b, p. 202, Proposition, and Laube and Pfanzagl 1971, p. 241, Theorem, in connection with Pfanzagl 1970, p. 139, Corollary).

Still, these results hold under the artificial assumption that (2.8.5) holds everywhere. They are statistically relevant under conditions on φ and S that imply that a relation (2.8.5) holding μ^2-a.e. holds, in fact, everywhere. This is the case if S fulfills a condition stronger than continuity. Denny (1970, p. 405, Theorem 3.3) proves the following result.

Let X be an arcwise connected Hausdorff space, and let $\varphi : X \to \mathbb{R}$ and $S : X \to \mathbb{R}$ continuous functions. If

$$\varphi(x) = \psi(S(x)) \quad \text{for } \mu\text{-a.a. } x \in X,$$

then there exists a function ψ_* *such that*

$$\varphi(x) = \psi_*(S(x)) \quad \text{for all } x \in X,$$

provided S *preserves ample sets. (The latter condition is guaranteed if* S *is locally Lipschitz.)*

See Hipp (1974, p. 1291, Lemmas 3.3 and 3.4) It appears that Denny was unaware of a similar result by Barankin and Katz (1959, p. 224, Lemma 2.3) for k-dimensional sufficient statistics, which will be presented below as Lemma 2.8.1.

Summarizing what had been obtained by 1970, one could state that for families of probability measures on an arcwise connected Hausdorff space the existence of a one-dimensional sufficient statistic for some sample size $n > 1$ implies exponentiality if the sufficient statistic is locally Lipschitz and the derivatives are continuous. This is summarized in the somewhat minute Theorem 4.1 by Denny (1970, pp. 408/9).

These results should be seen in connection with the Theorem in Denny (1964, p. 95) which implies the existence of a uniformly continuous statistic $S_n : \mathbb{R}^n \to (0, 1)$ with increasing partial functions that is (trivially) sufficient for the family of all probability measures on $\mathbb{B}^n$ with λ^n-density, hence in particular sufficient for $\{P^n : P \in \mathfrak{P}\}$ for any family $\mathfrak{P}|\mathbb{B} \ll \lambda$. This demonstrates that more than the uniform continuity of the sufficient statistic is needed to infer exponentiality, and that the need for stronger conditions on the sufficient statistic cannot be compensated for by regularity conditions on the densities. Therefore an approach that imposes regularity conditions on the densities in order to obtain equality everywhere from equality μ^n-everywhere cannot lead to an optimal result.

Since there are exponential families without continuous densities (see Hipp 1974, p. 1284, Example 1.2) these results are not wholly satisfactory. An important improvement was achieved by Hipp (1972, p. 36, Corollary 3.12) who shows that a family of probability measures on a region of $\mathbb{R}^k$, with positive λ^k-densities, is exponential if it admits for some sample size $n > 1$ a sufficient statistic with continuous partial derivatives. For $k = 1$ this condition on the sufficient statistic can be weakened to "locally Lipschitz" (Hipp 1974, p. 1283, Theorem 1.1). An important point of the proof is that being locally Lipschitz for S already implies a property of φ which comes close to "continuity a.e." (see p. 1290, Proposition 2.5).

In this connection, we should mention a paper by Brown (1964). Its main result, Theorem 2.1, p. 1458, asserts exponentiality under some sort of continuity condition on the sufficient statistic without a condition on the densities. Still, the proof uses an incorrect lemma (see Pfanzagl 1971), and a valid proof has never been supplied. In spite of this, Brown's paper is mentioned in numerous papers and textbooks. In "Kendall's Advanced Theory of Statistics" (see Stuart and Ord 1991, vol. 2, p. 636) it is praised for its "rigorous treatment".

So far our considerations have been restricted to sufficient statistics attaining their values in $\mathbb{R}$, and for one simple reason: The results are easy to review. In fact, sufficient statistics attaining their values in a Euclidean space of higher dimension have been a subject of interest from the beginning, starting with the profound paper by Koopman. Koopman (1936, p. 400) explicitly criticized the fact that Fisher never gave a precise definition of "sufficiency". Isolated from the special framework, Koopman's definition reads as follows:

$S : (X, \mathscr{A}) \to (Y, \mathscr{B})$ is *sufficient* if, for all $x', x'' \in X$, the relation $S(x') = S(x'')$ implies that the function $P \to p(x')/p(x'')$ is constant for $P \in \mathfrak{P}$.

In this definition, p is a density of P, and it is clear that this definition makes sense only if there are distinguished versions of the densities, say continuous ones. Notice that his definition of sufficiency implicitly presumes that a relation like (2.8.5) holds everywhere.

In Koopman's paper, $\mathfrak{P}$ is a multiparametric family of i.i.d. products, and S_n maps from $(\mathbb{R}^n, \mathbb{B}^n)$ to $(\mathbb{R}^k, \mathbb{B}^k)$. Under the assumptions that $n > k$ and that S_n is continuous and $(x, \vartheta) \to p(x, \vartheta)$ is analytic, he shows (p. 402, Theorem I) that $p(\cdot, \vartheta)$ is exponential of order less than or equal to k. In addition to the strong regularity condition on $p(x, \vartheta)$, his definition of sufficiency does not allow what corresponds to the occurrence of exceptional sets of measure zero in the Factorization Theorem. Moreover, he presumes that the number of components of the sufficient statistic is the same as the number of parameters. Yet, his paper does contain the essential idea, taken up in the papers by Barankin (and coauthors) more than twenty years later: that the minimal dimension of the exponential family is determined by the ranks of the matrices (2.8.7).

Not aware of Koopman's paper, Dynkin (1951) arrives at a comparable conclusion by pursuing a different approach. Substituting a general family $\mathfrak{P}$ for Dynkin's parametric family on an open subset of $\mathbb{R}$, his procedure can be described as follows: Let

$$\ell(x, P) := \log p(x)/p_0(x),$$

where P_0 is arbitrarily fixed. Again, one has to presume a distinguished version of q_P. Familiar with the Factorization Theorem, Dynkin certainly has in the back of his mind the fact that $\ell(x, P)$ depends on x through the sufficient statistic only. His result:

If the linear space generated by 1 and the functions $\ell(\cdot, P)$, $P \in \mathfrak{P}$, is of finite dimension $k + 1$, say, then there exist functions $T_i : X \to \mathbb{R}$ such that, for every $P \in \mathfrak{P}$,

$$\ell(x, P) = a_0(P) + \sum_{i=1}^{k} a_i(P) T_i(x) \quad \text{for} \quad x \in X. \tag{2.8.6}$$

Example For $P = N(\mu, \sigma^2)$ and $P_0 = N(0, 1)$, the linear space generated by the functions

$$\ell(x, (\mu, \sigma^2)) = -\log \sigma - \frac{\mu^2}{2}\sigma^{-2} + \mu\sigma^{-2}x + \frac{1}{2}(1 - \sigma^{-2})x^2, \quad x \in \mathbb{R},$$

is of dimension 3, with $T_1(x) = x$ and $T_2(x) = x^2$. □

As a consequence of (2.8.6), the family $\mathfrak{P}$ is exponential with sufficient statistic $S(x) = (T_1(x), \dots, T_k(x))$, and the map $(x_1, \dots, x_n) \to (\sum_{\nu=1}^n T_1(x_\nu), \dots, \sum_{\nu=1}^n T_k(x_\nu))$ is sufficient for $\{P^n : P \in \mathfrak{P}\}$.

In Dynkin's paper it remains unclear where the dimension $(k+1)$ really comes from and why the functions T_i can be chosen as $\ell(\cdot, P_i)$ (see p. 24: "Actually we have..."). It is just the case $k \geq n$ where the local properties of the functions $\ell(\cdot, P_i)$ are used to show that the sufficient statistic for the sample size n is locally bijective if the functions $\ell(\cdot, P_i)$ have continuous derivatives (Dynkin 1951, p. 24, Theorem 2). Perhaps the reader will get on with what Schmetterer (1966, Satz 7.4, pp. 257/8 and 1974, pp. 215/6) has to say about Dynkin's Theorem 2. A variant of Dynkin's Theorem 2 is Theorem A in Brown (1964, p. 1461).

An approach that derives the minimal dimension of "regular" sufficient statistics from local properties of the densities q_P is more enlightening. It was first explored by Barankin and Katz (1959), and continued by Barankin (1961), and later Barankin and Maitra (1963). The paper from Barankin and Katz (1959), original in its approach compared with Dynkin, is of poor technical quality and Barankin's paper (1961) is mainly a correction note. Hence readers interested in this approach should start with Barankin and Maitra (1963). Yet, even this paper is outdated. Written for parametric families it uses derivatives of the densities with respect to the parameters. According to Shimizu (1966), this is avoidable.

The essence of these papers becomes more transparent if (i) we restrict the considerations to i.i.d. products, and (ii) generalize the framework from parametric to general families. The basic assumption is that the densities q_P have continuous derivatives.

Let $\mathbf{x}_n = (x_1, \dots, x_n) \in X^n$, where $X \subset \mathbb{R}$ is an interval. For arbitrary $m \in \mathbb{N}$ and arbitrary $P_i \in \mathfrak{P}$, $i = 1, \dots, m$, let $M(\mathbf{x}_n; P_1, \dots, P_m)$ denote the rank of the matrix

$$(\ell'(x_\nu, P_i))_{i=1,\dots,m;\nu=1,\dots,n} \tag{2.8.7}$$

and let

$$\rho(\mathbf{x}_n) := \max\{M(\mathbf{x}_n; P_1, \dots, P_m) : m \in \mathbb{N}, P_i \in \mathfrak{P}, i = 1, \dots, m\}.$$

To simplify the presentation, we assume that $\rho(\mathbf{x}_n)$ (which will turn out to be the minimal dimension of "regular" sufficient statistics) is the same for every $\mathbf{x}_n \in X^n$, say r. By definition, $r \leq n$. For arbitrary $P_i \in \mathfrak{P}$, $i = 1, \dots, r$, let

$$U(P_1, \dots, P_r) := \{\mathbf{x}_n \in X^n : M(\mathbf{x}_n); P_1, \dots, P_r) = r\}.$$

Since $\ell'(\cdot, P)$ is continuous, the set $U(P_1, \dots, P_r)$ is open. The main results of this approach are

(i) *The map* $\mathbf{x}_n \to (\sum_{\nu=1}^n \ell(x_\nu, P_1), \dots, \sum_{\nu=1}^n \ell(x_\nu, P_r))$ *is minimal sufficient for* $\{P^n : P \in \mathfrak{P}\}$ *in some neighbourhood of every* $\mathbf{x}_n \in U(P_1, \dots, P_n)$. *(Sufficiency on a subset of* X^n *means in this connection that the factorization holds for all* $\mathbf{x}_n$ *in this subset.)*

(ii) *If* $\mathbf{x}_n \to (S_1(\mathbf{x}_n), \ldots, S_k(\mathbf{x}_n))$ *is sufficient for* $\{P^n : P \in \mathfrak{P}\}$, *then* $k \geq r$, *provided every* S_i *has continuous partial derivatives: The rank* r *is a lower bound for the "dimension" of any sufficient statistic with continuous partial derivatives.*

Remark When some statisticians speak of the "dimension" of a sufficient statistic, they simply mean that $S(x)$ can be written as $(S_1(x), \ldots, S_k(x)) \in \mathbb{R}^k$. Since there exists a continuous map from $\mathbb{R}^k$ to $\mathbb{R}$, say T, which is bijective λ^k-a.e. (see, Denny 1964, p. 95, Theorem), it is clear that the continuity of the functions $S_i : X \to \mathbb{R}$ is not enough to define the "dimension" of S: If S is a continuous k-dimensional statistic, $T \circ S$ is a continuous one-dimensional sufficient statistic, provided $P \circ S \ll \lambda^k$ for $P \in \mathfrak{P}$. A meaningful concept for the dimension can, therefore, be defined only for sufficient statistics subject to a condition stronger than continuity.

What has been stated under (i) and (ii) is essentially Lemma 3, pp. 50/1 in Shimizu (1966). A forerunner of this result is Theorem 3.2 in Barankin and Katz (1959, p. 228), repeated as Theorem 2.1 in Barankin and Maitra (1963, p. 222). All of these results are based on the assumption that the Factorization Theorem holds everywhere. This is justified by a Lemma of Barankin and Katz (see Lemma 2.8.1 below) which seems to have been overlooked by Shimizu.

The arrangement of Shimizu's proof is not very lucid. Perhaps one could argue as follows. For $\bar{\mathbf{x}}_n \in U(P_1, \ldots, P_r)$, the rank of the matrix (2.8.7) is r.

To simplify our notations, we assume that $(\ell'(\overline{x}_\nu, P_i))_{i=1,\ldots,\nu=1,\ldots,n}$ is nonsingular. By definition of r, the following matrix is singular for every $x \in X$ and every $P \in \mathfrak{P}$:

$$\begin{pmatrix} \ell'(\overline{x}_1, P_1) & \ldots & \ell'(\overline{x}_1, P_r) & \ell'(\overline{x}_1, P) \\ \vdots & & & \vdots \\ \ell'(\overline{x}_r, P_1) & \ldots & \ell'(\overline{x}_r, P_r) & \ell'(\overline{x}_r, P) \\ \ell'(x, P_1) & \ldots & \ell'(x, P_r) & \ell'(x, P) \end{pmatrix}.$$

In particular, there are $a_1(P), \ldots, a_r(P)$ such that

$$\ell'(x, P) = \sum_{i=1}^{r} a_i(P)\ell'(x, P_i)$$

and therefore

$$\ell(x, P) = a_0(P) + \sum_{i=1}^{r} a_i(P)\ell(x, P_i).$$

Hence, for every $P \in \mathfrak{P}$,

$$\sum_{\nu=1}^{n} \ell(x_\nu, P) = \psi_P\left(\sum_{\nu=1}^{n} \ell(x_\nu, P_1), \ldots, \sum_{\nu=1}^{n} \ell(x_\nu, P_r)\right)$$

with

$$\psi_P(y_1, \ldots, y_r) = a_0(P) + \sum_{i=1}^{r} a_i(P) y_i,$$

and this proves the "local" sufficiency claimed under (i).

To show that a sufficient statistic with the minimal dimension r does exist, we still need to piece the locally sufficient statistics $(\ell(\cdot, P_1), \ldots, \ell(\cdot, P_r))$ together. This is done in Theorem 1, p. 52. While Shimizu's definition of S is more elegant than the corresponding definition in the papers by Barankin and Katz (1959, p. 227) or Barankin and Maitra (1963, p. 222), in the construction of the globally sufficient statistic he closely follows the cumbersome procedure of Barankin and Katz in the proof of their Theorem 4.1, p. 235. The basic idea of this procedure is to cover X by using a countable family of bounded, pairwise disjoint sets V_ℓ, $\ell \in \mathbb{N}$, such that each V_ℓ is contained in some $U(P_1, \ldots, P_r)$. This implies that for each ℓ there exists a statistic S_ℓ, $S_\ell = (\ell(\cdot, P_1), \ldots, \ell(\cdot, P_r))$ which is locally sufficient on V_ℓ. The statistics S_ℓ are shifted by an amount c_ℓ such that the sets $\{S_\ell(x) + c_\ell : x \in V_\ell\}$, $\ell \in \mathbb{N}$, are pairwise disjoint. Then we can define the globally sufficient statistic S by $S(x) = S_\ell(x) + c_\ell$ if $x \in V_\ell$. In this way we arrive at the statement that $S(x') = S(x'')$ implies that x', x'' are in the same V_ℓ, so that $q_P(x')/q_P(x'')$ is independent of P.

An additional result (or, perhaps, the main result) of this approach is the following: If there exists for some sample size n a sufficient statistic with continuous partial derivatives and dimension $k < n$, then the family is exponential of dimension $\leq k$. The fact that this result presumes densities with a continuous derivative is a substantial drawback compared with the case of a real-valued sufficient statistic.

In this connection we should also mention a result put forward by (1969, p. 198, (ii)), who were not aware of Shimizu's paper. They prove for $X = \mathbb{R}$ (with hints for a generalization to $X = \mathbb{R}^m$) the exponentiality for densities with continuous derivatives. Their condition on the sufficient statistic is "continuity" only, but they assume that the factorization holds everywhere, a condition which these authors (in view of Denny's 1964 result) consider "indispensable" (p. 198). With the Lemma of Barankin and Katz (1959) this assumption is fulfilled if the sufficient statistic has continuous derivatives, but as far as exponentiality is concerned, this does not take us any farther than does Shimizu (1966).

Barankin and Katz (1959, Lemma 2.3, p. 224; repeated in, Barankin 1960, Lemma 2.2, p. 97 and, Barankin and Maitra 1963, Lemma 2.2, p. 221) show that a relation like (2.8.4) holds everywhere if it holds λ^n-a.e., provided the components $S_i(x_1, \ldots, x_n)$ admit continuous partial derivatives, and the densities are continuous.

Since Barankin and Katz need all in all more than five papers for the proof of their fundamental Lemma 2.3, we offer an independent proof. Isolated from the special context, this lemma reads as follows.

Lemma 2.8.1 *Let $X \subset \mathbb{R}^n$ be an open subset. Assume that the relation*

$$\varphi(x) = \psi(S_1(x), \ldots, S_k(x)) \tag{2.8.8}$$

holds for λ^n*-a.a.* $x \in X$*. Assume that every* S_i*,* $i = 1, \ldots, k$ *has continuous partial derivatives, and that the matrix*

$$\left(\partial S_i / \partial x_j\right)_{i=1,\ldots,j=1,\ldots,n}$$

has rank $k \le n$ *for every* $x \in X$*. Then relation (2.8.8) holds for every* $x \in X$ *if* φ *is continuous on* X*.*

Addendum. *If* φ *has continuous derivatives, then* ψ *has continuous derivatives, too.* The Addendum follows from the Implicit Function Theorem as in Barankin and Katz (1959, p. 226).

Proof Let $x_0, y_0 \in X$ be such that $S(x_0) = S(y_0)$ ($= s_0$, say). We have to show that $\varphi(x_0) = \varphi(y_0)$. Let $N \subset \mathbb{B}^n$ denote the exceptional λ^n-nullset for relation (2.8.8). We shall show that for any open $U \ni x_0$ and $V \ni y_0$ there exist $x \in U \cap N^c$ and $y \in V \cap N^c$ such that $S(x) = S(y)$, whence $\varphi(x) = \varphi(y)$. Since U and V are arbitrary and φ continuous, this implies $\varphi(x_0) = \varphi(y_0)$.

Using the Implicit Function Theorem, we may assume w.l.g. that $S(U)$ and $S(V)$ are open in $\mathbb{R}^k$. Since $S(U) \cap S(V) \neq \emptyset$, we have $\lambda^k(S(U) \cap S(V)) > 0$. Since S_i has continuous derivatives, S fulfills Lusin's condition. Hence $\lambda^n(N) = 0$ implies $\lambda^k(S(N)) = 0$ and therefore

$$\lambda^k(S(U \cap N^c) \cap S(V \cap N^c)) = \lambda^k(S(U) \cap S(V)) > 0.$$

Hence there exists $s \in S(U \cap N^c) \cap S(V \cap N^c)$, $s \neq s_0$, and therefore $x \in U \cap N^c$ and $y \in V \cap N^c$, such that $S(x) = s = S(y)$. □

2.9 Characterizations of Sufficiency

If $S : (X, \mathscr{A}) \to (Y, \mathscr{B})$ is sufficient, then for any statistical procedure based on the observation x, there exists a statistical procedure depending on x through $S(x)$ only which is "at least as good". This holds true for many kinds of statistical procedures (decisions, tests, estimators,...), and any of these can be used to characterize a statistic S with these properties as "sufficient". The characterization by measures of information is of a different nature, since it lacks operational significance.

Decision-Theoretic Characterization

Let T be a decision function mapping $(X, \mathscr{A})$ into a Polish space $(Z, \mathscr{C})$. Given a loss function $\ell(\cdot, P)$, the performance of T may be evaluated by the risk function

$$R(T, P) := \int \ell(z, P) P \circ T(dz).$$

If $S : (X, \mathscr{A}) \to (Y, \mathscr{B})$ is sufficient, there exists a Markov kernel $M|Y \times \mathscr{C}$ such that

$$\int M(y, C) P \circ S(dy) = P(T^{-1}C) \quad \text{for } C \in \mathscr{C}.$$

This implies in particular that the statistical procedure based on M has, for any loss function $\ell(\cdot, P)$, the same risk as the original procedure based on T, i.e.,

$$\int \ell(z, P) M(y, dz) P \circ S(dy) = \int \ell(z, P) P \circ T(dz) \quad \text{for } P \in \mathfrak{P}.$$

Bahadur (1955a) has the following converse in mind.

Let $S : (X, \mathscr{A}) \to (Y, \mathscr{B})$ be some measurable statistic. If for every $T : (X, \mathscr{A}) \to (Z, \mathscr{C})$ there exists a Markov kernel $M|Y \times \mathscr{C}$ such that

$$\int \ell(z, P) M(y, dz) P \circ S(dy) \leq \int \ell(z, P) P \circ T(dz) \quad \text{for } P \in \mathfrak{P}, \tag{2.9.1}$$

then S is sufficient.

Bahadur (1955a) proves this result (p. 288, Theorem) for dominated families $\mathfrak{P}$ containing more than one probability measure, and for $(Z, \mathscr{C}) = (\mathbb{R}^k, \mathbb{B}^k)$ (including the case $k = \infty$). The essential point for this converse is for which loss functions relation (2.9.1) is required. Bahadur gets along with a single loss function fulfilling a not very appealing condition (see p. 287), namely: For arbitrary $P_i \in \mathfrak{P}$, $i = 0, 1$, $\inf_z\{(1 - u)\ell(z, P_0) + u\ell(z, P_1)\}$ is attained for a unique value $z = z(u)$, and $z(u') = z(u'')$ implies $u' = u''$ (a condition not fulfilled for $\ell(z, P_\vartheta) = (z - \vartheta)^2$).

At first glance, results of this kind would seem to satisfy all needs, at least if the conditions on the loss function could be somehow improved. Yet, not all statistical problems fit easily into this framework. This holds, in particular, if equivariance or unbiasedness of estimators has to be taken into account.

This is trivial for unbiasedness: Let $\hat{\kappa} : (X, \mathscr{A}) \to (\mathbb{R}, \mathbb{B})$ be unbiased and of minimal convex risk for $\kappa(P) := \int \hat{\kappa}(x) P(dx)$. That means: For any unbiased estimator there exists an equivalent or better unbiased estimator depending on x through $\hat{\kappa}(x)$ only, namely $\hat{\kappa}$ itself. Yet, $\hat{\kappa}$ is not necessarily sufficient: This can happen if there exists a sufficient statistic $S : (X, \mathscr{A}) \to (Y, \mathscr{B})$ with $\mathfrak{P} \circ S|\mathscr{B}$ is complete, and one chooses a functional contraction of S for $\hat{\kappa}$. (See also Chap. 4.)

In connection with equivariant estimators, we mention the following result of Fieger (1978, p. 39, Satz 1). For $\vartheta \in \mathbb{R}$ let $P_\vartheta|\mathbb{B}$ be the probability measure with λ-density $x \to p(x - \vartheta)$. A shift equivariant function $S_n : X^n \to \mathbb{R}^k$ is sufficient iff for every convex loss function ℓ there exists a shift equivariant function $k_\ell : \mathbb{R}^k \to \mathbb{R}$ such that

$$\int \ell(k_\ell(S_n(x)) - \vartheta) P_\vartheta^n(dx) \leq \int \ell(\vartheta^{(n)}(x) - \vartheta) P_\vartheta^n(dx) \tag{2.9.2}$$

for every shift equivariant function $\vartheta^{(n)} : X^n \to \mathbb{R}$. (No randomization!)

Note that this theorem depends on the fact that (2.9.2) is required for a large class of loss functions. If $p(x) = 1_{[-1/2,1/2]}(x)$, the statistic $S_n(\mathbf{x}_n) = (x_{1:n}, x_{n:n})$ is sufficient for the pertaining location parameter family. The function $k(y_1, y_2) = (y_1 + y_2)/2$ fulfills (2.9.2) for every loss function ℓ which is convex and symmetric about 0; yet $\frac{1}{2}(x_{1:n} + x_{n:n})$ is not sufficient (Fieger 1978, p. 39).

Characterization by the Power of Tests

If $S : (X, \mathscr{A}) \to (Y, \mathscr{B})$ is sufficient, then for any critical function $\varphi : (X, \mathscr{A}) \to ([0, 1], \mathbb{B} \cap [0, 1])$ there exists a critical function $\psi : (Y, \mathscr{B}) \to ([0, 1], \mathbb{B} \cap [0, 1])$, i.e., the conditional expectation of φ, given S, such that

$$\int \psi(y) P \circ S(dy) = \int \varphi(x) P(dx) \text{ for every } P \in \mathfrak{P}.$$

The converse can be obtained as follows.

Let $S : (X, \mathscr{A}) \to (Y, \mathscr{B})$ be an arbitrary statistic. Assume that for every $A \in \mathscr{A}$ and any pair P_i, $i = 1, 2$ in a dominated family $\mathfrak{P}$ there exists $\psi_A : (Y, \mathscr{B}) \to ([0, 1], \mathbb{B} \cap [0, 1])$ such that

$$\int \psi_A(y) P_1 \circ S(dy) \le P_1(A),$$

$$\int \psi_A(y) P_2 \circ S(dy) \ge P_2(A).$$

Then S is sufficient for $\mathfrak{P}$. (See Pfanzagl 1974, Theorem and p. 198. See also Pfanzagl 1994, p. 10, Theorem 1.3.9.)

Characterization by Concentration of Mean Unbiased Estimators

If $S : (X, \mathscr{A}) \to (Y, \mathscr{B})$ is sufficient, then for any unbiased estimator $\hat{\kappa}: (X, \mathscr{A}) \to (\mathbb{R}, \mathbb{B})$ there exists a function $k : (Y, \mathscr{B}) \to (\mathbb{R}, \mathbb{B})$ (which is the conditional expectation of $\hat{\kappa}$, given S) such that

$$\int k(y) P \circ S(dy) = \int \hat{\kappa}(x) P(dx), \tag{2.9.3}$$

$$\int C(k(y)) P \circ S(dy) \le \int C(\hat{\kappa}(x)) P(dx), \tag{2.9.4}$$

for every $P \in \mathfrak{P}$ and every convex function $C \ge 0$.

Assume now that S is an arbitrary statistic such that (2.9.3) and (2.9.4) hold for every bounded unbiased estimator $\hat{\kappa}$. Since (2.9.4) holds for a large class of functions C, one can easily infer from (2.9.3) and (2.9.4) that k attains only values in $[0, 1]$ if $\hat{\kappa} = 1_A$. Hence the sufficiency of S follows from the result on critical functions presented above. (Hint: Use the loss function $u \to |u| 1_{(-\infty,0)}(u)$ to conclude that $k \ge 0$, and the loss function $u \to u 1_{(0,1]}(u) + u^2 1_{(1,\infty)}(u)$ to conclude that $k \le 1$). Note that the function k occurring in relations (2.9.3) and (2.9.4) depends on $\hat{\kappa}$. If condition (2.9.4) is strengthened in the sense that for every bounded $\hat{\kappa}$ there exists k such that (2.9.4) holds for *every* $\hat{\hat{\kappa}}$ fulfilling (2.9.3), then (2.9.4) with $C(u) = u^2$ is enough to infer the sufficiency of S. Observe, however, that this condition is somewhat artificial: It requires the existence of an optimal unbiased estimator whenever an unbiased estimator exists at all. (We mention this just as a variant of a theorem of Bahadur discussed in Sect. 4.2.) Observe that the extended condition mentioned above does not imply that $\mathfrak{P} \circ S$ is complete. Completeness of $\mathfrak{P} \circ S$ follows if $k \circ S \in K$ for *every* $k : Y \to \mathbb{R}$.

Characterization by Measures of Information

Let $P_i, i = 1, 2$ be probability measures with μ-densities p_i. For a convex function $C : [0, \infty) \to \mathbb{R}$, the C-divergence between P_1 and P_2, introduced by Csiszár (1963), p. 86, relation (4), is

$$I_C(P_1, P_2) := \int C(p_1(x)/p_2(x))P_2(dx).$$

Csiszár (1963, p. 90, Satz 1 and Ergänzung) implies the following assertions. For any statistic $S : (X, \mathscr{A}) \to (Y, \mathscr{B})$,

$$I_C(P_1 * S, P_2 * S) \leq I_C(P_1, P_2). \tag{2.9.5}$$

If S is sufficient for $\{P_1, P_2\}$, equality holds in (2.9.5) for every convex function C. Conversely, equality in (2.9.5) for some strictly convex C, implies that S is sufficient for $\{P_1, P_2\}$, provided $I_C(P_1, P_2) < \infty$.

This generalizes an earlier result of Kullback and Leibler (1951) for the special case $C(u) = u \log u$.

References

Ash, R. B. (1972). *Real Analysis and Probability*. New York: Academic Press.

Bahadur, R. R. (1954). Sufficiency and statistical decision functions. *The Annals of Mathematical Statistics*, *25*, 423–462.

Bahadur, R. R. (1955a). A characterization of sufficiency. *The Annals of Mathematical Statistics*, *26*, 289–293.

Bahadur, R. R. (1955b). Statistics and subfields. *The Annals of Mathematical Statistics*, *26*, 490–497.

Bahadur, R. R. (1957). On unbiased estimates of uniformly minimum variance. *Sankhyā*, *18*, 211–224.

Bahadur, R. R., & Lehmann, E. L. (1955). Two comments on "sufficiency and statistical decision functions". *The Annals of Mathematical Statistics*, *26*, 139–142.

Barankin, E. W. (1960). Sufficient parameters: solution of the minimal dimensionality problem. *Annals of the Institute of Statistical Mathematics*, *12*, 91–118.

Barankin, E. W. (1961). A note on functional minimality of sufficient statistics. *Sankhyā Series A*, *23*, 401–404.

Barankin, E. W., & Katz, M, Jr. (1959). *Sufficient statistics of minimal dimension. Sankhyā*, *21*, 217–246.

Barankin, E. W., & Maitra, A. P. (1963). Generalization of the Fisher-Darmois-Koopman-Pitman theorem on suficient statistics. *Sankhyā Series A*, *25*, 217–244.
Bar-Lev, S. K., & Plachky, D. (1989). Boundedly complete families which are not complete. *Metrika*, *36*, 331–336.
Barndorff-Nielsen, O., & Pedersen, K. (1969). Sufficient data reduction and exponential families. *Mathematica Scandinavica*, *22*, 197–202.
Bartlett, M. S. (1937). Properties of sufficiency and statistical tests. *Proceedings of the Royal Society of London. Series A*, *160*, 268–282.
Blackwell, D. (1951). Comparison of experiments. In *Proceedings of the Second Berkeley Symposium on Mathematical Statistics and Probability* (pp. 93–102). University of California Press.
Blackwell, D. (1953). Equivalent comparisons of experiments. *The Annals of Mathematical Statistics*, *24*, 265–272.
Blackwell, D., & Dubins, L. E. (1975). On existence and non-existence of proper, regular, conditional distributions. *Annals of Probability*, *3*, 741–752.
Blackwell, D., & Ryll-Nardzewski, C. (1963). Non-existence of everywhere proper conditional distributions. *The Annals of Mathematical Statistics*, *34*, 223–225.
Bomze, I. M. (1990). *A functional analytic approach to statistical experiments* (Vol. 237). Pitman research notes in mathematics series. Harlow: Longman.
Brown, L. D. (1964). Sufficient statistics in the case of independent random variables. *The Annals of Mathematical Statistics*, *35*, 1456–1474.
Burkholder, D. L. (1961). Sufficiency in the undominated case. *The Annals of Mathematical Statistics*, *12*, 1191–1200.
Csiszár, I. (1963). Eine informationstheoretische Ungleichung und ihre Anwendung auf den Beweis der Ergodizität von Markoff'schen Ketten. Magyar Tud. *Akad. Mat. Kutató Int. Közl.*, *8*, 85–108.
Denny, J. L. (1964). A continuous real-valued function on E^n almost everywhere 1–1. *Fundamenta Mathematicae*, *55*, 95–99.
Denny, J. L. (1970). Cauchy's equation and sufficient statistics on arcwise connected spaces. *The Annals of Mathematical Statistics*, *41*, 401–411.
Dieudonné, J. (1948). Sur le théorème de Lebesgue-Nikodym (III). *Annales de l'Institut Fourier (Grenoble)*, *23*, 25–53.
Doob, J. L. (1938). Stochastic processes with an integral-valued parameter. *Transactions of the American Mathematical Society*, *44*, 87–150.
Doob, J. L. (1953). *Stochastic processes*. New York: Wiley.
Dudley, R. M. (1989). *Real analysis and probability*. Pacific Grove: Wadsworth & Brooks/Cole.
Dudley, R. M. (1999). *Uniform central limit theorems* (Vol. 63). Cambridge studies in advanced mathematics. Cambridge: Cambridge University Press.
Dynkin, E. B. (1951). Necessary and sufficient statistics for a family of probability distributions. *Uspeh. Matem. Nauk*, *6* (Russian); *Selected Translations in Mathematical Statistics and Probability*, *1*, 23–41 (1961).
Fieger, W. (1978). Suffizienz bei verschiebungsinvarianten Schätzproblemen. *Metrika*, *25*, 37–48.
Fisher, R. A. (1920). A mathematical examination of the methods of determining the accuracy of an observation by the mean error, and by the mean square error. *Monthly Notices of the Royal Astronomical Society*, *80*, 758–770.
Fisher, R. A. (1922). On the mathematical foundation of theoretical statistics. Philosophical Transactions of the Royal Society of London. *Series A*, *222*, 309–368.
Fisher, R. A. (1934). Two new properties of mathematical likelihood. *Proceedings of the Royal Society of London. Series A*, *144*, 285–307.
Fraser, D. A. S. (1954). Completeness of order statistics. *Canadian Journal of Mathematics*, *6*, 42–45.
Hald, A. (1998). *A history of mathematical statistics from 1750 to 1930*. New York: Wiley.
Halmos, P. R. (1941). The decomposition of measures. *Duke Mathematical Journal*, *8*, 386–392.
Halmos, P. R. (1946). The theory of unbiased estimation. *The Annals of Mathematical Statistics*, *17*, 34–43.
Halmos, P. R. (1950). *Measure Theory*. New York: Van Nostrand.

Halmos, P. R., & Savage, L. J. (1949). Application of the Radon-Nikodym theorem to the theory of sufficient statistics. *The Annals of Mathematical Statistics*, *20*, 225–241.
Hasegawa, M., & Perlman, M. D. (1974). On the existence of a minimal sufficient subfield. *Annals of Statistics*, *2*, 1049–1055; *Correction Note*, *3*, 1371/2 (1975).
Heyer, H. (1969). Erschöpftheit und Invarianz beim Vergleich von Experimenten. *Z. Wahrscheinlichkeitstheorie verw. Gebiete*, *12*, 21–55.
Heyer, H. (1972). *Zum Erschöpftheitsbegriff von D. Blackwell. Metrika*, *19*, 54–67.
Heyer, H. (1982). *Theory of statistical experiments. Springer series in statistics*. Berlin: Springer.
Hipp, Ch. (1972). *Characterization of exponential families by existence of sufficient statistics*. Thesis: University of Cologne.
Hipp, Ch. (1974). Sufficient statistics and exponential families. *Annals of Statistics*, *2*, 1283–1292.
Hoeffding, W. (1977). Some incomplete and boundedly complete families of distributions. *Annals of Statistics*, *5*, 278–291.
Hoel, P. G. (1951). Conditional expectation and the efficiency of estimates. *The Annals of Mathematical Statistics*, *22*, 299–301.
Isenbeck, M., & Rüschendorf, L. (1992). Completeness in location families. *Probability and Mathematical Statistics*, *13*, 321–343.
Kolmogorov, A. N. (1933). *Grundbegriffe der Wahrscheinlichkeitstheorie*. Berlin: Springer.
Koopman, B. O. (1936). On distributions admitting a sufficient statistic. *Transactions of the American Mathematical Society*, *39*, 399–409.
Kullback, S., & Leibler, R. A. (1951). On information and sufficiency. *The Annals of Mathematical Statistics*, *22*, 79–86.
Kumar, A., & Pathak, P. K. (1977). Two applications of Basu's lemma. *Scandinavian Journal of Statistics*, *4*, 37–38.
Landers, D., & Rogge, L. (1972). Minimal sufficient σ-fields and minimal sufficient statistics. Two counterexamples. *The Annals of Mathematical Statistics*, *43*, 2045–2049.
Landers, D., & Rogge, L. (1976). A note on completeness. *Scandinavian Journal of Statistics*, *3*, 139.
Laplace, P. S. (1818). Deuxiéme supplément à la theórie analytique des probabilités. Paris: Courcier.
Laube, G., & Pfanzagl, J. (1971). A remark on the functional equation $\sum_{i=1}^{n} \varphi(x_i) = \psi(T(x_1, \ldots, x_n))$. *Aequationes Mathematicae*, *6*, 241–242.
Le Cam, L. (1986). *Asymptotic methods in statistical decision theory*. New York: Springer.
Lehmann, E. L. (1947). On families of admissible tests. *The Annals of Mathematical Statistics*, *18*, 97–104.
Lehmann, E. L. (1955). Ordered families of distributions. *The Annals of Mathematical Statistics*, *26*, 399–419.
Lehmann, E. L. (1959). *Testing statistical hypotheses*. New York: Wiley.
Lehmann, E. L., & Casella, G. (1998). *Theory of point estimation* (2nd ed.). Berlin: Springer.
Lehmann, E. L., & Scheffé, H. (1950, 1955, 1956). Completeness, similar regions and unbiased estimation. Sankhyā, **10**, 305–340; *15*, 219–236; *Correction*, *17*, 250.
Mandelbaum, A., & Rüschendorf, L. (1987). Complete and symmetrically complete families of distributions. *Annals of Statistics*, *15*, 1229–1244.
Mattner, L. (1996). Complete order statistics in parametric models. *Annals of Statistics*, *24*, 1265–1282.
Mattner, L. (1999). Sufficiency, exponential families, and algebraically independent numbers. *Mathematical Methods of Statistics*, *8*, 397–406.
Mattner, L. (2000). Minimal sufficient statistics in location-scale parameter models. *Bernoulli*, *6*, 1121–1134.
Mattner, L. (2001). Minimal sufficient order statistics in convex models. *Proceedings of the American Mathematical Society*, *129*, 3401–3411.
Messig, M. A., & Strawderman, W. E. (1993). Minimal sufficiency and completeness for dichotomous quantal response models. *Annals of Statistics*, *21*, 2149–2157.
Neyman, J. (1935). Sur un teorema concernente le cosiddette statistiche sufficienti. *Giorn. Ist. Ital. Attuari*, *6*, 320–334.

Neyman, J. (1937). Outline of a theory of statistical estimation based on the classical theory of probability. Philosophical Transactions of the Royal Society of London. *Series A*, *236*, 333–380.
Pfanzagl, J. (1964). On the topological structure of some ordered families of distributions. *The Annals of Mathematical Statistics*, *35*, 1216–1228.
Pfanzagl, J. (1970). On a functional equation related to families of exponential probability measures. *Aequationes Mathematicae*, *4*, 139–142; *Correction Note*, *6*, 120.
Pfanzagl, J. (1971a). A counterexample to a lemma of L. D. *Brown*. *The Annals of Mathematical Statistics*, *42*, 373–375.
Pfanzagl, J. (1971b). On the functional equation $\varphi(x) + \varphi(y) = \psi(T(x, y))$. *Aequationes Mathematicae*, *6*, 202–205.
Pfanzagl, J. (1974). A characterization of sufficiency by power functions. *Metrika*, *21*, 197–199.
Pfanzagl, J. (1981). A special representation of a sufficient randomization kernel. *Metrika*, *28*, 79–81.
Pfanzagl, J. (1994). *Parametric statistical theory*. Berlin: De Gruyter.
Pitcher, T. S. (1957). Sets of measures not admitting necessary and sufficient statistics or subfields. *The Annals of Mathematical Statistics*, *28*, 267–268.
Pitcher, T. S. (1965). A more general property than domination for sets of probability measures. *Pacific Journal of Mathematics*, *15*, 597–611.
Pitman, E. J. G. (1936). Sufficient statistics and intrinsic accuracy. *Proceedings of the Cambridge Philosophical Society*, *32*, 567–579.
Plachky, D. (1977). A characterization of bounded completeness in the undominated case. In *Transactions of the Seventh Prague Conference on Information Theory, Statistical Decision Functions, Random Processes A* (pp. 477–480). Reidel.
Rogge, L. (1972). The relations between minimal sufficient statistics and minimal sufficient σ-fields. *Z. Wahrscheinlichkeitstheorie verw. Gebiete*, *23*, 208–215.
Romano, J. P., & Siegel, A. F. (1986). *Counterexamples in probability and statistics*. Monterey: Wadsworth and Brooks.
Roy, K. K., & Ramamoorthi, R. V. (1979). Relationship between Bayes, classical and decision theoretic sufficiency. *Sankhyā Series A*, *41*, 48–58.
Sacksteder, R. (1967). A note on statistical equivalence. *The Annals of Mathematical Statistics*, *38*, 787–794.
Sato, M. (1996). A minimal sufficient statistic and representations of the densities. *Scandinavian Journal of Statistics*, *23*, 381–384.
Schmetterer, L. (1966). On the asymptotic efficiency of estimates. In Research Papers in Statistics. Festschrift for J. Neyman & F. N. David (Eds.) (pp. 301–317). New York: Wiley.
Schmetterer, L. (1974). *Introduction to mathematical statistics*. Translation of the 2nd German edition 1966. Berlin: Springer.
Shimizu, R. (1966). Remarks on sufficient statistics. *Annals of the Institute of Statistical Mathematics*, *18*, 49–55.
Stigler, S. M. (1973). Laplace, Fisher, and the discovery of the concept of sufficiency. *Biometrika*, *60*, 439–445. Reprinted in Kendall and Plackett (1977).
Strasser, H. (1985). *Mathematical theory of statistics*. Berlin: De Gruyter.
Stuart, A., & Ord, K. (1991). *Kendall's advanced theory of statistics. Classical inference and relationship* (5th ed., Vol. 2). London: Edward Arnold.
Sverdrup, E. (1953). Similarity, unbiasedness, minimaxity and admissibility of statistical test procedures. *Skand. Aktuarietidskr.*, *36*, 64–86.
Torgersen, E. N. (1991). *Comparison of statistical experiments* (Vol. 36). Encyclopedia of mathematics and its applications. Cambridge: Cambridge University Press.
Wijsman, R. A. (1958). Incomplete sufficient statistics and similar tests. *The Annals of Mathematical Statistics*, *29*, 1028–1045.
Witting, H. (1966). *Mathematische Statistik*. Eine Einführung in Theorie und Methoden: Teubner.
Zacks, S. (1971). *The theory of statistical inference*. New York: Wiley.

Chapter 3
Descriptive Statistics

3.1 Introduction

Descriptive statistics is an important link between mathematical constructs and reality. Its purpose is to transform inexact, prescientific concepts (like location, concentration, independence ...) into exact concepts which are operationally meaningful and suitable for mathematical treatment. Since the mathematical tools used in this area are elementary, it is surprising that there are no results worth mentioning prior to 1950. This explains why none of the textbooks which were on the market in the fifties (like Wilks 1943; Cramér 1946a, or Schmetterer 1956) has a chapter on descriptive statistics. But why do recent books that try to paint a broad picture of mathematical statistics (like "Kendall's Advanced Theory of Statistics") fail to introduce concepts like "unimodality" or "spread order"? A remarkable exception is Witting and Müller-Funk (1995) which contains a large Sect. 7.1, on descriptive functionals. Otherwise, readers interested in descriptive statistics must use more recent monographs like Bertin et al. (1997), or Müller and Stoyan (2002), and, above all, the papers on descriptive statistics by Bickel and Lehmann (1975–1979).

The following remarks are restricted to those topics in descriptive statistics which are relevant for the subsequent essays. In Sects. 3.4 and 3.7 we discuss concepts of *location* and *concentration*. The papers by Bickel and Lehmann Bickel and Lehmann (1975–1979) on location and Bickel and Lehmann Bickel and Lehmann (1975–1979) on dispersion serve as the basis for this discussion. Important papers on descriptive concepts not discussed here are Lehmann (1966) on dependence and van Zwet (1964) on inverse function orderings.

Though the emphasis of the following sections is on (partial) order relations between probability measures with respect to location and concentration, a quantification of these properties may be desirable for some purposes, in particular if no distribution is optimal with respect to the partial order. Quantification means defining a functional $\kappa : \mathfrak{P} \to \mathbb{R}$ which "measures" the property in question. It is essential that this measure be compatible with the given order relation. In general, the functional is not uniquely determined by the condition of compatibility, nor does

J. Pfanzagl, *Mathematical Statistics*, Springer Series in Statistics,
DOI 10.1007/978-3-642-31084-3_3

compatibility with a meaningful (partial) order relation guarantee that the functional itself is meaningful. Moreover, a comparison based on the functional κ will be less informative than a comparison with respect to the partial order, assuming such a comparison is possible. In particular: the comparison of estimators should not be restricted to the comparison of variances only if they are comparable with respect to their concentration on intervals. In their papers on descriptive functionals of location and dispersion, Bickel and Lehmann introduce another aspect for the selection of a descriptive functional: its robustness and, in this context, its estimability.

In its simplest form, estimation theory starts from a family $\mathfrak{P}$ of probability measures P defined on a measurable space $(X, \mathscr{A})$, and a functional $\kappa|\mathfrak{P}$, taking its values in a measurable space $(Y, \mathscr{B})$. In the period following 1950, a typical case was $Y = \mathbb{R}^k$, perhaps Y a function space if the problem was to estimate the density of $P \in \mathfrak{P}$. Estimators $\kappa^{(n)} : X^n \to Y$ were obtained from i.i.d. samples guided by P^n, for some (unknown) $P \in \mathfrak{P}$.

The big problem is the choice of the basic family. If the $\mathfrak{P}$ chosen is larger than necessary, then an estimator which is optimal for this family will be suboptimal for a subfamily $\mathfrak{P}_0 \subset \mathfrak{P}$. If the $\mathfrak{P}$ chosen is too small, then an estimator, though reasonable for every $P \in \mathfrak{P}$, might go wild if the true probability measure is not in $\mathfrak{P}$.

There are few situations where the knowledge of the random process generating the observations $x_1, x_2, \ldots$ suggests for $\mathfrak{P}$ a particular parametric family. Radioactive decay is one of these exceptions. Usually, there will be not more than a vague general experience leaving the choice between a variety of parametric models. The situation is no better for fully "nonparametric" theory. Even if $\mathfrak{P}$ is intended to be a "general" family, say a large family on $\mathbb{B}$ with Lebesgue densities, any mathematical treatment depends on certain nontestable smoothness assumptions about the densities, and smoothness assumptions which are roughly equivalent from an intuitive point of view (such as Hölder- or Sobolev-classes) produce different results. These are important problems which can only be addressed using asymptotic techniques.

There are some obvious conditions on the estimators like measurability of $(x_1, \ldots, x_n) \to \kappa^{(n)}(x_1, \ldots, x_n)$ which is indispensable, or continuity of this map which is natural and not really restrictive. Other obvious requirements are permutation invariance in the case of an i.i.d. sample, or equivariance if the family $\mathfrak{P}$ is closed under certain transformations.

The requirement that the estimators should be concentrated about the estimand as closely as possible is usually subject to the further condition that the estimators should be properly centered.

3.2 Parameters and Functionals

Statisticians take it as a matter of course that one has to estimate "functionals" in the case of general ("nonparametric") families, and "parameters" in the case of parametric families. A purist could object that this is not really true. Consider the distribution of lifetime with density $x \to \vartheta \exp[-\vartheta x]$, $x > 0$. The task to estimate the parameter is not fully specified unless the intended use of the estimator is known. The parameter ϑ has an interpretation as decay time, and one might be interested in an estimator which is median unbiased. If the task is to estimate the expected lifetime, this is $1/\vartheta$, and one will be interested in an estimator with expectation $1/\vartheta$. Perhaps the problem is to obtain an (unbiased) estimator of the density itself, or an (unbiased) estimator of $\exp[-t/\vartheta]$ $(= P_\vartheta(t, \infty))$?

Even in the case of a parametric family, it is a functional (expressed in terms of the parameter) which is the estimand. Regrettably, many statisticians use the term "parameter" though they really mean "functional".

In the opinion of prominent statisticians (like von Mises 1912, p. 16), parameters should have an intuitive interpretation (as functionals). Even if this is impossible, as is the case of the shape parameter of the Gamma-distribution (what, after all, is an operational definition of shape?), there are functions of the parameters, like $\int_0^\infty x\Gamma_{a,b}(dx) = ab$, that have operational significance.

If we take a sensible view, the real problem will always be to estimate the functional of an unknown distribution, and the parametric family enters the problem just as an approximation to the unknown distribution (hopefully supported by some knowledge of the process generating the unknown distribution).

In many (nonparametric) applications, the functional $\kappa : \mathfrak{P} \to \mathbb{R}$ is not uniquely determined: κ should, perhaps, be some "measure of location". It appears doubtful whether the use of refined mathematical techniques is adequate in such a case. Not all scholars share this opinion. Lehmann (1983, p. 365) claims that: "If each [functional κ] is an equally valued measure of location, what matters is how efficient the quantity $[\kappa(P)]$ can be estimated." In this connection, Bickel and Lehmann speak of the "robustness" of the functional, which means: Small changes in the distribution imply only small changes in the value of the functional. In other words: The functional should be continuous (with respect to some metric on $\mathfrak{P}$). Such a requirement is supported by the fact that κ cannot be estimated locally uniformly at P unless κ is continuous at P. Yet there continue to be some difficulties with functionals that are distinguished by being accurately estimable rather than by the nature of the problem, and one could even ask how important the use of particularly accurate estimators for only vaguely defined functionals really is. To say that an estimator estimates what it estimates, but does so efficiently, is tantamount to saying, "anything goes".

3.3 Estimands and Estimators

Sometimes the question arises: Given a function $\hat{\kappa} : \mathfrak{P} \to \mathbb{R}^k$, which value $\kappa(P)$ does it estimate? In some cases the question even arises: What does $\hat{\kappa}$ estimate? Consider for example estimators that estimate the center of symmetry of a symmetric distribution, like the Hodges–Lehmann estimator or the Huber estimators. What do they really estimate if P fails to be symmetric?

The problem "Given an estimator—identify the estimand" could perhaps be solved as follows. Suppose using $\hat{\kappa}(x)$ as a surrogate for the unknown $\kappa(P)$ causes the loss $\ell(\hat{\kappa}(x) - \kappa(P))$. Then $\hat{\kappa}_0$ will be preferred over $\hat{\kappa}$ if $\int \ell(\hat{\kappa}_0 - \kappa(P))dP \leq \int \ell(\hat{\kappa} - \kappa(P))dP$ for every $P \in \mathfrak{P}$, or, preferably, if $\ell(\hat{\kappa}_0 - \kappa(P))$ is stochastically smaller than $\ell(\hat{\kappa} - \kappa(P))$ (as defined in Sect. 3.4) for every $P \in \mathfrak{P}$. Since $\kappa(P)$ is usually not given, one might define a functional $\kappa(P)$ by the relation

$$\int \ell(\hat{\kappa} - \kappa(P))dP = \min_{\mu \in \mathbb{R}^p} \int \ell(\hat{\kappa} - \mu)dP. \tag{3.3.1}$$

If the functional $\kappa(P)$ is given, the relation (3.3.1) is a condition on the estimator. It confirms that $\hat{\kappa}$ does, in fact, what it should: It estimates $\kappa(P)$ (and not some other functional). Applied in asymptotic theory, this would also lead to the unpleasant result that the estimand is different for different sample sizes.

Relation (3.3.1) raises various problems: Unless loss is measured on a cardinal scale, the functional κ defined by (3.3.1) will depend on the particular scale. Moreover, different functions $\hat{\kappa}$ will define different estimands. Hence the definition of $\kappa(P)$ by (3.3.1) will usually only make sense in asymptotic considerations. (See Chap. 5.)

Beyond the formal definition of the estimand using (3.3.1), there may be various intuitive requirements: If P is symmetric about $\kappa(P)$, and $\hat{\kappa}$ is an estimator symmetric about $\kappa(P)$, one will be willing to accept $\hat{\kappa}$ as an estimator of $\kappa(P)$ even if condition (3.3.1) is not fulfilled with $\mu = \kappa(P)$. Applied to families $\mathfrak{P}|\mathbb{B}$, the loss function $\ell(u) = u^2$ defines $\kappa(P)$ as the mean of P, and $\ell(u) - |u|$ defines $\kappa(P)$ as the median. Depending on the basic model, conditions of mean unbiasedness or median unbiasedness might have an intuitive appeal of their own.

For *multivariate* functionals $\kappa : \mathfrak{P} \to \mathbb{R}^k$, mean unbiasedness of $\kappa^{(n)} : X^n \to \mathbb{R}^k$ poses no problem. It is defined componentwise,

$$\int \kappa_i(x)P(dx) = \int \kappa_i(P) \quad \text{for } i = 1, \ldots, k.$$

However, there is no operationally significant generalization of "median unbiasedness" to multivariate estimators. Median unbiasedness for each component $\kappa_i^{(n)}$, $i = 1, \ldots, k$ is too weak. With application to asymptotic considerations in mind, we suggest requiring of median unbiasedness of $\sum_{i=1}^k \alpha_i \kappa_i^{(n)}$ for $\sum_{i=1}^k \alpha_i \kappa_i(P)$ for every $(\alpha_1, \ldots, \alpha_k) \in \mathbb{R}^k$.

In the following we try to explore the domain in which mean unbiasedness is appropriate. For some problems, mean unbiasedness' is mandatory. This presumes in particular that the quantity to be estimated is measured on a cardinal scale. Typical examples are weights, costs, durations and probabilities. In such cases, the estimator $\kappa^{(n)}(x_1, \ldots, x_n)$ should be "correct" in the long run, i.e., should be mean unbiased. In such a case, one could say that the estimator $\kappa^{(n)}$ estimates $\int \kappa^{(n)} dP^n$.

Assume that $\mathfrak{P} = \{P_\vartheta;\ \vartheta \in \mathbb{R}\}$. In most elementary textbooks, estimators of ϑ, say $\vartheta^{(n)}$, are required to be unbiased. Without knowing the potential uses of this estimator, the requirement of unbiasedness lacks any justification. Unbiasedness of $\vartheta^{(n)}$ as an estimator for ϑ is irrelevant if the purpose is to use $P_{\vartheta^{(n)}}(A)$ as an estimator for $P_\vartheta(A)$. If P_ϑ has a μ-density $p(\cdot, \vartheta)$, then one needs an estimator $p^{(n)}(\cdot, x_1, \dots, x_n)$ such that $\xi \to p^{(n)}(\xi, x_1, \dots, x_n)$ is unbiased for $p(\xi, \vartheta)$ to have $\int 1_A(\xi) p^{(n)}(\xi, x_1, \dots, x_n) d\xi$ as an unbiased estimator of $P_\vartheta(A)$, for every $A \in \mathscr{A}$. This will, in general, not be $p(\cdot, \vartheta^{(n)})$.

Let $t_\alpha(\vartheta) \in \mathbb{R}$ denote the α-quantile of P_ϑ, defined by $P_\vartheta[t_\alpha(\vartheta), \infty) = \alpha$. If the intention is to use the estimator $(x_1, \dots, x_n) \to t_\alpha^{(n)}(x_1, \dots, x_n)$ in order to obtain a tolerance region $[t_\alpha^{(n)}, \infty)$ with average covering probability α, then the requirement will be

$$\int P_\vartheta[t_\alpha^{(n)}(x_1, \dots, x_n), \infty) P_\vartheta^n(d(x_1, \dots, x_n)) = \alpha \quad \text{for every } \vartheta \in \Theta. \qquad (3.3.2)$$

This, too, is a sort of unbiasedness, but the unbiasedness of $\vartheta^{(n)}$ as an estimator for ϑ has nothing to do with condition (3.3.2).

Unbiasedness of an estimator for a parameter ϑ can be justified only if this parameter can be interpreted as a functional for which, by its nature, unbiasedness is desirable. This certainly applies to location parameters. It is doubtful whether unbiasedness is adequate for scale parameters. One could argue that a deviation of $\hat{\sigma}/\sigma$ from 1 by the factor 2 is of the same "weight" as a deviation by the factor $1/2$—and this does not easily go together with unbiasedness.

Even statisticians who admire the beautiful result of Olkin and Pratt (1958, p. 202, relation (2.3)) on the existence of unbiased estimators for the correlation coefficient of a two-dimensional normal distribution will agree that, for this parameter, unbiasedness is without operational significance.

In spite of the fact that convincing arguments for mean unbiasedness as a universal requirement for every estimator have never been brought forward, estimation theory was almost exclusively concerned with mean unbiased estimators in the years between 1950 and 1970. This can, perhaps, be traced back to the fact that the mathematical tools for dealing with mean unbiasedness, like integrals and conditional expectations, were available at this time.

Compared with the mathematical appeal of the theory of mean unbiased estimators (see Chap. 4), the obvious shortcomings of this concept were neglected. The doubts raised by prominent statisticians concerning mean unbiasedness had no effect on the predominance of these concepts in textbooks—elementary ones and others. Here are some critical voices:

C.R. Rao (1945, p. 82): "... the inevitable arbitrariness of these postulates of unbiasedness and minimum variance needs no emphasis."
L.J. Savage (1954, Chapter 7, p. 244): "... it is now widely agreed that a serious reason to prefer [mean] unbiased estimators seems never to have been proposed."
Barnard (1974, p. 4): "Often one begins with the concept of an unbiased estimator—in spite of the fifty year old condemnation of this idea, as a foundation for theory, on the part of R.A. Fisher."
Fraser (1957, p. 49): "median unbiasedness has found little application in estimation theory

primarily because it does not lend itself to the mathematical analysis needed to find minimum risk estimates."

In some cases, unbiasedness comes as a deus ex machina unmotivated by applications. If the purpose is to estimate a quantile, median unbiasedness is cogent. What would be the interpretation of a median unbiased estimator of the mean?

Authors who feel obliged to motivate the requirement of unbiasedness—there aren't that many—present arguments which seem to be more shaky than they are convincing. They never admit that it is merely the availability of some nice-looking easy results that justifies the emphasis given to mean unbiasedness. Here are some examples.

Cramér (1946b, p. 87): "In order to avoid unnecessary [!] complications, we shall suppose throughout that [the estimators are unbiased]."
Heyer (1982, p. 116): "The obvious [!] aim of an optimal decision process will be the search for estimators ... of vanishing distortion [i.e., unbiasedness] and uniformly minimal variance..."
Strasser (1985, p. 160): "... mean unbiasedness is a natural condition if the problem has a linear structure." [?]
Witting (1985, p. 300): "Die Beschränkung auf erwartungstreue Schätzer ist zur Auszeichnung optimaler Elemente häufig zweckmäßig".

The opinion put forward by Aitken and Silverstone (1942, p. 189) on unbiasedness is downright absurd: "... To obtain an unbiased estimator with minimum variance we must in most cases estimate not ϑ but some function of ϑ." Similarly, Barton (1956) suggests a method "which may only be reasonably applied when the property of unbiasedness is of more importance than the functional form of the parameter estimated" (see p. 202). In spite of the absurdity of this idea: What can one do in the case of a parametric family where *no* function of the parameter admits an unbiased estimator?

A basic shortcoming of mean unbiased estimators: As opposed to median unbiased estimators, they are not invariant under arbitrary monotone transformations of the functional: If $\mathfrak{P} = \{N(\mu, \sigma^2)^n : \mu \in \mathbb{R}, \sigma^2 > 0\}$, the square roots of unbiased estimators for σ^2 are not unbiased for σ.

The bad news is: Even if the model requires an unbiased estimator for some functional, unbiased estimators may not exist at all, or they may show certain disturbing properties (e.g. by not being proper). We will abstain from presenting an example; a plethora of them can be found in textbooks like Lehmann and Casella (1998).

Even if unbiasedness is not required by the nature of the model, one might surmise that unbiasedness somehow ensures that the estimator is "impartial", that it does not favour some members of $\mathfrak{P}$ to the disadvantage of others. Yet, there are numerous examples available in the literature where mean unbiased estimators spectacularly fail to meet this requirement. The first example of this phenomenon was provided by D. Basu (1955, p. 346), a variant of which can be found in Pfanzagl (1994, p. 71, Example 2.2.9). There are estimators $\vartheta^{(n)}$ of ϑ which are mean unbiased in the family $\{N(\vartheta, 1)^n : \vartheta \in \mathbb{R}\}$, but $N(\vartheta, 1)^n\{\vartheta^{(n)} = 0\} > 0$ for every $\vartheta \in \mathbb{R}$. Zacks (1971, p.

119, Example 3.9) presents an unbiased estimator for ϑ in the family with density $(x_1, x_2) \to \vartheta^{-2} 1_{(\vartheta,2\vartheta)}(x_1) 1_{(\vartheta,2\vartheta)}(x_2)$ with variance 0 for $\vartheta \in \{2^k : k = 0, \pm 1, \ldots\}$.

Historical Remark

Let $\mathfrak{P}$ be a family of probability measures on $(X, \mathscr{A})$, and $\kappa : \mathfrak{P} \to \mathbb{R}$ a functional. One of the basic qualities of an estimator $\hat{\kappa} : X \to \mathbb{R}$ is its concentration about $\kappa(P)$. One aspect of this is that it should be properly centered, i.e., without a systematic error which over- or underestimates $\kappa(P)$ for every $P \in \mathfrak{P}$. The question is how these ideas could be made precise.

It appears that the concept of "bias" originally referred to the observations, not to estimators. Working on the assumption that there is a "true" value and that the observations differ from this true value by a degree of error, Bowley (1897, p. 859) distinguishes between "biased errors" going all in the same direction, and "unbiased errors", which are equally likely to be positive or negative. His interest is in the influence of these errors of the observations on the error of an estimator (taken as an average of the observations), and his conclusion (based on some elementary computations) is: "Unbiassed errors can be neglected in comparison with biassed". Basically the same attitude can be found in Markov (1912, p. 202) who requires as a starting point for his computations the absence of a systematic error (of the individual observations).

A precise definition of an "unbiased estimator" (in the sense of $\int \hat{\vartheta}(x) P(dx) = \vartheta$) first occurs in David and Neyman (1938, p. 106, Definition 1). One could, perhaps, say that the concept of an unbiased estimator grew out of a compelling condition formulated as the "absence of a systematic error", but meaning, in fact, that the model underlying the analysis should be correct. There is no conclusive argument leading from this condition to the requirement of "unbiasedness" for estimators. In the paper by David and Neyman, dealing with the Markov Theorem on least squares, unbiasedness is an essential ingredient for this particular theory. It is not meant as a general condition to be imposed on all estimators.

The idea of the median as an important descriptive functional is as old as probability theory (see Huygens 1657, de Moivre 1756, Problem 5), and median unbiasedness as a requirement for the location of estimators was suggested as early as 1774 (Laplace, p. 363).

3.4 Stochastic Order

For probability measures on $\mathbb{B}$, various order relations have been suggested to express that the random variable corresponding to P_1 is in a stochastic sense larger than the random variable corresponding to P_0. To define the following order relations, let F_i denote the distribution function corresponding to P_i.

Order A. P_1 *is* stochastically larger *than* P_0 *iff* $F_1(t) \leq F_0(t)$ *for every* $t \in \mathbb{R}$.

Order B.

$$t \to F_1(t)/F_0(t) \quad \textit{is nondecreasing}, \tag{3.4.1}$$

$$t \to (1 - F_1(t))/(1 - F_0(t)) \quad \textit{is nonincreasing}. \tag{3.4.2}$$

Order C. $(F_1(t'') - F_1(t'))/(F_0(t'') - F_0(t'))$ *is nondecreasing in both variables whenever* $F_0(t') < F_0(t'')$.

Order A was introduced by Mann and Whitney (1947). It was used by Lehmann (see 1951a, 1952, and, in particular, 1955, pp. 399/400) in various equivalent versions, such as:

$$\text{There exists a function } f \text{ fulfilling } f(x) \geq x \text{ such that } P_1 = P_0 \circ f$$

or

$$\int m(u) P_0(du) \leq \int m(u) P_1(du) \quad \text{for every nondecreasing function } m.$$

Here is a natural example of order A: For any location parameter family $\{P_\vartheta : \vartheta \in \mathbb{R}\}$, the family $\{P_\vartheta^n \circ T_n : \vartheta \in \mathbb{R}\}$ is ordered A if $(x_1, \ldots, x_n) \to T_n(x_1, \ldots, x_n)$ is equivariant under shifts.

Measures κ of location compatible with the stochastic order are studied in Bickel and Lehmann (1975–1979). In addition to the compatibility with the stochastic order, i.e., $\kappa(P_0) \leq \kappa(P_1)$ if $P_0 \leq P_1$, they require that

$$\kappa(P \circ (x \to ax + b)) = a\kappa(P) + b \quad \text{for } a > 0 \text{ an } b \in \mathbb{R} \tag{3.4.3}$$

and perhaps, in addition, $\kappa(P \circ (x \to -x)) = -\kappa(P)$. Compatibility with stochastic order implies in particular that

$$\kappa(P \circ f) \geq \kappa(P) \text{ if } f(x) \geq x$$

and that

$$\kappa(P_0) \leq \kappa(P_1) \text{ implies } \kappa(P_0 \circ m) \leq \kappa(P_1 \circ m) \text{ if } m \text{ is nondecreasing.}$$

If relation (3.4.3) is strengthened to $\kappa(P \circ m) = m(\kappa(P))$ for every nondecreasing function m, this implies that $\kappa(P)$ is a quantile of P (Fraser 1954b, p. 51).

If P_2 is dominated by P_1 ($P_2 \ll P_1$), order C is equivalent to the existence of a nondecreasing density dP_2/dP_1, usually called the *m.l.r.* property ("m.l.r." for *monotone likelihood ratios*).

It was Rubin who, in an abstract (1951, p. 608), pointed out that it is the monotonicity of likelihood ratios that accounts for certain attractive properties of exponential families. The theory of m.l.r. families was fully developed by Karlin and Rubin (1956, p. 279, Theorem 1) and Lehmann (1959). Various complete class- and optimality results by Allen (1953), Sobel (1953) and, in particular, Blackwell and Girshick (1954, p. 182, Lemma 7.4.1) which hold for m.l.r. families are formulated for the special case of an exponential family only.

It can easily be seen that order C implies order B, and that (3.4.1) as well as (3.4.2) imply A. Orders A and C are useful in statistical theory. Order B was introduced in Pfanzagl (1964, p. 217) because it is strong enough to imply certain topological properties of $\mathfrak{P}$. If $\mathfrak{P}$ is B-ordered, pointwise convergence of the distribution functions implies uniform convergence on $\mathbb{B}$. If the probability measures have continuous Lebesgue densities, pointwise convergence of the distribution functions implies pointwise convergence a.e. of the densities. (See Pfanzagl 1964, p. 1220, Theorem 1, and 1969, p. 61, Theorem 2.12.) This result applies in particular to exponential families (which have monotone likelihood ratios and continuous densities). As a consequence: $P_{\vartheta_n} \Rightarrow P_{\vartheta_0}$ weakly implies $\vartheta_n \to \vartheta_0$. (See in this connection also Barndorff-Nielsen 1969.)

A Side Remark on m.l.r. Families and the Power of Tests

For m.l.r. families, Karlin and Rubin (1956) state, among many other results, that the class of all critical functions φ fulfilling

$$\varphi(t) = \begin{cases} 1 \\ 0 \end{cases} \text{for } t \begin{matrix} > \\ < \end{matrix} t_0 \tag{3.4.4}$$

is complete. (See p. 282, Theorem 4, and Lehmann 1959, p. 72, Theorem 3 for a more accessible version.)

This is a special case of a more general result by Karlin and Rubin (1956, p. 279, Theorem 1) which asserts that the class of all monotone procedures is complete with respect to any subconvex loss function. An improved version of this result is Theorem 2.1, p. 714, in Brown et al. (1976).

Lehmann (1959, p. 68, Theorem 2) expresses for a parametric family $\{P_\vartheta : \vartheta \in \Theta\}$, $\Theta \in \mathbb{R}$, what is characteristic for the optimality of tests in m.l.r. families. To come closer to applications, assume now the existence of a function $T : X \to \mathbb{R}$ such that $\{P_\vartheta : \vartheta \in \Theta\}$ has "monotone likelihood ratios in T", i.e., that for any ϑ_1, ϑ_2 there exists a nondecreasing function $H_{1,2}$ such that

$$dP_{\vartheta_2}/dP_{\vartheta_1}(x) = H_{1,2}(T(x)) \quad \mu\text{-a.e.} \tag{3.4.5}$$

Under this assumption, for any critical function φ of type (3.4.4), $x \to \varphi(T(x))$ is most powerful for any of the test problems $P_{\vartheta_1} : P_{\vartheta_2}$. Or, in other words:

If

$$\int \psi(x) P_{\vartheta_0}(dx) = \int \varphi(T(x)) P_{\vartheta_0}(dx),$$

then

$$\int \psi(x) P_\vartheta(dx) \underset{\geq}{\leq} \int \varphi(T(x)) P_\vartheta(dx) \quad \text{for } \vartheta \underset{<}{>} \vartheta_0. \tag{3.4.6}$$

Warning: The optimality of the critical functions $\varphi \circ T$ depends on the fact that the family $\{P_\vartheta : \vartheta \in \Theta\}$ has m.l.r. in T in the sense of (3.4.5), which implies that T is sufficient. It is not sufficient for the family $\{P_\vartheta \circ T : \vartheta \in \Theta\}$ to have m.l.r.

M.l.r. families provide tests which not only maximize the power for every $\vartheta > \vartheta_0$. They also minimize the power for every $\vartheta < \vartheta_0$. This is more than is usually required for testing the hypothesis $\vartheta \leq \vartheta_0$ against alternatives $\vartheta > \vartheta_0$. The usual requirement on the critical function is $P_\vartheta(\varphi) \leq \alpha$ for $\vartheta \leq \vartheta_0$ and $P_\vartheta(\varphi)$ as large as possible for $\vartheta > \vartheta_0$. The existence of an optimal critical function in this class does not imply that the family has m.l.r.—not even if such a critical function exists for every $\vartheta_0 \in \Theta$ and every $\alpha \in (0, 1)$. (For an example see Pfanzagl 1960 and 1962, p. 112.)

The existence of tests with the strong optimum property indicated under (3.4.6) is more or less confined to m.l.r. families. Assume that a dominated family $\mathfrak{P}$ is ordered by a relation $\leq$ with the following property: For every $P_0 \in \mathfrak{P}$ and every $\alpha \in (0, 1)$ there exists a critical function φ such that

$$\int \varphi(x) P_0(dx) = \alpha$$

and

$$\int \psi(x) P(dx) \underset{\geq}{\leq} \int \varphi(x) P(dx) \quad \text{if } P \underset{<}{>} P_0$$

for any critical function ψ fulfilling $\int \psi(x) P(dx) = \alpha$. Then there exists a function $T : X \to \mathbb{R}$ and for any pair $P_i \in \mathfrak{P}$, $i = 1, 2$, a nondecreasing function $H_{1,2}$ such that

$$dP_2/dP_1(x) = H_{1,2}(T(x)) \quad (P_1 + P_2)\text{-a.e.}$$

(See Pfanzagl 1962, p. 110, Satz. For a generalization see Mussmann 1987.)

Exponential families with density $C(\vartheta)h(x)\exp[a(\vartheta)T(x)]$ have m.l.r. if the function a is increasing. For such families the existence of most powerful critical functions of the type (3.4.4) was already remarked by Lehmann (1947, p. 99, Theorem 1). Relevant for applications are families with m.l.r. for every sample size. Exponential families are obviously of this type (with m.l.r. in $\sum_{\nu=1}^n T(x_\nu)$). According to Borges and Pfanzagl (1963, p. 112, Theorem 1), a family of mutually absolutely continuous probability measures which has m.l.r. for every sample size is necessarily exponential.

The results concerning the relationship between m.l.r. families and the existence of most powerful tests, obtained in the fifties, are not yet familiar to all statisticians. Berger (1980, p. 369) writes:

> The most important class of distributions for which [uniformly most powerful tests] sometimes [!] exist is the class of distributions with monotone likelihood ratio.

3.5 Spread

A construct expressing concentration (or rather dispersion) without reference to a particular center is *spread*. If Q_0 is less spread out than Q_1 (in symbols: $Q_0 \preceq Q_1$), this is much more than greater concentration about a given center μ like in (3.7.1). It means, in an intuitive sense, greater inherent concentration "everywhere".

The intuitive meaning of this concept has been expressed by different authors in different ways, which afterwards proved mathematically equivalent.

Doksum (1969, p. 1169) introduced the *tail order:* Q_1 has *heavier tails* than Q_0 if

$$F_1^{-1} - F_0^{-1} \quad \text{is nondecreasing on } (0, 1). \tag{3.5.1}$$

The same relation, written slightly differently as

$$x \to F_1^{-1}(F_0(x)) - x \quad \text{is nondecreasing,}$$

had earlier been used by Fraser (1957).

Bickel and Lehmann (1975–1979, p. 34, relation 1.3) introduced the *spread order:* Q_1 is *more spread out* than Q_0 (i.e., $Q_0 \preceq Q_1$) if

$$F_0^{-1}(\beta) - F_0^{-1}(\alpha) \leq F_1^{-1}(\beta) - F_1^{-1}(\alpha) \quad \text{for } 0 < \alpha < \beta < 1, \tag{3.5.2}$$

and they show that this is equivalent to

$$q_0(F_0^{-1}(\alpha)) \geq q_1(F_1^{-1}(\alpha)) \quad \text{for } \lambda\text{-a.a. } \alpha \in (0, 1)$$

if Q_i admits a Lebesgue density q_i.

Relation (3.5.2) has been used earlier by Saunders and Moran (1978), who prove for the Gamma-distribution $\Gamma(a, b)$ with scale parameter a and shape parameter b that $a_0 \leq a_1$ and $b_0 \leq b_1$ imply $\Gamma(a_0, b_0) \preceq \Gamma(a_1, b_1)$ (1978, p. 429, Theorem 1).

The merits of the paper by Saunders and Moran lie in their results for the Gamma-distribution. The priority concerning the construct (3.5.2) is hard to decide. After all, Saunders and Moran used an unpublished paper by Lewis which presumably contains the result of the paper by Lewis and Thompson (1981). Lewis and Thompson (1981, pp. 78/79), unaware of the paper by Bickel and Lehmann Bickel and Lehmann (1975–1979), define a *dispersion order* as follows.

Q_1 is *more dispersed* than Q_0 if for every $\alpha \in (0,1)$,

$$F_0(F_0^{-1}(\alpha)+x) \overset{\geq}{\underset{\leq}{}} F_1(F_1^{-1}(\alpha)+x) \quad \text{for } x \overset{>}{\underset{<}{}} 0. \tag{3.5.3}$$

The equivalence between (3.5.2) and (3.5.3) is already indicated in a somewhat vague version in Saunders and Moran 1978, p. 427, relation 1.3.

According to Shaked (1982, p. 312, Theorem 2.1), relation (3.5.3) is equivalent to the following:

For every $c \in \mathbb{R}$, $x \to F_0(x) - F_1(x+c)$ changes sign at most once, from $-$ to $+$.

It speaks for the power of this construct that it was discovered in different versions by so many authors. Though the equivalence of the different definitions can easily be seen, the authors were not always aware of the other related papers. It appears that the equivalence between (3.5.1) and (3.5.2) remained unnoticed until 1983 (see Deshpande and Kochar 1983, p. 686). One might even say that the spread order was discovered once more: The property of estimator sequences asserted in Hájek (1970, p. 329, Corollary 2), followed by Roussas (1972, p. 145, Proposition 4.1) is just a version of (3.5.3).

Probability measures which are comparable in the spread order emerge from the Convolution Theorem. According to the results presented in Sect. 5.13, $Q_1 = Q_0 * R$ is "more" spread out than Q_0 if Q_0 is logconcave. That regularly attainable limit distributions on $\mathbb{B}$ are comparable with the optimal limit distribution in the spread order may also be shown directly (see Sect. 5.11).

To obtain from $Q_0 \preceq Q_1$ an assertion about the concentration on intervals, one has to distinguish a certain center μ. The following result is straightforward:

Lemma 3.5.1 *If μ is a common quantile of Q_i, $i = 0, 1$, then $Q_0 \preceq Q_1$ implies that Q_0 is "more" concentrated than Q_1 on all intervals about μ.*

$Q_0 \preceq Q_1$ implies $Q_0 \circ m \preceq Q_1 \circ m$ if $m(x) = ax + b$ with $a > 0$, but not for an arbitrary increasing function m. Yet, an important consequence of $Q_0 \preceq Q_1$ is preserved under monotone transformations: If μ is a common quantile of Q_i, then $m(\mu)$ is a common quantile of $Q_i \circ m$, and $Q_0 \circ m$ is more concentrated than $Q_1 \circ m$ is on all intervals about $m(\mu)$.

In order to stress that comparability with respect to spread is much more than comparability in the sense of (3.7.1), we mention the example of the Beta distributions: For $\alpha > \beta$, $B_{\alpha,\alpha}$ is more concentrated than $B_{\beta,\beta}$ on all intervals containing the common median $\mu = 1/2$. Yet $B_{\alpha,\alpha}$ and $B_{\beta,\beta}$ are not comparable in the spread order, since they have the same bounded support. (See Pfanzagl 1994, p. 99, Example 2.7.7.)

Another relation based on logconcavity (and useful for the evaluation of confidence intervals) is

Proposition 3.5.2 *If P has a logconcave density, then*

$$\alpha \to F(F^{-1}(\alpha)+t) \text{ is } \begin{matrix}\text{concave}\\ \text{convex}\end{matrix} \text{ if } t \begin{matrix}>\\ <\end{matrix} 0.$$

(See Pfanzagl 1994, p. 274, Lemma 8.2.14 for $F = \Phi$).

It is surprising that an intuitively convincing and mathematically useful concept like "spread" is neglected in almost all textbooks. Many highly qualified mathematical statisticians have never heard of "spread".

3.6 Unimodality; Logconcave Distributions

A probability measure on $\mathbb{B}$ with distribution function F is unimodal at 0 if F is convex on $(-\infty, 0)$ and concave on $(0, \infty)$. For probability measures with Lebesgue density this is equivalent to the existence of a density which is nondecreasing on $(-\infty, 0]$ and nonincreasing on $[0, \infty)$. This concept is usually attributed to de Helguero (1904) and Khintchine (1938), but it was Gauss (1823) who established a bound for the concentration of unimodal distributions far earlier (see Hald 1998, pp. 462–465). In spite of its intuitive appeal, unimodality was not considered worthwhile for theoretic analysis until the middle of the 20th century. The first such step was taken by two mathematical giants, Gnedenko and Kolmogorov (1954), and it was a slip. In Sect. 32 they assert that the convolution product of two probability measures unimodal at 0 is unimodal at 0, with a proof adapted from a thesis of A.I. Lapin, 1947. This assertion is disproved by the example of Chung (1953, pp. 583/4) of a probability measure $P|\mathbb{B}$ unimodal at 0 such that $P * P$ is not unimodal. (See Dharmadhikari and Joag-Dev 1988, pp. 11/13, for more examples.) (According to Wintner 1938, $P * P$ is unimodal if $P|\mathbb{B}$ is unimodal and symmetric. The example in Davidovič (1969, p. 480) shows that this does not extend to probability measures on $\mathbb{B}^k$ with $k > 1$.)

There are various ways to generalize the concept of unimodality from probability measures on $\mathbb{B}$ to probability measures on $\mathbb{B}^k$. Convexity of $\{x \in \mathbb{R}^k : p(x) > r\}$ for every $r \geq 0$ is the most natural and the most useful concept. Another possibility is to require that $\{x \in \mathbb{R}^k : p(x) > r\}$ is *star-shaped* for every $r \geq 0$, i.e. that there exists a center μ such that $p(x) > r$ implies $p((1-\alpha)\mu+\alpha x) > r$ for every $\alpha \in [0, 1]$. The most general concept of unimodality, i.e., that $\{x \in \mathbb{R}^k : p(x) > r\}$ is a connected set, has found no use. To distinguish between the two concepts introduced above, we speak of *convex-unimodality* and *star-unimodality,* if necessary.

In (1956), Ibragimov introduces in connection with this problem the concept of *strong unimodality.* A probability measure P is strongly unimodal if $P * Q$ is unimodal for every unimodal Q. Ibragimov (p. 255, Theorem) establishes that a non-degenerate probability measure is strongly unimodal iff it admits a logconcave density. This density is continuous and positive on the support of P. (Klaassen 1985, p. 906, Lemma 2.1.) Ibragimov's result preserves the assertion of Gnedenko and Kolmogorov in a modified version: The assertion is true if at least one of the unimodal distributions has a logconcave density. A side result of Ibragimov: $P_n \Rightarrow P_0$ implies $d(P_n, P_0) \to 0$ if the measures have a unimodal density.

The relevance of logconcave densities goes far beyond the special context of strong unimodality where it was first recognized by Ibragimov. According to Lehmann (1955, p. 406, Example 3.2), P has a logconcave density iff the location parameter family generated by P has monotone likelihood ratios. Observe also the relation between logconcavity and logconcavity of the order statistics, and the monotonicity of failure rates. (See in this connection also the monograph by Reiss 1989.)

Logconcavity extends to Euclidean spaces of any dimension, where the relation to strong unimodality no longer applies. In particular, the following holds true for any logconcave probability measure $P|\mathbb{B}^k$.

(i) $\{x \in \mathbb{R}^k : p(x) > r\}$ is convex for any $r \geq 0$. Hence any logconcave probability measure is unimodal.

(ii) $P \circ (u \to Au)$ is logconcave for any $m \times k$-matrix A (Dharmadhikari and Joag-Dev 1988, p. 47, Lemma 2.1).

The implications of unimodality and logconcavity for convolution products will be discussed in Sect. 3.9.

3.7 Concentration

Though the intention of estimation theory is to find estimators that are concentrated about the estimand as closely as possible, there were no efforts prior to 1950, say, to find a suitable mathematical construct corresponding to the intuitive concept of "concentration". The statisticians settled for comparing estimators on the basis of their quadratic risk, even though Fisher had already expressed his doubts. L.J. Savage (1954, p. 224) made the obvious suggestion to consider the concentration of probability measures Q_0, Q_1 on $\mathbb{B}$ about a given center μ, and to define $Q_0 \geq Q_1$ if

$$Q_0(\mu - t', \mu + t'') \geq Q_1(\mu - t', \mu + t'') \quad \text{for } t', t'' \geq 0. \tag{3.7.1}$$

In the case $Q_0\{\mu\} = 0$, relation (3.7.1) implies that μ is a common quantile of Q_0 and Q_1. The consequence: If μ is a median of Q_0, a strong comparison as in (3.7.1) is possible only with distributions Q_1, having the same median.

What makes the spread order particularly important is that $Q_0 \preceq Q_1$ implies (3.7.1) if μ is a common quantile of Q_0 and Q_1 (see Lemma 3.5.1).

If one considers unbiased estimators, one will be interested in the concentration about μ as the common expectation $\int u Q_i(du), i = 0, 1$. Unless the distributions are symmetric about μ, one cannot expect that μ is, at the same time, a common quantile of Q_i. Hence, for mean unbiased estimators, a strong comparison as in (3.7.1) will be possible in exceptional cases only.

The Peak Order

The natural basis for the comparison of estimators is their concentration about the estimand. For estimators on $\mathbb{B}$ the obvious concept is the concentration on intervals containing the estimand, as suggested by Pitman (1939, p. 401), Z.W. Birnbaum (1948, p. 76, who suggested the term "peakedness") and L.J. Savage (1954, p. 224). Pitman and Birnbaum consider intervals symmetric about the estimand only; presumably they did not expect situations where a comparison on arbitrary intervals about 0 is possible. To see the need for these more general comparisons, consider the case of a scale parameter, say a, where the concentration on intervals $(t^{-1}a, ta)$ is more relevant than the concentration on $(a-t, a+t)$.

Extending Z.W. Birnbaum's peakedness-concept from $\mathbb{B}$ to $\mathbb{B}^k$, Sherman (1955, p. 765, relation 5) defines a *peak order* for probability measures on $\mathbb{B}^k$

$$Q_1 \geq Q_2 \quad \text{if } Q_1(C) \geq Q_2(C) \tag{3.7.2}$$

for all $C \in \mathbb{B}^k$ which are convex and symmetric about 0. Half a century later, the same order relation appears in Witting and Müller-Funk (1995, p. 439) as the "Anderson-Halbordnung".

As can easily be seen, the peak order is equivalent to

$$Q_1 \geq Q_2 \quad \text{if} \int g dQ_1 \geq \int g dQ_2$$

for all *gain functions* $g : \mathbb{R}^k \to [0, \infty)$ which are symmetric and unimodal. (See Witting and Müller-Funk 1995, p. 439, Hilfssatz 6.214.)

Equivalent to (3.7.2) is the statement that the distribution of subconvex loss is stochastically smaller under Q_1 than under Q_2.

Notice that $Q_1 \geq Q_2$ in the peak order on $\mathbb{B}^k$ implies that

$$Q_1 circ(u \to a^\top u) \geq Q_2 \circ (u \to a^\top u) \quad \text{in the peak order on } \mathbb{R} \text{ for every } a \in \mathbb{B}^k.$$

There is an obvious but technically useful extension of this equivalence to the class, say $\mathscr{M}$, of all symmetric gain functions that are approximable (from below) by functions $\Sigma a_i g_i$ with $a_i > 0$ and $g_i \geq 0$, symmetric and unimodal.

Generalizing Birnbaum's Lemma 1 (p. 77), Sherman's Lemma 3, p. 766, asserts that $P_1 \geq P_2$ and $Q_1 \geq Q_2$ in the peak order imply $P_1 * Q_1 \geq P_2 * Q_2$, if all probability measures are symmetric and unimodal. In fact, a stronger result holds true:

Proposition 3.7.1 *If $Q|\mathbb{B}^k$ is symmetric and unimodal, then*

$$P_1 \geq P_2 \;\; \textit{implies} \;\; Q * P_1 \geq Q * P_2, \tag{3.7.3}$$

without any further condition on P_1 and P_2.

This follows directly from the extended version of Anderson's Theorem, which asserts that $y \to Q(C + y)$ is in $\mathscr{M}$ if C is convex and symmetric about 0.

In this connection one should also mention an early result of Z.W. Birnbaum (1948, p. 79, Theorem 1) on unimodal distributions with a Lebesgue density symmetric about 0: If Q_1 is more concentrated than Q_2 on symmetric intervals about 0, then the same is true of all n-fold convolution products. (Notice that for symmetric distributions, "more concentrated on all intervals symmetric about 0" is the same as saying "more concentrated on all intervals containing 0".)

The Löwner Order

The family $N(0, \Sigma)|\mathbb{B}^k$ is an instance where the peak order applies. Löwner (1934) introduced an order relation between positive definite matrices Σ_i by $\Sigma_1 \leq_L \Sigma_2$ if $\Sigma_2 - \Sigma_1$ is positive semidefinite. Since $N(0, \Sigma_2) = N(0, \Sigma_1) * N(0, \Sigma_2 - \Sigma_1)$, Anderson's Theorem implies that

$$N(0, \Sigma_1) \geq N(0, \Sigma_2) \text{ in the peak order iff } \Sigma_2 - \Sigma_1 \text{ is positive semidefinite.} \tag{3.7.4}$$

In other words, $N(0, \Sigma_1) \geq N(0, \Sigma_2)$ in the peak order iff $\Sigma_1 \leq \Sigma_2$ in the Löwner order. Amazingly, the basic relation (3.7.4) seems to have been unknown prior to Anderson's Theorem.

In Das Gupta (1972, p. 254, Theorem 3.3), relation (3.7.4) is generalized from $N(0, \Sigma)$ to distributions P_Σ with density $x \to |\Sigma|^{-1/2} p(x^\top \Sigma^{-1} x)$.

Rao (1945, p. 85, and 1947, pp. 281/2) remarks that $\Sigma_1 \geq_L \Sigma_2$ implies

$$(\Sigma_2)_{ii} \geq (\Sigma_1)_{ii} \quad \text{for } i = 1, \ldots, p, \tag{3.7.5}$$

and that $\Sigma_2 = \Sigma_1$ if equality holds in (3.7.5) for every $i = 1, \ldots, p$. Since Anderson's Theorem was not yet available, he gives no further interpretation of it in terms of concentration.

Cramér (1946a, Sect. 2.7 and 1946b, p. 85 ff) tries to illustrate the difference between $N(0, \Sigma_1)$ and $N(0, \Sigma_2)$ using Markov's concentration ellipsoid, a comparison without operational significance.

Some statisticians, now that Anderson's Theorem has become available, still express their final result in a formal way as "$\Sigma_2 - \Sigma_1$ is positive semidefinite", without using the operational expression (3.7.4). (See e.g. Bahadur 1964, p. 1550; Roussas 1968, p. 257 and 1972, p. 161; Serfling 1980, p. 142.)

The use of the ellipsoid of concentration in Schmetterer (1974, p. 292, Theorem 1.13) almost thirty years after Cramér, and twenty years after Anderson's Theorem, is quite surprising.

If Q_0 is unimodal and symmetric about μ, then $Q_1 \preceq Q_2$ implies that Q_1 is more concentrated than Q_2 (whether μ is a common quantile or not) on all *symmetric* intervals, i.e., (3.7.1) holds with $t' = t''$. (See Pfanzagl 1995, p. 78, Theorem 2.3.17.) One could imagine that this holds with arbitrary $t', t'' \geq 0$ if Q_1 is not just more spread out than Q_1, but is of the special type $Q_2 = Q_1 * R$. This is, however, not the case. Example 6.1 in Pfanzagl (2000b, p. 7) shows that for any $t', t'' > 0$, $t' \neq t''$, there exists R with expectation 0 such that

$$N(0,1)(-t', t'') < N(0,1) * R(-t', t'').$$

(See also Lynch et al. 1983, p. 890, Theorem 2.)

There is yet another condition which implies (3.7.1): If Q_i has a continuous Lebesgue density q_i such that $\{x \in \mathbb{R} : q_1(x) \geq q_2(x)\}$ is an interval containing a common quantile μ, then (3.7.1) holds true (see Pfanzagl 1994, p. 75, Lemma 2.3.2).

3.8 Anderson's Theorem

For the interpretation of the Convolution Theorem (Sect. 5.13) with convolution kernel Q, one needs conditions on $Q|\mathbb{B}^k$ which imply

$$Q * R(B) \leq Q(B) \tag{3.8.1}$$

for a relevant family $\mathscr{B}$ of sets $B|\mathbb{R}^k$. Since relation (3.8.1) is required to hold for any $R|\mathbb{B}^k$, this amounts to finding conditions on Q and $\mathscr{B}$ such that

$$Q(B+y) \leq Q(B) \quad \text{for every } y \in \mathbb{R}^k \text{ and } B \in \mathscr{B}. \tag{3.8.2}$$

The usual textbooks provide for this purpose a shortened version of Anderson's Theorem, which asserts (3.8.1) under the condition that Q is unimodal and symmetric about 0, and B is convex and symmetric about 0. In fact, Anderson's Theorem offers a stronger result: Rewritten in our notation it asserts the following.

Anderson's Theorem. *If $Q|\mathbb{B}^k$ is unimodal and symmetric about 0, then*

$$Q(C+ry) \geq Q(C+y) \quad \textit{for every } y \in \mathbb{R}^k \textit{ and every } r \in [0,1],$$

provided that C is convex and symmetric about 0.

Remark. A conjecture by Sherman (1955, p. 766) claims that symmetry and convex-unimodality of Q might even be necessary for the validity of (3.8.2). This conjecture was disproved by Wells (1978), using a suggestion by Dharmadhikari and Jogdeo (1976, p. 612, Example 4.1): There is a symmetric Q, star-unimodal but not convex-unimodal, such that $y \to Q(B+y)$ is star-unimodal for every symmetric convex set B. An example by Wefelmeyer (1985) presents a star-unimodal symmetric Q such that $y \to Q(B+y)$ is not star-unimodal for every symmetric convex set B.

The shortened version of Anderson's Theorem suffices for the interpretation of the Convolution Theorem by means of relation (3.8.1). Yet, it might be of interest to have a closer look at the function $y \to Q(C+y)$. Anderson's Theorem asserts that it is star-down, starting from $y = 0$. For $k = 1$, this is the same as unimodality.

As already remarked by Andersen (1955, p. 171; see Sherman 1955, p. 764 for a counterexample), this function is not necessarily unimodal if $k > 1$. (Sherman's example is reproduced in Das Gupta 1976, Example 1, pp. 90/1 and in Dharmadhikari and Joag-Dev 1988, p. 65.)

Though $y \to Q(C + y)$ is not unimodal in general, it can be approximated by a sequence of unimodal functions, which serves nearly the same purpose. To see this, we may go back to an alternative proof of Anderson's Theorem offered by Sherman (1955). According to his Lemma 1, p. 764, the function

$$y \to \int 1_C(x + y) 1_K(x) \lambda^k(dx) \tag{3.8.3}$$

is unimodal if C and K are bounded convex sets in $\mathbb{B}^k$. (No symmetry!) In fact, Sherman's result is just a weaker version of Satz 3 in Fáry and Rédei (1950, 207) which implies that the function (3.8.3) is logconcave.

Consider now the function

$$y \to \int 1_C(x + y) q(x) \lambda^k(dx), \tag{3.8.4}$$

where q is a unimodal density. Approximating q by a sequence of functions of the form $\sum_{i=1}^N a_i 1_{K_i}$ with $a_i \geq 0$ and K_i convex, one obtains that (3.8.4) is approximable by functions $y \to \sum_i^n a_i \int 1_C(x+y) 1_{K_i}(x) \lambda^k(dx)$. Hence for unimodal distributions Q, the function $y \to Q(C + y)$ is in $\mathscr{M}$. If we now add the assumption that Q and C are symmetric about 0, we obtain that the function $y \to Q(C + y)$ is star-down from $y = 0$ (which is Anderson's Theorem). Since it is, in addition, in $\mathscr{M}$, relation (3.8.1) can be extended to

$$Q * P_1 \leq Q * P_2 \quad \text{if } P_1 \leq P_2. \tag{3.8.5}$$

Since relation (3.8.3) already occurs in Fáry and Rédei (1950), Anderson's Theorem could have been obtained five years earlier, but nobody was interested. Even so, Anderson's Theorem had to wait more than ten years to find its pivotal role in the interpretation of the Convolution Theorem.

In connection with the characterization of maximum likelihood (ML) sequences as "asymptotically Bayes", Le Cam (1953, p. 315) introduces the concept of gain functions g such that

$$y \to \int g(x) N(y, \Sigma)(dx) \tag{3.8.6}$$

attains its maximum at $y = 0$. He abstained from identifying these gain functions, and Anderson, looking for statistical applications of his theorem, did not realise that he had provided an answer to Le Cam's problem: Relation (3.8.6) holds for all gain functions which are unimodal and symmetric about 0. Le Cam, who loved to surprise his readers by unusual references, had missed the chance to cite Satz 3 in Fáry and Rédei (1950, p. 207), from which Anderson's Theorem easily follows.

The unimodality and symmetry of Q are the basic conditions of Anderson's Theorem. For interpretations of the Convolution Theorem, it is only the case $Q = N(0, \Sigma)$ where symmetry applies. Yet, $N(0, \Sigma)$ has an important feature going beyond unimodality: Its density is logconcave. Though logconcavity had already turned out to be an important property in Ibragimov's paper from 1956, the question was never taken up whether there is a stronger version of Anderson's Theorem for the particular case $Q = N(0, \Sigma)$. The key for such a possible improvement is the following theorem[1] (Prékopa 1973, p. 342, Theorem 6).

Prékopa's Theorem. *If $f : \mathbb{R}^m \times \mathbb{R}^k \to [0, \infty)$ is logconcave, then*

$$y \longrightarrow \int f(x, y)\lambda^m(dx) \quad \textit{is logconcave.}$$

If $q \in dQ/d\lambda^k$ is logconcave and C is convex, then the map $(x, y) \to 1_C(x+y)q(x)$ is logconcave, too. Hence

$$y \to \int 1_{C+y}(x)q(x)\lambda^k(dx) = Q(C + y) \quad \text{is logconcave.}$$

If Q and C are symmetric about 0, this implies that the function $y \to Q(C+y)$, too, is symmetric about 0. Being logconcave, it is, therefore, unimodal about 0. This is a property which does not apply, in general, if Q is merely unimodal and symmetric. (Recall the example in Sherman 1955, p. 764.)

Remark. Obviously Prékopa was unaware of the relation between his result and both Anderson's Theorem and the Convolution Theorem. He also seems to have overlooked a forerunner of his result in the paper by Davidovič et al. (1969), which states that $y \to \int f_1(x + y) f_2(x)\lambda^k(dx)$ is logconcave if the functions f_i are logconcave. These authors, in turn, were unaware of a just slightly weaker result by Lekkerkerker (1953, p. 505/6, Theorem 1), which asserts that $y \to \int f_1(x + y) f_2(x)dx$ is decreasing and logconcave on $y \in (0, \infty)$ if the functions f_i have this property. Readers interested in this field, with a surprising number of multiple discoveries, would be well advised to consult Dharmadhikari and Joag-Dev (1988), Pečarić et al. (1992), and Das Gupta (1980). More may be found in Bertin et al. (1997), but even readers ready to follow the abstract approach of these authors will be daunted by their forbidding notation. □

---

[1]According to Pfanzagl (1994, p. 86, Corollary 2.4.10), the function $y \to \int f(x, y)\lambda^k(dx)$ is concave if f is unimodal. This stronger assertion is obviously wrong: It would imply that any subconvex function is concave, since $(x, y) \to 1_{[0,1]^k}(x)\ell(y)$ is subconvex if ℓ is subconvex, and $\ell(y) = \int 1_{[0,1]^k}(x)\ell(y)\lambda^k(dx)$. The proof uses that $y \to \lambda^k(C_y)$ is concave, which is true only on $\{y \in \mathbb{R}^m : \lambda^k(C_y) > 0\}$, not on $\mathbb{R}^m$. None of the reviewers mentioned this blunder.

Does the logconcavity of $y \to Q(C + y)$ contribute to a refined interpretation of the Convolution Theorem? It is certainly a welcome property that $y \to Q(C + y)$ varies "regularly", that the set $\{y \in \mathbb{R}^k : Q(C + y) \geq r\}$ is not "fuzzy" in some intuitive sense. The more basic requirement, that $Q(C + y)$ is "decreasing" along rays $y = ty_0$, is already fulfilled in the general case of symmetric and unimodal Q.

To what extent do results like (3.8.2) extend from 1_C to general gain functions g, i.e., what are the properties of

$$y \to \int g(x + y)Q(dx)?$$

The extension of the most fundamental result is immediate: If $y \to \int 1_C(x + y)Q(dx)$ is star-down for every convex and symmetric set C, then $y \to \int g(x + y)Q(dx)$ is star-down for every symmetric and unimodal function g. More complex properties of $y \to \int 1_C(x+y)Q(dx)$ like unimodality or logconcavity require more subtle arguments. If $q \in dQ/d\lambda^k$ and g are logconcave, the function

$$y \to \int g(x + y)Q(dx) \tag{3.8.7}$$

is logconcave and therefore unimodal according to Prékopa's Theorem, applied with $f(x, y) = g(x + y)q(x)$. Of course, one would like to have unimodality (3.8.7) under conditions on g weaker than logconcavity (though logconcavity is not an unreasonable property of gain functions.) For $k = 1$, the function $y \to \int g(x + y)Q(dx)$ is, indeed, unimodal if g is unimodal and Q logconcave (see Pfanzagl 2000b, p. 12, Theorem 9.1.) Sherman's example shows that this does not extend to dimensions $k > 1$.

3.9 The Spread of Convolution Products

The results on the concentration of convolution products for measures on $\mathbb{B}^k$, presented in Sect. 3.8, are confined to the concentration on convex sets which are symmetric about 0. For probability measures on $\mathbb{B}$, the concept of "spread" offers the possibility to characterize the concentration of convolution products in a different way. For this purpose, a new theorem was required (see Lewis and Thompson 1981, pp. 88/9).

Theorem of Lewis and Thompson. *If $Q|\mathbb{B}$ has a logconcave density, then $P_1 \succeq P_2$ implies*

$$Q * P_1 \succeq Q * P_2. \tag{3.9.1}$$

Observe the analogy to (3.7.3) for Q and P_i on $\mathbb{B}^k$.

For the interpretation of the Convolution Theorem, the application of (3.9.1) with $R_2\{0\} = 1$ suffices:

If $Q|\mathbb{B}$ has a logconcave density, then

$$Q \preceq Q * P \quad \textit{for every } P|\mathbb{B}. \tag{3.9.2}$$

Of course, one would like to have (3.9.2) for a larger class of distributions Q. This is ruled out by a result of Droste and Wefelmeyer (1985, p. 237, Proposition 2): The validity of (3.9.2) for arbitrary R already characterizes Q as logconcave. (See also Klaassen 1985, p. 905, Theorem 1.1.) The fact that relation (3.9.1) is, with respect to the spread order, restricted to logconcave Q, does not limit its usefulness for the interpretation of the Convolution Theorem.

Now we will discuss the consequences which spread order has for the concentration on intervals. Since spread is shift invariant, the relation $Q_0 \preceq Q_1$ leads to a comparison of the respective concentrations only if Q_0 and Q_1 are comparable with respect to their location. Lemma 3.5.1 implies the following.

If $Q|\mathbb{B}$ is logconcave, then

$$Q * R(I) \le Q(I) \quad \text{for every interval } I \text{ containing } 0, \tag{3.9.3}$$

if Q and $Q * R$ have median 0.

If Q is normal, the inequality (3.9.2) is strict unless $R\{0\} = 1$. This is not necessarily true for arbitrary logconcave distributions Q. If Q is exponential, there are non-degenerate R such that Q and $Q * R$ have median 0, and yet $Q * R(-t', t'') = Q(-t', t'')$ for some $t', t'' > 0$. (See Pfanzagl 2000b, p. 6, Remark 5.2.)

Relation (3.9.3) refers to regularly attainable limit distributions of median unbiased estimator sequences. If the Convolution Theorem holds for such a limit distribution with a logconcave factor Q, then Q is in this class of limit distributions maximally concentrated on all intervals containing the median. A straightforward application is to limit distributions which are convolution products with an exponential factor, evaluated by means of an arbitrary (i.e., not necessarily symmetric) loss function.

If the convolution factor is $N(0, \sigma^2)$, then both Anderson's Theorem and the Theorem of Lewis and Thompson apply. Hence $N(0, \sigma^2)$ maximizes the probability of $(-t, t)$ in the class of all regularly attainable limit distributions, and it maximizes the probability of $(-t', t'')$ in the class of all regularly attainable limit distribution with median 0. Expressed in terms of loss functions this occurs in Witting and Müller-Funk (1995, pp. 440/1), the only textbook on mathematical statistics taking notice of the concept of "spread" and the Theorem of Lewis and Thompson (see p. 447, Satz 6.221).

Remark. Given a convolution product $Q = N(0, \sigma^2) * R$, let Q_0 be the shifted version of Q with median 0. Then

$$Q_0(I) \le N(0, \sigma^2)(I) \quad \text{for every interval } I \text{ containing } 0. \tag{3.9.4}$$

This leads to the question whether for $k > 1$, too, there is always a shifted version Q_0 of $Q = N(0, \Sigma) * R$ such that $Q_0(B) \leq N(0, \Sigma)(B)$ for a class of sets B more general than the convex and symmetric ones, say for all convex sets containing 0.

Kaufman (1966, Sect. 6, pp. 176–178) presents for $k = 2$ the example of a distribution $Q = N(0, \Sigma) * R$ (in fact a regularly attainable limit distribution) with the following property. For every shifted version Q_0 there are rectangles $I_1 \times I_2$ containing 0 such that

$$Q_0(I_1 \times I_2) > N(0, \Sigma)(I_I \times I_2).$$

What could be considered as a generalization for not necessarily symmetric intervals from $k = 1$ to arbitrary k is that the distribution of $a^\top u$ is more spread out under $N(0, \Sigma) * R$ than under $N(0, \Sigma)$, so that (3.9.4) holds for the distribution of $a^\top u$. In a somewhat disguised form this occurs in Hájek (1970, p. 329, Corollary 2). (See Sect. 5.13 for details.)

3.10 Interpretation of Convolution Products

When Kaufman presented his fundamental result (1966, p. 157, Theorem 2.1) he simply said: A uniformly attainable limit distribution cannot be more concentrated—on convex sets symmetric about zero—than the limit distribution of the ML-sequence. It was an idea of Inagaki (1970, p. 10, Theorem 3.1), followed by Hájek (1970, p. 324, Theorem), to express limit distributions as convolutions with a factor $N(0, \Sigma)$, say $N(0, \Sigma) * R$. Even though they supply their result with an interpretation based on Anderson's Theorem, one could question whether it was a good idea to use the convolution form to express an optimality result. It appears that certain authors consider the convolution form as some sort of a final result. This happens already in LeCam's paper (1972, p. 259, Examples 2 and 3), where he presents an abstract Convolution Theorem. Though he points to convolutions with an exponential factor as a special case, he feels no need for an interpretation in terms of probabilities.

Ibragimov and Has'minskii present a Convolution Theorem containing the factor $N(0, \Sigma)$ (1981, Theorem II,9,1, p. 154). Then they use four pages (155–158) to prove Anderson's Theorem, which they need for the interpretation of this result (see Theorem II.11.2, p. 160). In Theorem V.5.2, p. 278, they present a Convolution Theorem containing an exponential factor which

> is analogous to Theorem II.9.1 with the normal distribution replaced by an exponential one.

No further comment on the interpretation of convolutions with an exponential factor where Anderson's Theorem does not apply.

Section 3.9 presents conditions under which the relation $Q = \overline{Q} * R$ admits an interpretation in terms of probabilities, conditions which justify the assertion that "estimator sequences with limit distribution $\overline{Q}$ are preferable over estimator sequences with limit distribution Q" or, more precisely, "$\overline{Q}$ is more concentrated than Q".

Are such considerations really necessary? Some scholars accept the statement that "$\overline{Q}$ is better than $\overline{Q} * R$" at face value. Their argument: $Q = \overline{Q} * R$ is a disturbed version of $\overline{Q}$; the convolution with R spreads out mass. An argument of this kind already occurs in Andersen (1955, p. 173):

> $\overline{Q} * R$ is "in a certain sense...more spread out than $\overline{Q} * R_t$ for $t \in (0, 1)$", where $R_t := R \circ (v \to tv)$.

Millar (1983, p. 155, Remark 4.3) says so explicitly.

> Since convolution "spreads out mass", it is natural to define [!] a sequence of regular estimators to be *efficient* if [... R ...] is unit mass at $\{0\}$.

A similar argument occurs in Bickel et al. (1993, p. 24).

What could be the intuitive background of the "spreading out mass" idea? By convolution, each point u is replaced by a probability measure, which distributes the point-mass of u over $\mathbb{R}^k$. Hence any fixed set $B \subset \mathbb{R}^k$ loses part of its mass. Could it be that the proponents of the "spread out mass" idea have overlooked the fact that, at the same time, B gains mass from points $u \in B^c$, which ruins the idea of the set B losing mass, an idea which, obviously, cannot apply to B and B^c at the same time. For which sets B should Millar's argument apply then? It is, in fact, surprising that this back and forth—with an unknown (!) R—leads to a predictable result in certain cases.

The representation as a convolution product is not unique. If $Q = \overline{Q} * R$, then $Q = \overline{Q}_c * R_{-c}$ for every $c \in \mathbb{R}^k$ (where P_c is a shifted version of P). Hence, which one of the probability measures $\overline{Q}_c$ is better than Q? If Q is a normal distribution, $N(0, \Sigma)$ is, in some sense, distinguished among the convolution kernels $N(c, \Sigma)$, $c \in \mathbb{R}^k$. If $\overline{Q}$ is the exponential distribution E, such a distinguished version does not exist. How can we, then, express the superiority of E over $E * R$?

If a given class $\mathfrak{Q}$ of limit distributions contains an element $\overline{Q}$ which is minimal in the convolution order, i.e., every $Q \in \mathfrak{Q}$ can be represented as $Q = \overline{Q} * R$, then $\overline{Q}$ is unique up to a shift, provided $\mathfrak{Q}$ is closed under shifts. (For a proof see Pfanzagl 2000b, p. 3, Proposition 3.1.)

Millar's argument does not refer to any property of $\overline{Q}$. Even in a favourable case with symmetric sets B and a symmetric $\overline{Q}$, it depends on more subtle properties of $\overline{Q}$ (like unimodality) whether $\overline{Q}$ loses mass on B. If $\overline{Q} = \frac{1}{2}N(-c, 1) + \frac{1}{2}N(c, 1)$ and $R = N(0, 1)$, then $\hat{Q} * R(-t, t) > \overline{Q}(-t, t)$ if t is sufficiently small.

Since Millar's argument refers to any kind of randomization, it should also apply to randomization using a Markov kernel: $\int M(x, B)Q(dx)$ ought to be smaller than $Q(B)$—for which B? Even in a situation which seems to be particularly favourable for Millar's claim, it can easily be refuted: For Q symmetric about 0, and B convex and symmetric about 0. If $\hat{Q} = N(0, 1)$ and $M(x, \cdot) = N(0, x^{-2})$, then

$$\int M(x, (-t, t))N(0, 1)(dx) > N(0, 1)(-t, t) \quad \text{for every } t > 0.$$

3.11 Loss Functions

Let $\hat{\kappa} : X \to \mathbb{R}^k$ be an estimator of the functional $\kappa : \mathfrak{P} \to \mathbb{R}^k$. The task of the loss function $\ell_P : \mathbb{R}^k \to [0, \infty)$ is to express the loss resulting from the deviation between $\hat{\kappa}(x)$ and $\kappa(P)$. This requires that $\ell_P(u) = 0$ if $u = \kappa(P)$, and that $\ell_P(u)$ is nondecreasing as u moves away from $\kappa(P)$. For $k = 1$, this means that $u \to \ell_P(u)$ is nondecreasing as u moves away from $\kappa(P)$ in either direction. The obvious extension to arbitrary dimensions k is that ℓ_P is monotone with center $\kappa(P)$, in other words: that $\{u \in \mathbb{R}^k : \ell_P(u) \leq r\}$ is for every $r > 0$ star-shaped with center $\kappa(P)$. The generally accepted concept of a loss function is more restrictive: $\{u \in \mathbb{R}^k : \ell_P(u) \leq r\}$ is for every $r > 0$ a convex set containing $\kappa(P)$. In the usual terminology, such loss functions are called "subconvex" about $\kappa(P)$. Warning: Some authors include in these terms that ℓ_P is symmetric about $\kappa(P)$. Recall that the restriction to symmetric loss functions is not required if $k = 1$.

The example of the binomial distribution demonstrates the need for loss functions other than $\ell_P(u) = \ell(u - \kappa(P))$. A deviation of 0.1, say, from $p = 0.5$ has a weight smaller than the same deviation from $p = 0.8$. $|\log(\hat{p}/(1-\hat{p})) - \log(p/(1-p))|$ might be an expression for the difference between $\hat{p}$ and p more adequate than $|\hat{p} - p|$. After all,

$$u \to |\log(u/(1-u)) - \log(p/(1-p))|, \quad u \in (0, 1)$$

is subconvex and attains its minimum for $u = p$.

If one believes in the possibility of measuring the loss caused by the deviation of an estimate from the estimand, one would expect something like a "law of diminishing returns", hence a loss function with decreasing derivative. This excludes a priori all convex loss functions. One could even argue that a loss function should be bounded. With unbounded loss functions, the expected loss might be determined by values the estimators hardly ever attain, and one could question whether that's what really counts.

So far, the comparison between estimators $\hat{\kappa}_i$, $i = 1, 2$, has been based on the concentration of $P \circ \hat{\kappa}_i$ on certain sets. To focus on the essentials, we simplify the notations by omitting P, which is fixed throughout. The problem is now to evaluate probability measures $Q|\mathbb{B}^k$, expressing the distribution of an estimator $\hat{\kappa}$ under P, by means of a loss function $\ell : \mathbb{R}^k \to [0, \infty)$ with the properties indicated above. For asymptotic considerations, Q denotes the (non-degenerate) limit distribution of an appropriately standardized estimator sequence, say $Q = \lim_{n\to\infty} P^n \circ c_n(\kappa^{(n)} - \kappa(P))$.

In Sects. 3.7 and 3.9 the criterion was the concentration on certain interesting subsets of $\mathbb{B}^k$. In the present section we shall investigate whether a comparison based on loss functions offers additional insights. Most authors accept an expression like $\int \ell(\hat{\kappa} - \kappa(P))dP$ or $\int \ell(c_n(\kappa^{(n)} - \kappa(P)))dP^n$ as a measure of the accuracy of the estimator $\hat{\kappa}$. (See e.g. Ibragimov and Has'minskii 1981, p. 16, Lehmann and Casella 1998, p. 5, and many more.) Already used by Laplace (1820) and Gauss (1821), it was revived by Wald, who introduced (1939, p. 304) the misleading term "risk". We prefer to speak of "expected loss" or "average loss". The use of "gain functions" by Le Cam (1953), which is more convenient since "gain" is in direct relation to "concentration", was not accepted by the statistical community.

Remark. Some authors accept the evaluation of estimator sequences by $\ell(c_n(\kappa^{(n)} - \kappa(P)))$ only with certain reservations. After all, $u \to \int \ell(c_n(u - \kappa(P)))$ evaluates the loss using a different loss function for every $n \in \mathbb{N}$. Arguments brought forward by some scholars to justify the application of the loss function to the standardized estimator sequence are spurious. As an example we might mention Millar (1983, p. 145):

> [Since $n^{1/2}$ is the best possible rate of convergence] it is reasonable to specify the loss by $\ell(n^{1/2}(\vartheta^{(n)} - \vartheta))$.

As J.K. Ghosh (1985, p. 318) succinctly states,

> the scaling by $\sqrt{n}$ remains, to some extent, more a matter of tradition and mathematical convenience than good statistical common sense.

The justification for the use of $\ell(c_n(\kappa^{(n)} - \kappa(\vartheta)))$ is more convincing in the special case $\ell(u) = 1_{[-t,t]}(u)$: The concentration of estimator sequences $\kappa^{(n)}$ is then compared on sets $(\kappa(P) - c_n t, \kappa(P) + c_n t)$ containing these estimator sequences with a probability which is neither negligible nor close to 1.

With ℓ fixed, $\int \ell dQ$ defines a total order between all probability measures $Q|\mathbb{B}^k$. Applied with an arbitrarily chosen ℓ, this order says nothing about the reality. One way out is to consider a large family $\mathscr{L}_0$ of all potential loss functions (including the unknown true one). Yet, is there any chance that the order based on the expected loss will be the same for every $\ell \in \mathscr{L}_0$? Since $\int \ell dQ = \int_0^\infty Q\{u \in \mathbb{R}^k : \ell(u) > r\}dr$, it seems unlikely that

$$\int \ell dQ_0 \leq \int \ell dQ_1 \tag{3.11.1}$$

for a large number of loss functions, unless there is an inherent relationship between the probability measures Q_i and the loss functions ℓ, which implies that

$$Q_0\{u \in \mathbb{R}^k : \ell(u) \geq r\} \leq Q_1\{u \in \mathbb{R}^k : \ell(u) \geq r\} \quad \text{for every } r > 0, \tag{3.11.2}$$

or that there is a large family $\mathscr{B}_0$ of subsets $B \subset \mathbb{B}^k$ such that

$$Q_0(B) \leq Q_1(B) \quad \text{for } B \in \mathscr{B}_0.$$

If this is the case, relation (3.11.1) will be true for all loss functions ℓ such that

$$\{u \in \mathbb{R}^k : \ell(u) \leq r\} \in \mathscr{B}_0 \quad \text{for every } r > 0.$$

Here is the relevant example, with two variants.

(i) If $Q_0|\mathbb{B}^k$ is symmetric and unimodal, then (3.11.2) holds for every set B which is convex and symmetric about zero (see Sect. 3.8).

(ii) If $Q_0|\mathbb{B}$ is logconcave, and if both Q_0 and Q_1 have the same median, then relation (3.11.2) holds for every interval containing the median (see Sect. 3.9).

In either case, $Q_0 \circ \ell$ is stochastically smaller than $Q_1 \circ \ell$, and, therefore, $\int \ell dQ_0 \leq \int \ell dQ_1$ for $\ell \in \mathscr{L}$, the set of all loss functions which are subconvex and—in case (i)—also symmetric. This implies in particular that the optimal limit distribution is comparable with every regularly attainable limit distribution (which are not mutually comparable in general).

Remark. In a context where unimodality of Q is not available (say probability measures on $C[0,1]$) a comparable result might be obtained under more restrictive conditions on the loss function.

If ℓ is convex, then $y \to \int \ell(u+y)Q(du)$ is convex. If Q and ℓ are symmetric, then $y \to \int \ell(u+y)Q(du)$ is symmetric. Symmetry and convexity together imply

$$\int \ell(u+y)Q(du) \geq \int \ell(u)Q(du) \quad \text{for every } y \in \mathbb{R}^k,$$

whence

$$\int \ell dQ * P \geq \int \ell dQ \quad \text{for every } P|\mathbb{B}^k.$$

This argument was used by Dvoretzky et al. (1956, p. 666) and Beran (1977, p. 402).

If ℓ is subconvex, then $m \circ \ell$ is subconvex for any nondecreasing function m with $m(0) = 0$. Apart from this special type of loss functions, it is a reasonable requirement on any class of loss functions to be closed under monotone transformations.

Some authors think of "loss" measured on a cardinal scale. Yet convincing examples are lacking. A more realistic assumption is that the loss is measured on an ordinal scale only, and this, too, requires that $m \circ \ell$ be a possible loss function if ℓ is such a one.

Proposition 3.11.1 *For families $\mathscr{L}$ of loss functions which are closed under monotone transformations, the following relations are equivalent.*

(i) $\int \ell dQ_0 \leq \int \ell dQ_1$ for every $\ell \in \mathscr{L}$.
(ii) $Q_0 \circ \ell$ is stochastically smaller than $Q_1 \circ \ell$ for every $\ell \in \mathscr{L}$.

Proof If $\int m \circ \ell dQ_0 \leq \int m \circ \ell dQ_1$ for any m on decreasing m with $m(0) = 0$, then this relation holds in particular with $m_s(u) = 1_{[s,\infty)}(u)$ for any $s > 0$. Since

$$\int m_s \circ \ell dQ = \int_0^\infty Q\{u \in \mathbb{R}^k : m_s(\ell(u)) \geq r\}, dr = Q\{u \in \mathbb{R}^k : \ell(u) \geq s\},$$

the equality $\int m_s \circ \ell dQ_0 \leq \int m_s \circ \ell dQ_1$ for every $s > 0$ implies that $Q_0 \circ \ell$ is stochastically smaller than $Q_1 \circ \ell$. □

If $\mathscr{L}$ is restricted to loss functions fulfilling certain additional conditions (such as smoothness, semicontinuity or $\ell(u) > 0$ for $u \neq 0$), the function m_s may be approximated by $\bar{m}_s$ such that $\bar{m}_s \circ \ell \in \mathscr{L}$ if $\ell \in \mathscr{L}$.

The operational significance of expected loss is called into question if the relation between $\int \ell(au)Q_0(du)$ and $\int \ell(au)Q_1(du)$ depends on a. This problem arises necessarily in asymptotic comparisons which are on standardized estimator sequences $c_n(\kappa^{(n)} - \kappa(P)$, where c_n is chosen such that $P^n \circ c_n(\kappa^{(n)} - \kappa(P))$ converges to a non-degenerate limit distribution. Since the rate c_n is not unique, standardization by $\bar{c}_n = ac_n$ is equally justified. (If the rate $c_n = n^{1/2}$ leads to the limit distribution $N(0, \sigma^2(P))$, it is not unusual to standardize by $\sigma(P)^{-1}n^{1/2}$ to obtain the limit distribution $N(0, 1)$.)

Let us assume, for the sake of illustration, that ℓ is bounded and continuous, so that $P^n \circ c_n(\kappa^{(n)} - \kappa(P)) \Rightarrow Q$ implies

$$\int \ell(c_n(\kappa^{(n)} - \kappa(P)))dP^n \to \int \ell(u)Q(du).$$

If the standardization uses $\bar{c}_n = ac_n$, then

$$\int \ell(\bar{c}_n(\kappa^{(n)} - \kappa(P)))dP^n \to \int \ell(au)Q(du).$$

If the asymptotic performance of two estimator sequences $(\kappa_i^{(n)})_{n\in\mathbb{N}}$ with

$$P^n \circ c_n(\kappa_i^{(n)} - \kappa(P)) \to Q_i$$

is evaluated by a comparison between $\lim_{n\to\infty} \int \ell(c_n(\kappa_i^{(n)} - \kappa(P)))dP^n$, $i = 0, 1$, it might turn out that, for large n,

$$\int \ell(c_n(\kappa_0^{(n)} - \kappa(P)))dP^n < \int \ell(c_n(\kappa_1^{(n)} - \kappa(P)))dP^n,$$

yet

$$\int \ell(\bar{c}_n(\kappa_0^{(n)} - \kappa(P)))dP^n > \int \ell(\bar{c}_n(\kappa_1^{(n)} - \kappa(P)))dP^n,$$

corresponding to

$$\int \ell(u)Q_0(du) < \int \ell(u)Q_1(du)$$

and

$$\int \ell(au)Q_0(du) > \int \ell(au)Q_1(du).$$

(To obtain an example with risks that can be computed explicitly, choose $Q_0 = N(0,1)$, $Q_1 = \frac{1}{2}N(-\mu, \sigma^2) + \frac{1}{2}N(\mu, \sigma^2)$, and $\ell(u) = 1 - \exp[-u^2]$.)

To summarize: Comparisons based on the expected loss under a single loss function are significant only if this is the true loss function. Consistent results for a large class of loss functions can be expected under special conditions only. In such cases, the comparison of losses leads to the same result as the comparison of the distribution of losses or the concentration of the estimators in certain sets. It therefore adds nothing to the results already known from the comparison of concentrations.

The mathematical argument for applying a loss function (attaining its minimum at 0) to $c_n(\kappa^{(n)} - \kappa(P))$ rather than to $\kappa^{(n)} - \kappa(P)$ is clear: Since $\kappa^{(n)} - \kappa(P) \to 0\ (P^n)$, the asymptotic performance of $\int \ell(\kappa^{(n)} - \kappa(P))dP^n$, $n \in \mathbb{N}$, depends on the local properties of ℓ at 0, and with a twice differentiable loss function one ends up with $\int (\kappa^{(n)} - \kappa(P))^2 dP^n$ as an approximation for the risk.

It is adequate to evaluate the accuracy of an estimator according to the length of an interval containing the estimate with high probability. (Recall the time-honoured comparison based on the "probable error".) This amounts to considering the concentration of $\kappa^{(n)}$ in intervals $(\kappa(P) - c_n^{-1}t', \kappa(P) + c_n^{-1}t'')$, or the concentration of $c_n(\kappa^{(n)} - \kappa(P))$ in $(-t', t'')$. The possibility of expressing this in terms of the loss function $\ell(u) = 1 - 1_{(-t',t'')}(u)$, applied to $c_n(\kappa^{(n)} - \kappa(P))$, does not imply that the evaluation of $c_n(\kappa^{(n)} - \kappa(P))$ by means of other loss functions yields a meaningful result.

If the evaluation using loss functions serves any purpose beyond comparing the asymptotic concentration on intervals, then that purpose is to obtain a global measure for this difference. It seems questionable whether this purpose is achieved. Depending on the loss function ℓ, the value of $\int \ell(\hat{\kappa}(x) - \kappa(P))P(dx)$ might depend mainly on the tails of $P \circ (\hat{\kappa} - \kappa(P))$. The objection against expressing concentration by means of loss functions gathers momentum if *asymptotic* concentration is the point: estimator sequences which are asymptotically equivalent (in terms of concentration in intervals) may widely diverge in terms of risks if the loss function is unbounded. There is yet another problem, resulting from the standardization by c_n. The purpose of this standardization is to ensure that $P^{(n)} \circ c_n(\kappa^{(n)} - \kappa(P))$, $n \in \mathbb{N}$, converges to a non-degenerate limit distribution. If this is achieved by means of standardization with the rate $(c_n)_{n\in\mathbb{N}}$, it is also achieved using standardization with the rate $\hat{c}_n = ac_n$, for any $a > 0$. The results are not necessarily the same (see Sect. 3.11.)

Few authors admit that comparisons of expected loss are operationally significant only if based on the *true* loss function. The "mathematical convenience" of a loss function has nothing to do with the very nature of the particular problem. Choosing a loss function on the basis of mathematical convenience, therefore, means —if judged on the basis of its suitability for a particular problem—choosing it ad libitum. Hence the optimality of an estimator with respect to such a loss function bears no relationship to reality. Here are a few important opinions on this problem.

C.R. Rao (1962, p. 74) is skeptical that the "criterion of minimum expected squared error", used—among many others—by Berkson (1956), is justified "unless he believes or makes us believe that the loss to society is proportional to the square of the error in his estimate".

The following statement by Stein (1964, p. 156) is even more important, as it occurs in a paper on the admissibility of the usual estimator for σ^2 if evaluated by the quadratic loss function: "I find it hard to take the problem of estimating σ^2 with quadratic loss function very seriously."

Usually one finds arguments of a purely formal nature.

Lehmann (1983, p. 8): "The choice of the squared error as loss has the twofold advantage of ease of computation and of leading to estimators that can be obtained explicitly", an argument already brought forward by Gauss.

Zellner (1986, p. 450) on the linex loss functions defined in (3.11.6) below: "The analytic ease with which results can be obtained makes them attractive for the use in applied problems..."

Another example of the casual approach to dealing with loss functions is (in our notations)

M.M. Rao (1965, p. 135): "It is more convenient [!] ... to consider the optimality of an (unbiased) estimator $\hat{\kappa}$ ... at P_0 as the minimum value of $N(\hat{\kappa} - \kappa(P_0))$ instead of that of $\int C(\hat{\kappa} - \kappa(P_0))dP_0$",

where

$$N(f) := \inf\{k > 0 : \int C(f/k)d\lambda \leq 1\}$$

and C is a convex and symmetric loss function with $C(0) = 0$.

Remark. Some authors (Shaked 1980, p. 193, Condition B, and Schweder 1982, p. 165, Definition 1) introduced for probability measures $Q_i|\mathbb{B}^k$ the concept of a *dilation order* $Q_0 \preceq Q_1$ by

$$\int C(u)Q_0(du) \leq \int C(u)Q_1(du) \quad \text{for every convex function } C|\mathbb{R}^k \tag{3.11.3}$$

(provided the integrals exist).

At first glance, this definition does not seem to refer to a particular center. Yet, relation (3.11.3), applied with $C(u) = u$ and $C(u) = -u$), implies

$$\int uQ_0(du) = \int uQ_1(du). \tag{3.11.4}$$

Theorem 2 in Muñoz-Perez and Sanchez-Gomes (1990, p. 442) implies that (3.11.3) is equivalent to (3.11.4) in combination with

$$\int |u - \mu|Q_0(du) \leq \int |u - \mu|Q_1(du) \quad \text{for every } \mu \in \mathbb{R}. \tag{3.11.5}$$

Since the dilation order is not very convincing from an intuitive point of view (observe that the convex function C is not required to be nonnegative), it is of interest to relate it to other order relations. According to Witting and Müller-Funk (1995, p.

493, Satz 7.24), relation (3.11.3) follows if the (necessary) condition (3.11.4) is fulfilled, and if (3.11.5) holds true for *some* μ.

Moreover, $Q_0 \preceq Q_1$ (in the spread order) implies

$$\int C(u - \mu_0) Q_0(du) \leq \int C(u - \mu_1) Q_1(du) \quad \text{with } \mu_i = \int u Q_i(du)$$

for every convex function C (since $Q_0 \preceq Q_1$ implies $Q_0^* \preceq Q_1^*$ and $\int u Q_i^*(du) = 0$, if $Q_i^* := Q_i - \mu_i$. From this, Q_0^* and Q_1^* have a common quantile, for which (3.11.5) applies). (See Shaked 1982, p. 313.)

Equality in (3.11.3) for $C(u) = |u|$ implies $Q_0 = Q_1$ (Pfanzagl 2000b, p. 8, Lemma 6.1).

The fact that relations like (3.11.3) are compatible with meaningful order relations (based on probabilities) does not entail that they are operationally significant in cases where an underlying meaningful order relation is not available. Moreover: In view of the enormous number of functions ℓ or C for which these relations hold true, it would need additional arguments to distinguish one of the expressions $\int \ell(u - \mu) Q(du)$ (say the one with $\ell(u) = |u|$ or $\ell(u) = u^2$) as "the" global measure of concentration.

Loss Functions and Unbiasedness

Lehmann (1951b, p. 587) suggests a general concept of unbiasedness, based on a given loss function. For the problem of estimation, this reads as follows:

Given a functional κ and a loss function $\ell(\cdot, Q)$ with minimum $\ell(\kappa(Q), Q)$, the estimator $\hat{\kappa}$ is *unbiased* for κ at P if the function $Q \to \int \ell(\hat{\kappa}(x), Q) P(dx)$ attains its minimum at $Q = P$. (Lehmann does not mention that $\kappa(Q)$ should minimize $\ell(\cdot, Q)$.

Specialized to $\ell_P(u) = (u - \kappa(P))^2$ and $\ell_P(u) = |u - \kappa(P)|$ this leads to mean unbiasedness and median unbiasedness, respectively. Yet, it appears that mean and median unbiasedness have a strong appeal of their own, and not because they come from some loss function. Loss functions are a rather artificial construct, and it seems questionable whether one should sacrifice concepts like mean or median unbiasedness with a clear operational significance for an unbiasedness concept based on a loss function whose only selling point is that it is "not totally unreasonable". For a discussion of various concepts of unbiasedness see also H.R. van der Vaart (1961).

Even if one abandons the idea of using an unbiasedness concept derived from a loss function, there should be no inherent conflict between, say, mean unbiasedness and the loss function. As a point of departure, let us assume that $\ell_P(u) = \ell(u - \kappa(P))$, where ℓ is subconvex, attaining its minimum at 0. Unbiasedness supposes, implicitly, that the deviation from $\kappa(P)$ by the amount Δ has the same weight, whether it is to $\kappa(P) + \Delta$ or to $\kappa(P) - \Delta$. Correspondingly, ℓ should be symmetric about 0.

Assume now that $\{P_a : a \in (0, \infty)\}$ is a scale parameter family, and that $\kappa(P_a) = a$. In this case it is natural to express the deviation of an estimate $\hat{a}$ from a by the deviation of $\hat{a}/a$ from 1, and to suppose that a deviation $\hat{a}/a$ from 1 by the factor 2, say, has the same weight as a deviation by the factor 1/2. If this is accepted, it makes no sense to require that $\hat{a}$ should be unbiased. If $\hat{a}$ is evaluated by $\ell(\hat{a}/a)$, with a loss function ℓ attaining its minimum at 1, then one would, if anything, require that $\ell(1/u) = \ell(u)$ rather than symmetry of $\ell(u)$ about $u = 1$. A loss function with this property is $\ell(u) = (\log u)^2$, suggested by Ferguson (1967, p. 179) as "more appropriate for scale parameters than the squared error loss". His opinion is, however, not incontrovertible. On p. 191 he also uses the squared error loss for a scale parameter only if doing so leads to a nicer result.

One would hesitate to write down such obvious remarks but for the fact that the textbooks are full of suggestions to measure the deviation of $\hat{a}(x)$ from a by means of loss functions like $(\hat{a}(x)/a - 1)^2$. A loss function not downright absurd for scale parameters is

$$\ell_0(u) = u - \log u - 1.$$

This loss function (suggested by James and Stein 1961, p. 376, relation (72) for matrix-valued estimators) is convex and attains its minimum at $u = 1$. Yet, it appears that this loss function harbours an inherent contradiction: ℓ_0 attributes different weights to $\hat{a}(x)/a = 1 + \Delta$ and $\hat{a}(x)/a = 1 - \Delta$. If this is the adequate description of a real situation, one would not think that—at the same time—mean unbiasedness is a natural requirement. Nevertheless, it is the mean unbiased estimator which minimizes the risk for the loss function ℓ_0. The function

$$a \to \int \ell_0(\hat{a}(x)/a) P_{a_0}(dx)$$

attains its minimum at $a = a_0$ iff $\int \hat{a}(x) P_{a_0}(dx) = a_0$. It follows from a result of Brown (1968, p. 35, Theorem 3.1) that ℓ_0 is, up to the transformation $\ell_0 \to \alpha\ell_0 + \beta$, the only loss function with this property (except for $u \to (u - a)^2$, of course). For real estate assessment, Varian (1975, p. 196) suggests evaluating the difference $\hat{\kappa}(x) - \kappa(P)$ by means of the so-called *linex* (for "linear-exponential") loss function

$$\ell_1(u) = \exp[au] - au - 1, \tag{3.11.6}$$

expressing—somehow—the idea that under-assessment results in an approximately linear loss of revenue, whereas over-assessment often results in substantial costs. It appears that Varian missed the relation between the linex loss function and the loss function ℓ_0, namely: $\ell_1(u) = \ell_0(\exp[au])$, and he also missed a characterization of ℓ_1 provided by Klebanov (1974). Given a family $\{P_\vartheta : \vartheta \in \Theta\}$, $\Theta \subset \mathbb{R}$, consider the following property of the loss function $u \to \ell(u - \vartheta)$: For any sufficient statistic $S|X$ and any ℓ-unbiased estimator there exists an ℓ-unbiased estimator which is a contraction of S and has "smaller" risk (with respect to this loss function). By the Rao–Blackwell Theorem, this statement holds true for $\ell(u) = u^2$. According to Klebanov (1974, p. 380, Theorem 1) there is just one more sufficiently regular loss function ℓ (convex with a continuous 2nd derivative and $\ell(0) = 0$) with this property: The linex.

3.12 Pitman Closeness

Given a concept of concentration (of a probability measure $Q|\mathbb{B}$ about 0) it is natural to evaluate the accuracy of an estimator $\hat{\kappa}$ by the concentration of $P \circ (\hat{\kappa} - \kappa(P))$ about 0, and to prefer $\hat{\kappa}_0$ over $\hat{\kappa}_1$ if $P \circ (\hat{\kappa}_0 - \kappa(P))$ is more concentrated about 0 than $P \circ (\hat{\kappa}_1 - \kappa(P))$. A new aspect is brought in if one considers the joint distribution $P \circ (\hat{\kappa}_0 - \kappa(P), \hat{\kappa}_1 - \kappa(P))$. Based on this joint distribution, Pitman (1937, p. 213) introduced the following concept of "closeness": $\hat{\kappa}_0$ is closer to $\kappa(P)$ than $\hat{\kappa}_1$ if

$$P\{|\hat{\kappa}_0 - \kappa(P)| < |\hat{\kappa}_1 - \kappa(P)|\} > 1/2.$$

Not all statisticians think that it is meaningful to compare estimators according to their closeness. L.J. Savage (1954, p. 225) insists on comparing estimators only by their concentration. Without a convincing argument, he states that "it makes no sense to consider the joint distribution of two estimators".

It would be an additional argument for the use of an estimator $\hat{\kappa}_0$ if it were not only more concentrated about $\kappa(P)$ than $\hat{\kappa}_1$, but, in addition, in Pitman's sense closer to $\kappa(P)$ than $\hat{\kappa}_1$; and it is of interest if an optimum property of an estimator is reflected in its Pitman closeness.

The concept of closeness would certainly have found general acceptance were it not for shortcomings which soon became evident. One important shortcoming was aired by Pitman himself (1937, p. 222): The order according to closeness is not necessarily transitive.

It is not surprising that a comparison based on a global measure like

$$\int |\hat{\kappa}_0 - \kappa(P)| dP < \int |\hat{\kappa}_1 - \kappa(P)| dP$$

does not tell us much about the joint distribution. One might, however, expect that $\hat{\kappa}_0$ is closer to $\kappa(P)$ than $\hat{\kappa}_1$ if, say, $|\hat{\kappa}_0 - \kappa(P)|$ is stochastically smaller than $|\hat{\kappa}_1 - \kappa(P)|$. Yet this is not the case. An example by Blyth (1972, p. 367, the "clocking paradox"), referring to a parametric family, shows that $P\{|\hat{\kappa}_0 - \kappa(P)| < |\hat{\kappa}_1 - \kappa(P)|\}$ might be close to zero, even though $\hat{\kappa}_0$ and $\hat{\kappa}_1$ have the same distribution. A slight modification of this example in Blyth and Pathak (1985, p. 46, Example 1) shows that the same effect may occur if $|\hat{\kappa}_0 - \kappa(P)|$ is stochastically smaller than $|\hat{\kappa}_1 - \kappa(P)|$.

Yet such examples are highly artificial, and the question arises whether the concept of closeness might be of some use in a more natural context. One such possibility is the following: If $\hat{\kappa}_0$ and $\hat{\kappa}_1 - \hat{\kappa}_0$ are stochastically independent, then $\hat{\kappa}_0$ is closer to $\kappa(P)$ than $\hat{\kappa}_1$, provided $\hat{\kappa}_0$ is median unbiased. This is (a special case of) Pitman's Comparison Theorem (see 1937, p. 214), proved under somewhat fuzzy conditions. The Convolution Theorem suggests that this Comparison Theorem could be used to show that an asymptotically optimal estimator sequence $\kappa_0^{(n)}$, $n \in \mathbb{N}$, is asymptotically closer to $\kappa(P)$ than any other "regular" estimator sequence.

We first consider a result for probability measures $Q|\mathbb{B}^k$, which will later be applied to limit distributions. If Q is a normal distribution, multidimensional median unbiasedness is a natural assumption.

Lemma 3.12.1 *Assume that $Q|\mathbb{B}^k$ is median unbiased in the sense that $Q \circ (u \to a^\top u)$ has median 0 for every $a \in \mathbb{R}^k$. Then the following holds true for every positive definite $k \times k$-matrix M and every probability measure $R|\mathbb{B}^k$.*

$$Q \times R\{(u, v) \in \mathbb{R}^k \times \mathbb{R}^k : u^\top M u \leq (u + v)^\top M(u + v)\} \geq 1/2. \qquad (3.12.1)$$

Proof Follows immediately from

$$u^\top M u \leq (u + v)^\top M(u + v) \quad \text{iff} \quad 2v^\top M u \geq -v^\top M v.$$

□

For $k = 1$, a relation equivalent to

$$Q \times R\{(u, v) \in \mathbb{R} \times \mathbb{R} : |u| \leq |u + v|\} \geq 1/2 \qquad (3.12.2)$$

occurs as Theorem 1 in M. Ghosh and Sen (1989, p. 1089). Relation (3.12.2) is, in fact, a special case of Pitman's "Comparison Theorem".

Ghosh and Sen mainly consider applications for parametric families and a median unbiased estimator $\hat{\vartheta}_0$ which is the function of a complete sufficient statistic. If $\hat{\vartheta}_1$ is another estimator such that $\hat{\vartheta}_1 - \hat{\vartheta}_0$ is ancillary, then $\hat{\vartheta}_0 - \vartheta$ and $\hat{\vartheta}_1 - \hat{\vartheta}_0$ are stochastically independent (according to Basu's Theorem), so that

$$P_\vartheta\{|\hat{\vartheta}_0(x) - \vartheta| \leq |\hat{\vartheta}_1(x) - \vartheta|\} \geq 1/2.$$

This rather special result is then illustrated by various examples.

Asymptotic results obtained from the application of (3.12.1) to limit distributions are of broader applicability. Assume now that $(\kappa^{(n)})_{n\in\mathbb{N}}$ is a regular estimator sequence with

$$P^{(n)} \circ c_n(\kappa^{(n)} - \kappa(P)) \Rightarrow Q|\mathbb{B}^k,$$

and $(\kappa_0^{(n)})_{n\in\mathbb{N}}$ an asymptotically optimal estimator sequence with

$$P^{(n)} \circ c_n(\kappa_0^{(n)} - \kappa(P)) \Rightarrow Q_0|\mathbb{B}^k,$$

where Q_0 is the optimal limit distribution. According to the Convolution Theorem (Sect. 5.13),

$$P^{(n)} \circ (c_n(\kappa_0^{(n)} - \kappa(P)), c_n(\kappa^{(n)} - \kappa_0^{(n)})) \Rightarrow Q_0 \times R.$$

Since the set in (3.12.1) is closed, Alexandrov's Theorem implies that

$$\limsup_{n\to\infty} P^{(n)}\big\{c_n(\kappa_0^{(n)} - \kappa(P))^\top M c_n(\kappa_0^{(n)} - \kappa(P)) \\ \le c_n(\kappa^{(n)} - \kappa(P))^\top M c_n(\kappa^{(n)} - \kappa(P))\big\} \ge 1/2. \tag{3.12.3}$$

In these relations, M is an arbitrary positive definite $k \times k$-matrix. The natural application is with M equal to the identity.

Relation (3.12.3) has a forerunner in Sen (1986, p. 54, Theorem 3.1). Though Sen refers (pp. 51, 56) to the papers by Hájek (1970), Inagaki (1970) and to Ibragimov and Has'minskii (1981), he does not fully exploit the force of the Convolution Theorem. His result refers to a k-dimensional regular parametric family $\{P_\vartheta : \vartheta \in \Theta\}, \Theta \subset \mathbb{R}^k$. Under the assumption that (in our notations) $P_\vartheta^n \circ (n^{1/2}(\vartheta_0^{(n)} - \vartheta), n^{1/2}(\vartheta^{(n)} - \vartheta))$, $n \in \mathbb{N}$, converges to a $2k$-dimensional normal distribution [!] with mean vector 0 and covariance matrix

$$\begin{pmatrix} \Sigma_0 & \Sigma_1 \\ \Sigma_1 & \Sigma \end{pmatrix},$$

he shows that

$$\lim_{n\to\infty} P_\vartheta^n\big\{n^{1/2}(\vartheta_0^{(n)} - \vartheta)^\top \Sigma_0^{-1} n^{1/2}(\vartheta_0^{(n)} - \vartheta) \\ \le n^{1/2}(\vartheta^{(n)} - \vartheta)^\top \Sigma_0^{-1} n^{1/2}(\vartheta^{(n)} - \vartheta)\big\} \ge 1/2. \tag{3.12.4}$$

Recall that $(u-v)^\top \Sigma_0^{-1}(u-v)$ has a traditional interpretation as the Mahalanobis distance between u and v.

Theorem 6.2.1 in Keating et al. (1993, p. 182), which asserts, in fact, that

$$\lim_{n\to\infty} P_\vartheta^n\big\{|\vartheta_0^{(n)} - \vartheta| \le |\vartheta^{(n)} - \vartheta|\big\} \ge 1/2$$

is the special case of (3.12.4) for $k = 1$.

The more general assertion (3.12.3) also does not establish Pitman closeness as a concept on which the comparison between estimators can be based. Pitman closeness is, however, a welcome additional property of an estimator which is considered to be "good" on the basis of other criteria.

There is another field where one would possibly expect to find a use for Pitman closeness: That an unbiased estimator of minimal variance is Pitman closer than any other unbiased estimator. This, however, is not true.

References

Aitken, A. C., & Silverstone, H. (1942). On the estimation of statistical parameters. *Proceedings of the Royal Society of Edinburgh Section A, 61*, 186–194.
Allen, S. G. (1953). A class of minimax tests for one-sided composite hypotheses. *The Annals of Mathematical Statistics, 24*, 295–298.
Anderson, T. W. (1955). The integral of a symmetric unimodal function over a symmetric convex set and some probability inequalities. *Proceedings of the American Mathematical Society, 6*, 170–176.
Bahadur, R. R. (1964). On Fisher's bound for asymptotic variances. *The Annals of Mathematical Statistics, 35*, 1545–1552.
Barnard, G. A. (1974). Can we all agree on what we mean by estimation? Util. *Mathematics, 6*, 3–22.
Barndorff-Nielsen, O. (1969). Lévy homeomorphic parametrization and exponential families. *Z. Wahrscheinlichkeitstheorie verw. Gebiete, 12*, 56–58.
Barton, D. E. (1956). A class of distributions for which the maximum-likelihood estimator is unbiased and of minimum variance for all sample sizes. *Biometrika, 43*, 200–202.
Basu, D. (1955). A note on the theory of unbiased estimation. *The Annals of Mathematical Statistics, 26*, 345–348.
Beran, R. (1977). Robust location estimates. *The Annals of Statistics, 5*, 431–444.
Berger, J. O. (1980). *Statistical decision theory: Foundations, concepts, and methods*. Springer, New York: Springer Series. in Statistics.
Berkson, J. (1956). Estimation by least squares and maximum likelihood. In *Proceedings of the Berkeley symposium on statistics and probability I* (pp. 1–11), California: University California Press.
Bertin, E. M. J., Cuculescu, I., & Theodorescu, R. (1997). *Unimodality of probability measures*. Dordrecht: Kluwer Academic Publishers.
Bickel, P. J., & Lehmann, E. L. (1975–1979). Descriptive statistics for nonparametric models. I: *The Annals of Statistics, 3*, 1038–1045; II: *The Annals of Statistics, 3*, 1045–1069; III: *The Annals of Statistics, 4*, 1139–1159. IV: In Contributions to Statistics, Hájek Memorial Volume (pp. 33–40). Academia, Prague.
Bickel, P.J., Klaassen, C.A.J., Ritov, Y., & Wellner, J. A. (1993). *Efficient and adaptive estimation for semiparametric models*. Johns Hopkins University Press. (1998 Springer Paperback)
Birnbaum, Z. W. (1948). On random variables with comparable peakedness. *The Annals of Mathematical Statistics, 19*, 76–81.
Blackwell, D., & Girshick, M. A. (1954). *Theory of games and statistical decisions*. New Jersey: Wiley.
Blyth, C. R.,& Pathak, P. K. (1985). *Does an estimator's distribution suffice?* In L. M. Le Cam & R. A. Olshen (Eds.), *Proceedings of the Berkeley conference in honor of Jerzy Neyman and Jack Kiefer* (Vol. 1, pp. 45–52). Wadsworth.
Blyth, C. R. (1972). Some probability paradoxes in choice from among random alternatives. With discussion. *Journal of the American Statistical Association, 67*, 366–381.
Borges, R., & Pfanzagl, J. (1963). A characterization of the one-parameter exponential family of distributions by monotonicity of likelihood ratios. Z. *Wahrscheinlichkeitstheorie verw. Gebiete, 2*, 111–117.
Bowley, A. L. (1897). Relations between the accuracy of an average and that of its constituent parts. *Journal of the Royal Statistical Society, 60*, 855–866.
Brown, L. D. (1968). Inadmissibility of the usual estimators of scale parameters in problems with unknown location and scale parameters. *The Annals of Mathematical Statistics, 39*, 29–48.
Brown, L. D., Cohen, A., & Strawderman, W. E. (1976). A complete class theorem for strict monotone likelihood ratio with applications. *Annals of Statistics, 4*, 712–722.
Chung, K. L. (1953). Sur les lois de probabilité unimodales. *C. R. Acad. Sci. Paris, 236*, 583–584.

Cramér, H. (1946b). A contribution to the theory of statistical estimation. *Skand. Aktuarietidskr.*, *29*, 85–94.

Cramér, H. (1946a). *Mathematical methods of statistics*. Princeton: Princeton University.

Das Gupta, S., Eaton, M. L., Olkin, I., Perlman, M., Savage, L. J., & Sobel, M. (1972). Inequalities on the probability content of convex regions for elliptically contoured distributions. In *Proceedings of the sixth Berkeley symposium on mathematical statistics and probability II* (pp. 241–265). Berkeley: University of California Press.

Das Gupta, S. (1976). A generalization of Anderson's theorem on unimodal functions. *Proceedings of the American Mathematical Society*, *60*, 85–91.

Das Gupta, S. (1980). Brunn-Minkowski inequality and its aftermath. *Journal of Multivariate Analysis*, *10*, 296–318.

Davidovič, Ju S, Korenbljum, B. I., & Hacet, B. I. (1969). A property of logarithmically concave functions. *Soviet Mathematics Doklady*, *10*, 477–480.

de Helguero, F. (1904). Sui massimi delle curve dimorfiche. *Biometrika*, *3*, 84–98.

de Moivre, A. (1756). *The doctrine of chances* (3rd ed.). London: Pearson.

Deshpande, J. V., & Kochar, S. C. (1983). Dispersive ordering is the same as tail-ordering. *Advances in Applied Probability*, *15*, 686–687.

Dharmadhikari, S., & Joag-Dev, K. (1988). *Unimodality, Convexity and Applications*. Cambridge: Academic Press.

Dharmadhikari, S. W., & Jogdeo, K. (1976). Multivariate unimodality. *The Annals of Statistics*, *4*, 607–613.

Doksum, K. (1969). Starshaped transformations and the power of rank tests. *The Annals of Mathematical Statistics*, *40*, 1167–1176.

Droste, W. (1985). *Lokale asymptotische Normalität und asymptotische Likelihood-Schätzer*. Thesis: University of Cologne.

Dvoretzky, A., Kiefer, J., & Wolfowitz, J. (1956). Asymptotic minimaxcharacter of the sample distribution function and of the classical multinomial estimator. *The Annals of Mathematical Statistics*, *27*, 642–669.

Fáry, I., & Rédei, L. (1950). Der zentralsymmetrische Kern und die zentralsymmetrische Hülle von konvexen Körpern. *Mathematische Annalen*, *122*, 205–220.

Ferguson, T. S. (1967). *Mathematical Statistics. A Decision Theoretic Approach*. Cambridge: Academic Press.

Fraser, D. A. S. (1957). *Nonparametric methods in statistics*. New Jersey: Wiley.

Fraser, D. A. S. (1954b). Nonparametric theory: Scale and location parameters. *The Canadian Journal of Mathematics*, *6*, 46–68.

Gauss, C. F. (1821). *Theoria combinationis observationum erroribus minimis obnoxiae, pars prior.* Commentationes Societatis Regiae Scientiarum Gottingensis Recentiores.

Gauss, C. F. (1823). *Theoria combinationis observationum erroribus minimis obnoxiae, pars posterior.* Commentationes Societatis Regiae Scientiarum Gottingensis Recentiores.

Ghosh, J. K. (1985). Efficiency of estimates. *Part I. Sankhyā Series A*, *47*, 310–325.

Ghosh, M., & Sen, P. K. (1989). Median-unbiasedness and Pitman closeness. *Journal of the American Statistical Association*, *84*, 1089–1091.

Gnedenko, B. W., & Kolmogorov, A. N. (1954). *Limit Distributions for Sums of Independent Random Variables* Cambridge, Mass. (Revised version of the Russian original, 1949).

Hájek, J. (1970). A characterization of limiting distributions of regular estimates. *Z. Wahrscheinlichkeitsheorie verw. Gebiete*, *14*, 323–330.

Hald, A. (1998). *A History of Mathematical Statistics from 1750 to 1930*. New York: Wiley.

Heyer, H. (1982). *Theory of statistical experiments. Springer Series in Statistics*. Berlin: Springer.

Huygens, C. (1657). *De ratiociniis in ludo aleae.* English translation: Woodward, London 1714.

Ibragimov, I. A. (1956). On the composition of unimodal distributions. *Theory of Probability and Its Applications*, *1*, 255–260.

Ibragimov, I. A., & Has'minskii, R. Z. (1981). *Statistical estimation: Asymptotic theory.* Springer, New York. Translation of the Russian original (1979).

Inagaki, N. (1970). On the limiting distribution of a sequence of estimators with uniformity property. *Annals of the Institute of Statistical Mathematics, 22*, 1–13.
James, W., & Stein, C. (1961). Estimation with quadratic loss. In *Proceedings of the 4th Berkeley symposium on mathematical statistics and probability I* (pp. 361–379). Berkeley: University California Press.
Karlin, S., & Rubin, H. (1956). The theory of decision procedures for distributions with monotone likelihood ratios. *The Annals of Mathematical Statistics, 27*, 272–291.
Kaufman, S. (1966). Asymptotic efficiency of the maximum likelihood estimator. *Annals of the Institute of Statistical Mathematics, 18*, 155–178. See also Abstract. *The Annals of Mathematical Statistics, 36*, 1084 (1965).
Keating, J. P., Mason, R. L., & Sen, P. K. (1993). *Pitman's measure of closeness*. A comparison of statistical estimators: SIAM.
Khintchine, A. Y. (1938). On unimodal distributions. *Izv. Nauchno. Issled. Inst. Mat. Mech. Tomsk. Gos. Univ., 2*, 1–7.
Klaassen, C. A. J. (1985). Strong unimodality. *Advances in Applied Probability, 17*, 905–907.
Klebanov, L. B. (1974). Unbiased estimators and sufficient statistics. *Theory of Probability and Its Applications, 19*, 379–383.
Laplace, P. S. (1774). Mémoire sur les suites récurro-récurrentes et sur leurs usages dans la théorie des hasards. *Mémories of the Academia Royal Science Paris, 6*, 353–371.
Laplace, P. S. (1820). *Théorie analytique des probabilités*. Courcier, Paris: Troisième édition.
Le Cam, L. (1972). Limits of experiments. In *Sixth Berkeley symposium on mathematical statistics and probability I* (pp. 249–261).
Le Cam, L. (1953). On some asymptotic properties of maximum likelihood estimates and related Bayes' estimates. *University of California Publications in Statistics, 1*, 277–330.
Lehmann, E. L. (1959). *Testing statistical hypotheses*. New Jersey: Wiley.
Lehmann, E. L. (1983). *Theory of point estimation*. New Jersey: Wiley.
Lehmann, E. L., & Casella, G. (1998). *Theory of point estimation* (2nd ed.). Berlin: Springer.
Lehmann, E. L. (1947). On families of admissible tests. *The Annals of Mathematical Statistics, 18*, 97–104.
Lehmann, E. L. (1951b). A general concept of unbiasedness. *The Annals of Mathematical Statistics, 22*, 587–592.
Lehmann, E. L. (1951a). Consistency and unbiasedness of certain nonparametric tests. *The Annals of Mathematical Statistics, 22*, 165–179.
Lehmann, E. L. (1952). Testing multiparameter hypotheses. *The Annals of Mathematical Statistics, 23*, 541–552.
Lehmann, E. L. (1955). Ordered families of distributions. *The Annals of Mathematical Statistics, 26*, 399–419.
Lehmann, E. L. (1966). Some concepts of dependence. *The Annals of Mathematical Statistics, 37*, 1137–1153.
Lekkerkerker, C. G. (1953). A property of logarithmic concave functions. I and II. *Indagationes Mathematics, 15*, 505–513 and 514–521.
Lewis, T., & Thompson, J. W. (1981). Dispersive distributions, and the connection between dispersivity and strong unimodality. *Journal of Applied Probability, 18*, 76–90.
Löwner, K. (1934). Über monotone Matrixfunktionen. *Mathematische Zeitschrift, 38*, 177–216.
Lynch, J., Mimmack, G., & Proschan, F. (1983). Dispersive ordering results. *Advances in Applied Probability, 15*, 889–891.
Mann, H. B., & Whitney, D. R. (1947). On tests whether one of two random variables is stochastically larger than the other. *The Annals of Mathematical Statistics, 18*, 50–60.
Markov, A. A. (1912). *Wahrscheinlichkeitsrechnung*. Translation by Liebmann, H. of the 2nd ed. of the Russian orginal. Teubner.
Millar, P. W. (1983). The minimax principle in asymptotic statistical theory. In Ecole d'Eté & de Probabilités de Saint-Flour XI-1981, Hennequin, P.L. (Eds.), *Lecture Notes in Mathematics 976* (pp. 75–265). New York: Springer.

Mises, R. von (1912). Über die Grundbegriffe der Kollektivmaßlehre. *Jahresbericht der Deutschen Mathematiker-Vereinigung, 21*, 9–20.

Müller, A., & Stoyan, D. (2002). *Comparison methods for stochastic models and risks*. Sussex: Wiley.

Muños-Perez, J., & Sanchez-Gomez, A. (1990). Dispersive ordering by dilation. *Journal of Applied Probability, 27*, 440–444.

Mussmann, D. (1987). On a characerization of monotone likelihood ratio experiments. *Annals of the Institute of Statistical Mathematics, 39*, 263–274.

Neyman, J. (1938). L'estimation statistique traitée comme un probléme classique de probabilité. *Actualités Sci. Indust., 739*, 25–57.

Olkin, I., & Pratt, J. W. (1958). Unbiased estimation of certain correlation coefficients. *The Annals of Mathematical Statistics, 29*, 201–211.

Pečarić, J. E., Proschan, F., & Tong, Y. L. (1992). *Convex functions, partial orderings, and statistical applications*. Cambridge: Academic Press.

Pfanzagl, J. (1960). Über die Existenz überall trennscharfer Tests. Metrika 3, 169–176. *Correction Note, 4*, 105–106.

Pfanzagl, J. (1994). *Parametric statistical theory.* De Gruyter.

Pfanzagl, J. (1962). Überall trennscharfe Tests und monotone Dichtequotienten. *Z. Wahrscheinlichkeitstheorie verw. Gebiete, 1*, 109–115.

Pfanzagl, J. (1964). On the topological structure of some ordered families of distributions. *The Annals of Mathematical Statistics, 35*, 1216–1228.

Pfanzagl, J. (1969). Further remarks on topology and convergence in some ordered families of distributions. *The Annals of Mathematical Statistics, 40*, 51–65.

Pfanzagl, J. (1995). On local and global asymptotic normality. *Mathematical Methods of Statistics, 4*, 115–136.

Pfanzagl, J. (2000b). Subconvex loss functions, unimodal distributions, and the convolution theorem. *Mathematical Methods of Statistics, 9*, 1–18.

Pitman, E. J. G. (1937). The "closest" estimates of statistical parameters. *Mathematical Proceedings of the Cambridge Philosophical Society, 33*, 212–222.

Pitman, E. J. G. (1939). Tests of hypotheses concerning location and scale parameters. *Biometrika, 31*, 200–215.

Prékopa, A. (1973). On logarithmic concave measures and functions. *Acta Scientiarum Mathematicarum, 34*, 335–343.

Rao, C. R. (1945). Information and accuracy attainable in the estimation of statistical parameters. *Bulletin of Calcutta Mathematical Society, 37*, 81–91.

Rao, C. R. (1947). Minimum variance estimation of several parameters. *Proceedings of the Cambridge Philosophical Society, 43*, 280–283.

Rao, C. R. (1962). Efficient estimates and optimum inference procedures in large samples. *Journal of the Royal Statistical Society. Series B, 24*, 46–72.

Rao, M. M. (1965). Existence and determination of optimal estimators relative to convex loss. *Annals of the Institute of Statistical Mathematics, 17*, 133–147.

Reiss, R. D. (1989). *Approximate distributions of order statistics. With applications to nonparametric statistics*. Berlin: Springer.

Roussas, G. G. (1968). Some applications of asymptotic distribution of likelihood functions to the asymptotic efficiency of estimates. *Zeitschrift für Wahrscheinlichkeitsheorie und verwandte Gebiete, 10*, 252–260.

Roussas, G. G. (1972). *Contiguity of probability measures: Some applications in statistics*. Cambridge: Cambridge University Press.

Saunders, I. W., & Moran, P. A. P. (1978). On the quantiles of the gamma and the F distributions. *Journal of Applied Probability, 15*, 426–432.

Savage, L. J. (1954). *The foundations of statistics*. New Jersey: Wiley.

Schmetterer, L. (1974). *Introduction to mathematical statistics.* Translation of the 2nd German edition 1966. Berlin: Springer.

Schmetterer, L. (1956). *Einführung in die Mathematische Statistik* (2nd ed., p. 1966). Wien: Springer.
Schweder, T. (1982). On the dispersion of mixtures. *Scandinavian Journal of Statistics*, *9*, 165–169.
Sen, P. K. (1986). *Are BAN estimators the Pitman-closest ones too? Sankhyā Series A 48*, 51–58.
Serfling, R. J. (1980). *Approximation theorems of mathematical statistics*. New York: Wiley.
Shaked, M. (1980). On mixtures from exponential families. *The Journal of the Royal Statistical Society*, *42*, 192–198.
Shaked, M. (1982). Dispersive ordering of distributions. *Journal of Applied Probability*, *19*, 310–320.
Sherman, S. (1955). A theorem on convex sets with applications. *The Annals of Mathematical Statistics*, *26*, 763–766.
Sobel, M. (1953). An essentially complete class of decision functions for certain standard sequential problems. *The Annals of Mathematical Statistics*, *24*, 319–337.
Stein, Ch. (1964). Inadmissability of the usual estimator for the variance of a normal distribution with unknown mean. *Annals of the Institute of Statistical Mathematics*, *16*, 155–160.
Strasser, H. (1985). *Mathematical theory of statistics*. De Gruyter.
Vaart, H. R. van der (1961). Some extensions of the idea of bias. *The Annals of Mathematical Statistics, 32*, 436–447.
van Zwet, W. R. (1964). *Convex Transformations of Random Variables*. Mathematical Centre Tracts **7**. Mathematisch Centrum, Amsterdam.
Varian, H. R. (1975). A Bayesian approach to real estate assessment. In S. E. Fienberg & A. Zellner (Eds.), *Studies in Bayesian econometrics and statistics in honor of Leonard J. Savage* (pp. 195–208). North-Holland.
Wald, A. (1939). Contributions to the theory of statistical estimation and testing hypotheses. *The Annals of Mathematical Statistics, 10*, 299–326.
Wefelmeyer, W. (1985). A counterexample concerning monotone unimodality. *Statistics and Probability Letters*, *3*, 87–88.
Wells, D. R. (1978). A monotone unimodal distribution which is not central convex unimodal. *The Annals of Statistics*, *6*, 926–931.
Wilks, S. S. (1943). *Mathematical statistics*. Princeton University Press: Princeton.
Wintner, A. (1938). *Asymptotic distributions and infinite convolutions*. Ann Arbor: Edwards Brothers.
Witting, H. (1985). *Mathematische Statistik I*. Parametrische Verfahren bei festem Stichprobenumfang: Teubner.
Witting, H., & Müller-Funk, U. (1995). *Mathematische Statistik II*. Asymptotische Statistik: Parametrische Modelle und nichtparametrische Funktionale. Teubner.
Zacks, S. (1971). *The theory of statistical inference*. New Jersey: Wiley.
Zellner, A. (1986). Bayesian estimation and prediction using asymmetric loss functions. *Journal of the American Statistical Association*, *81*, 446–451.

Chapter 4
Optimality of Unbiased Estimators: Nonasymptotic Theory

4.1 Optimal Mean Unbiased Estimators

Given a family $\mathfrak{P}$ of probability measures on $(X, \mathscr{A})$, the estimator $\hat{\kappa}_0 : X \to \mathbb{R}$ is *ℓ-optimal* at P_0 among all mean unbiased estimators of the functional $\kappa(P)$ if $\hat{\kappa}_0$ minimizes

$$\int \ell(\hat{\kappa} - \kappa(P_0))dP_0$$

among all estimators $\hat{\kappa}$ fulfilling $\int \hat{\kappa}dP = \kappa(P)$ for $P \in \mathfrak{P}$. We speak of *convex optimality* if this relation holds for every convex loss function, and of *quadratic optimality* if it holds for $\ell(u) = u^2$.

Certain results become more transparent if we call $\hat{\kappa}_0$ quadratically [convex] optimal if it is an ℓ-optimal estimator of its expectation (without an explicit reference to the functional $P \to \int \hat{\kappa}dP$).

Unlike the case of median unbiased estimators, a comparison of mean unbiased estimators with respect to the concentration on intervals is impossible. Hence the following considerations are based on the comparison of estimators by means of certain loss functions only.

If a functional $\kappa : \mathfrak{P} \to \mathbb{R}$ admits a mean unbiased estimator, all that one could expect in general is to find an estimator that minimizes the risk for a given loss function at a certain $P_0 \in \mathfrak{P}$. It is of mainly mathematical interest whether the infimum of $\int \ell_0(\hat{\kappa} - \kappa(P_0))P_0(dx)$ over all unbiased estimators $\hat{\kappa}$ is attained or not.

For $\ell(u) = |u|^s$, $s > 1$, this was proved by Barankin (1949, p. 483, Theorem 2(iii)) under the assumption that

$$\int (p(x)/p_0(x))^{s/(s-1)} P_0(dx) < \infty \quad \text{for } P \in \mathfrak{P}.$$

For $\ell(u) = u^2$ see Stein (1950, pp. 407/8, Theorem 1). See Witting (1985, p. 306, Satz 2.119) for a detailed proof under the condition that $\int (p(x)/p_0(x))^2 P_0(dx) < \infty$ for $P \in \mathfrak{P}$.

J. Pfanzagl, *Mathematical Statistics*, Springer Series in Statistics,
DOI 10.1007/978-3-642-31084-3_4

Of some practical interest are criteria which guarantee that a given unbiased estimator, say $\hat{\kappa}_0$, does, in fact, minimize $\hat{\kappa} \to \int \ell_{P_0}(\hat{\kappa}(x))P_0(dx)$. (Recall that there is at most one such estimator if ℓ_P is strictly convex and bounded from below.) Starting with the quadratic loss function, i.e., $\ell_P(u) = (u - \kappa(P))^2$, Rao (1952) suggests the following

Criterion. *The estimator $\hat{\kappa}_0$ minimizes the quadratic risk among all mean unbiased estimators at P_0 iff $\int v(x)\hat{\kappa}_0(x)P_0(dx) = 0$ holds for every function $v \in \mathscr{V}_2$, where $\mathscr{V}_2$ is the set of all functions $v : X \to \mathbb{R}$ with*

$$\int v(x)P(dx) = 0 \quad \text{and} \quad \int v^2(x)P(dx) < \infty \quad \text{for } P \in \mathfrak{P}.$$

This criterion is more important for general considerations than for particular applications, since the class $\mathscr{V}_2$ is not always easy to characterize. (See the examples 2.112, p. 301 and 2.113, p. 302 in Witting 1985.)

Rao's criterion was generalized from the power $s = 2$ to higher powers $s \geq 2$: *The estimator $\hat{\kappa}_0$ minimizes the risk for $\ell(u) := |u|^s$ at P_0 iff*

$$\int v(x)|\hat{\kappa}_0(x) - \kappa(P_0)|^{s-1}\text{sgn}\,(\hat{\kappa}_0(x) - \kappa(P_0))P_0(dx) = 0$$

holds for every function $v \in \mathscr{V}_s$, where $\mathscr{V}_s$ now denotes all functions $v \in \mathscr{V}_2$ fulfilling $\int |v(x)|^s P(dx) < \infty$ for $P \in \mathfrak{P}$.

See Schmetterer (1960, p. 1155, Theorem 3). (Correct three misprints in the statement of this theorem.) For a more precise proof of this result see Heyer (1982), pp. 124/5, Theorem 17.3. This result was further generalized to convex functions ℓ: *The estimator $\hat{\kappa}_0$ minimizes the risk for a convex function ℓ* at *P_0 iff*

$$\int v(x)\ell'(\hat{\kappa}_0(x) - \kappa(P_0))P_0(dx) = 0$$

holds for every function $v \in \mathscr{V}_\ell$, it where *$\mathscr{V}_\ell$ is now denotes all functions v with $\int v(x)P(dx) = 0$ and $\int \ell(v(x))P(dx) < \infty$.*

See Linnik and Rukhin (1971, p. 839, Proposition). See also Schmetterer (1960, p. 1155, Theorem 3) and Krámli (1967, p. 160, Theorem). A presentation of these results can be found in Heyer (1982, Sect. 17). For an extension to arbitrary convex loss functions see Kozek (1977, p. 188, Theorem 4.4).

Optimality with respect to *some convex* loss function at a *particular* $P_0 \in \mathfrak{P}$ is neither important for applications, nor does it provide deeper insights from a methodological point of view. Yet, it had a certain appeal to mathematically minded statisticians. Here we may refer to the usual textbooks for examples in which the unbiased estimator minimizing the risk at P_0 does, in fact, just what one is afraid of: It depends on P_0. An early example of this kind is given by Lehmann and Scheffé (1950, p. 253, Example 5.3): For the family of uniform distributions with densities $\{1_{(\vartheta-1/2,\vartheta+1/2)} : \vartheta \in \mathbb{R}\}$, no function of ϑ admits an unbiased estimator based on a sample of size $n > 1$ which minimizes the quadratic risk simultaneously for all $\vartheta \in \mathbb{R}$. For an example of a parametric family in which an unbiased estimator exists for every sample size, but none of them minimizes the risk for any strictly convex loss function, simultaneously for every $\vartheta \in \Theta$, see Pfanzagl (1994, p. 103, Example 3.1.6).

In fact one would expect that the existence of an optimal unbiased estimator (i.e., one that minimizes the risk for a given loss function simultaneously for all $P_0 \in \mathfrak{P}$) is a rare exception. It might, therefore, come as a surprise that there is an important type of families in which for every functional admitting an unbiased estimator there exists an unbiased estimator that minimizes the risk, simultaneously for all convex loss functions and all $P_0 \in \mathfrak{P}$. These are the families $\mathfrak{P}$ admitting a sufficient statistic $S : X \to Y$ for which $\mathfrak{P} \circ S$ is complete. Historically, it was this finding that sparked interest in a theory of unbiased estimators around 1950, and which is still responsible for its presence in textbooks.

Now let $\hat{\kappa}$ be unbiased on $\mathfrak{P}$ for $\kappa : \mathfrak{P} \to \mathbb{R}^k$ in the family $\mathfrak{P}$. If $S|X$ is sufficient, there exists a conditional expectation of $\hat{\kappa}$, given S, say $k \circ S$, which is independent of P, hence an estimator again. Since $\int k(S(x))P(dx) = \int \hat{\kappa}(x)P(dx)$ for $P \in \mathfrak{P}$, the estimator $k \circ S$ is unbiased, too. According to Jensen's inequality for conditional expectations, for every convex function ℓ,

$$\ell(k(y)) \leq P(\ell(\hat{\kappa}) \mid S = y) \quad \text{for } P \circ S\text{-a.a. } y \in Y, \tag{4.1.1}$$

hence

$$\int \ell(k(S(x))P(dx) \leq \int \ell(\hat{\kappa}(x))P(dx). \tag{4.1.2}$$

(If ℓ is strictly convex, the inequality (4.1.1) is strict unless $\hat{\kappa} = k \circ S$ P-a.e.) Together with unbiasedness, (4.1.2) implies

$$\int \ell(k(S(x)) - \kappa(P))P(dx) \leq \int \ell(\hat{\kappa}(x) - \kappa(P))P(dx).$$

Rao (1945, p. 83) proves relation (4.1.2) for $\ell(u) = u^2$, using an argument specific to this quadratic loss function. His argument refers to a one-parameter family of probability measures P_ϑ over $\mathbb{R}^n$, with $\kappa(P_\vartheta) = \vartheta$. Presumably, he had a real-valued sufficient statistic in mind. Rewritten in our notations, his argument, given in his equation (3.8), reads as follows:

$$\begin{aligned} &\int (\hat{\kappa}(x) - \kappa(P))^2 P(dx) \\ &\quad \int (\hat{\kappa}(x) - k(S(x)))^2 P(dx) + \int (k(S(x) - \kappa(P))^2 P(dx) \\ &\geq \int (k(S(x)) - \kappa(P))^2 P(dx). \end{aligned}$$

In this derivation, he uses

$$\int (\hat{\kappa}(x) - k(S(x)))k(S(x))P(dx) = 0 \tag{4.1.3}$$

without further comment. In Rao (1947, pp. 280/1, Theorem 1) this argument is extended to k-parameter families.

Independently of C.R. Rao, the same result was obtained by Blackwell (1947, p. 106, Theorem 2). Relation (4.1.3), used by Rao without further ado, is found worth of a careful proof by Blackwell (see p. 105, Theorem 1).

The extension to convex loss functions is due to Hodges and Lehmann (1950, p. 188, Theorem 3.3). Their proof is based on Jensen's inequality for conditional expectations (p. 195, Lemma 3.1), corresponding to (4.1.1). The authors are aware of the fact that this inequality is straightforward if a "regular conditional probability" exists. Yet they take the trouble to give a proof which goes through without the conditions needed to ensure the existence of a regular conditional probability in general. The role of Barankin's paper (1950) which appeared in the same journal, but after the paper by Hodges and Lehmann remains unclear. Barankin's Theorem on p. 281 gives inequality (4.1.1) for $\ell(u) = |u|^s$, with a proof attributed to "the referee", and he applies this inequality in Corollary 1, p. 283 to obtain (4.1.2) with $\ell(u) = |u|^s$. Unclear, too, is the purpose of Barankin's paper (1951). His Theorem on p. 168 is just another proof of Jensen's inequality for conditional expectations, and its application repeats the result of Hodges and Lehmann.

The idea that unbiased estimators can be improved by taking the conditional expectation with respect to a sufficient statistic, met with some reserve. First of all, "improved" just means that the risk is decreased, simultaneously for every convex loss function; it does not mean that the improved estimator is more concentrated on intervals containing the estimand. Moreover, the improved estimator may have certain properties uncalled for (and not shared by the original estimator). It might, for instance, fail to be proper. Finally, there are cases where the conditional expectation, given S, cannot be expressed in closed form.

The role of the improvement procedure is accentuated by a result of Lehmann and Scheffé: The improved estimator is optimal (in the sense of minimizing the convex risk in the class of all unbiased estimators) if the conditioning is taken with respect to a sufficient statistic S for which $\mathfrak{P} \circ S$ is complete. Theorem 5.1 in Lehmann and Scheffé (1950, p. 321) asserts that an unbiased estimator is of minimal quadratic risk iff it is the contraction of a sufficient statistic S with $\mathfrak{P} \circ S$ complete. This formulation does not exhibit the core of the argument (which depends in by no means on the assumption that $\ell(u) = u^2$): If $\mathfrak{P} \circ S$ is complete, there is at most one unbiased estimator that is a contraction of S. Recall a forerunner of this result, due to Halmos (1946).

In their final form, these results are an inevitable topic in any textbook under the title

Theorem of Rao–Blackwell–Lehmann–Scheffé. *Let $\mathfrak{P}$ be a family admitting a sufficient statistic S such that $\mathfrak{P} \circ S$ is complete. Then for every mean unbiased*

estimator $\hat{\kappa}$, its conditional expectation $k \circ S$ is optimal in the sense that it minimizes the convex risk in the class of all mean unbiased estimators.

If we consider the steps leading to this result, we find that two essential points are already present in Rao (1945, in particular p. 83): (i) There is a bound for the quality of mean unbiased estimators, and (ii) this bound can be achieved by taking conditional expectations with respect to a certain sufficient statistic.

Yet it needed five authors (and five years) to bring these vague ideas to their final shape. The reason: Rao had difficulties to cope with the concept of a conditional expectation (see 1945, p. 83, relation 3.7, and 1947, p. 281). If he had used the stochastic independence between S and $\hat{\kappa} - k \circ S$ (rather than their being uncorrelated) he could have obtained the convex-optimality (rather than the quadratic optimality). Taking conditional expectations leads to an improvement, but not necessarily to optimality. The optimality follows from the completeness of $P \circ S$, introduced by Lehmann and Scheffé in 1950. Something close to "completeness" is foreshadowed in Rao (1945), p. 83_{4-6} and (1947), p. $281_{1,2}$.

The use of "uncorrelated" rather than "stochastically independent" leads to a serious disadvantage when Rao extends his results to k-parameter families (see 1945, p. 84–86). Instead of arriving at "minimal convex risk" he ends up with the result that the covariance matrix of the improved estimators is minimal in the Löwner-order among all covariance matrices of mean unbiased estimators—a result which follows immediately, since $(u_1, \ldots, u_k) \to \left(\sum_{i=1}^k \alpha_i u_i\right)^2$ is convex for every $(\alpha_1, \ldots, \alpha_k) \in \mathbb{R}^k$.

Remark Various optimality results for unbiased estimators are of the type: $\hat{\kappa}_0$ minimizes $\int \ell(\hat{\kappa}) dP$ for $\hat{\kappa}$ in a certain class of unbiased estimators, *simultaneously* for all convex functions $\ell : \mathbb{R} \to [0, \infty)$. This is not a convincing optimum property, since the location of $P \circ \hat{\kappa}$ enters through the condition $\int \hat{\kappa} dP = \kappa(P)$ only. The loss function ℓ itself makes no allowance for location (say by the property $\ell(u) = 0$ for $u = \kappa(P)$), nor does it distinguish between estimators that are optimal for the particular loss function, and estimators that are optimal for *every* convex loss function.

Recall that optimality with respect to all subconvex loss functions is equivalent to the following statements:

(i) The distribution of subconvex losses is minimal in the stochastic order.

(ii) the concentration is maximal on all intervals containing $\kappa(P)$.

As against that, minimality of $\hat{\kappa}$ with respect to every convex loss function says nothing about the distribution of the convex losses $\ell \circ \hat{\kappa}$. Moreover, the improvement of an estimator by taking a conditional expectation reduces the convex risk, but the distribution of the convex losses of the improved estimator is not necessarily stochastically smaller than the distribution of the original estimator. Finally, an estimator which is of minimal convex risk in the class of all mean unbiased estimators may be inferior to other mean unbiased estimators if evaluated by a subconvex loss function or by the concentration on intervals. An early example in this regard is due to Basu (1955, p. 347).

How insufficient "convex risk" as a measure of the quality of an estimator really is can be seen from examples of estimators with minimal convex risk in the class of all mean unbiased estimators that are inferior to other mean unbiased estimators if evaluated by the concentration on certain intervals.

Example For $\vartheta > 0$, let P_ϑ be the exponential distribution with density $x \to \vartheta^{-1}\exp[-x/\vartheta]$, $x > 0$. Since the family $\{P_\vartheta : \vartheta > 0\}$ is complete, and $\int x P_\vartheta(dx) = \vartheta$, the function x is mean unbiased for ϑ with minimal convex risk. If $Q|\mathbb{B}_+$ fulfills $\int_0^\infty u Q(du) = 1$, the (randomized) estimator $(u, x) \to ux$ is mean unbiased, too, and has, therefore, larger convex risk. Yet $Q \times P_\vartheta\{\vartheta t' < ux < \vartheta t''\} > P_\vartheta\{\vartheta t' < x < \vartheta t''\}$ for some $t' < 1 < t''$ if Q is chosen appropriately. (Hint: choose $Q\{1-\delta_1\} = \delta_2/(\delta_1+\delta_2)$ and $Q\{1+\delta_2\} = \delta_1/(\delta_1+\delta_2)$, with $\delta_i > 0$ and sufficiently small.) □

An essential aspect of this is that the higher concentration in intervals containing ϑ holds for *every* $\vartheta > 0$. Examples of mean unbiased estimators with higher concentration for *some* ϑ are much easier to obtain. In the family $\{N(\vartheta, 1)^n : \vartheta \in \mathbb{R}\}$, the estimator $\overline{x}_n$ is of minimal convex risk in the class of all mean unbiased estimators. The estimator $(x_1, \dots, x_n) \to 2\overline{x}_n 1_{\{x_1 \le x_n\}}(x_1, \dots, x_n)$ is mean unbiased and is in small intervals $(\vartheta - t, \vartheta + t)$ with higher probability than $\overline{x}_n$, provided ϑ is close to zero.

Considering the fact that (a) mean unbiasedness is not a natural condition to be imposed upon all estimators, (b) minimal convex risk is not a convincing criterion for the evaluation of estimators, and (c) mean unbiasedness is in no inherent relation to convex loss, one might ask why this theory flourished for some time. The answer: It was the foremost common way to prove one's mathematical skills.

4.2 Bahadur's Converse of the Rao–Blackwell–Lehmann–Scheffé Theorem

Consider the following statements:

(1) There exists a statistic $S : (X, \mathscr{A}) \to (Y, \mathscr{B})$ which is sufficient for $\mathfrak{P}$, and $\mathfrak{P} \circ S$ is complete.
(2) For every functional $\kappa : \mathfrak{P} \to \mathbb{R}$ admitting an unbiased estimator there exists an unbiased estimator that minimizes every convex risk.

The Rao–Blackwell–Lehmann–Scheffé Theorem asserts that (1) implies (2). In this Theorem, the sufficiency of S is needed in order to obtain from every unbiased estimator a "better" S-measurable unbiased estimator P; the completeness of $\mathfrak{P} \circ S$ is needed in order to show that this estimator minimizes every convex risk. Bahadur's intention (see 1957) was to prove that (2) implies (1).

Bahadur's paper is fairly poorly arranged, containing six Theorems and seven Propositions. The presentation of Bahadur's results in Schmetterer (1966, pp. 332–352) requires 6 Theorems, and it had not become much simpler twenty years later: The presentation in Strasser (1985, pp. 168–172) consists of 4 Lemmas, 3 Theorems and one Corollary. In other textbooks like Eberl and Moeschlin (1982), Heyer (1973, 1982) and Witting (1985) Bahadur's result is not even mentioned. In roughly a dozen of papers dealing with Bahadur's approach (see Eberl 1984, for further references) one is missing what I would consider the main result of Bahadur (1957): That a bounded quadratically optimal estimator is convex optimal. This result, based on Bahadur's approach, was explicitly put forward by Padmanabhan (1970, p. 109, Theorem 3.1). See also Schmetterer and Strasser (1974, p. 60). (For more examples and counterexamples see the papers by Bomze 1986, 1990; Eberl 1984; Heizmann 1989.)

To make a long story short: All these papers were based on an ingenious idea of Bahadur (1957, p. 218). Given a family $\mathfrak{P}$ of probability measures $P|(X, \mathscr{A})$, let $\mathscr{V}_2$ be the set of all functions $v : X \to \mathbb{R}$ fulfilling the conditions $\int v^2 dP < \infty$ and $\int v dP = 0$ for $P \in \mathfrak{P}$. Bahadur introduces

$$\mathscr{A}_{\mathfrak{P}} := \{A \in \mathscr{A} : \int 1_A v dP = 0 \text{ for } v \in \mathscr{V}_2 \text{ and } P \in \mathfrak{P}\}, \tag{4.2.1}$$

the σ-field of all subsets A of $\mathscr{A}$ for which 1_A is a quadratically optimal mean unbiased estimator of the functional $P \to P(A)$.

The following Proposition characterizes the $\mathscr{A}_{\mathfrak{P}}$-measurable functions.

Proposition 4.2.1 *(i) If $\hat{\kappa} : X \to \mathbb{R}$ with $\int \hat{\kappa}^2 dP < \infty$ is $\mathscr{A}_{\mathfrak{P}}$-measurable, then*

$$\int \hat{\kappa} \, v dP = 0 \quad \textit{for } v \in \mathscr{V}_2 \textit{ and } P \in \mathfrak{P}. \tag{4.2.2}$$

(ii) If $\hat{\kappa} : X \to \mathbb{R}$ is bounded, then relation (4.2.2) implies that $\hat{\kappa}$ is $\mathscr{A}_{\mathfrak{P}}$-measurable.

Proof (i) By definition, relation (4.2.2) holds for $\hat{\kappa} = 1_A$ if $A \in \mathscr{A}_{\mathfrak{P}}$. Relation (4.2.2) is extended to $\mathscr{A}_{\mathfrak{P}}$-measurable functions $\hat{\kappa}$ by approximation with $\mathscr{A}_{\mathfrak{P}}$-measurable functions.

(ii) For bounded $\hat{\kappa}$, relation (4.2.2) implies that $\hat{\kappa} v \in \mathscr{V}_2$ if $v \in \mathscr{V}_2$. Proceeding inductively, we obtain that for bounded $\hat{\kappa}$,

$$\hat{\kappa}^k v \in \mathscr{V}_2 \quad \text{for every } k \in \mathbb{N}.$$

Following the basic idea of Bahadur (1957, p. 218, proof of Theorem 5(i)), this implies that $\int 1_B(\hat{\kappa}) v dP = 0$ for $v \in \mathscr{V}_2$ and $P \in \mathfrak{P}$. Hence $\hat{\kappa}^{-1} B$ is $\mathscr{A}_{\mathfrak{P}}$-measurable for every $B \in \mathbb{B}$, which implies the $\mathscr{A}_{\mathfrak{P}}$-measurability of $\hat{\kappa}$. □

A detailed proof can be found in Strasser (1972, p. 110, Theorem 5.6). See also Strasser (1985, p. 170, Theorem 35.14) or Pfanzagl (1994, p. 121).

An example provided by Bahadur (1957, Sect. 6) shows that the boundedness in Proposition 4.2.2 is essential: Measurable functions of (unbounded) unbiased estimators of minimal quadratic risk are not necessarily estimators of minimal quadratic risk of their expectation. Bahadur's example has been used by Padmanabhan (1970) to show that an (unbounded) estimator of minimal quadratic risk may be inferior if evaluated by the loss function $u \to |u|^3$. (See Padmanabhan 1970, pp. 110/1 and 112/3.)

Bednarek-Kozek and Kozek (1978) present examples of unbiased estimators that minimize the risk for *some,* but not for *all* convex loss functions. The reader interested in a more systematic study of the mathematical aspects of this problem could consult Kozek (1980).

What remains of Bahadur's approach is that certain optimal estimators may be considered as conditional expectations, a side result that was never aspired to in Bahadur's paper Bahadur (1957).

Proposition 4.2.2 *Assume that $\hat{\kappa}_0$ fulfilling $\int \hat{\kappa}_0^2 dP < \infty$ is $\mathscr{A}_{\mathfrak{P}}$-measurable. If $\int \hat{\kappa} dP = \int \hat{\kappa}_0 dP$ for $P \in \mathfrak{P}$, then $\hat{\kappa}_0$ is for every $P \in \mathfrak{P}$ a conditional expectation of $\hat{\kappa}$, given $\mathscr{A}_{\mathfrak{P}}$, with respect to $\mathfrak{P}$. This implies $\int \ell(\hat{\kappa}_0) dP \le \int \ell(\kappa) dP$ for every convex loss function ℓ.*

Proof We have to show that

$$\int (\hat{\kappa} - \hat{\kappa}_0) 1_A dP = 0 \quad \text{for every } A \in \mathscr{A}_{\mathfrak{P}} \text{ and every } \ P \in \mathfrak{P}.$$

By definition of $\mathscr{A}_{\mathfrak{P}}$,

$$\int 1_A v dP = 0 \quad \text{for } v \in \mathscr{V}_2 \text{ and } P \in \mathfrak{P}. \tag{4.2.3}$$

Since $\hat{\kappa} - \hat{\kappa}_0 \in \mathscr{V}_2$, relation (4.2.3) applied with $v = \hat{\kappa} - \hat{\kappa}_0$ implies

$$\int 1_A (\hat{\kappa} - \hat{\kappa}_0) dP = 0 \quad \text{for } P \in \mathfrak{P}. \tag{4.2.4}$$

Since relation (4.2.4) holds for every $A \in \mathscr{A}_{\mathfrak{P}}$, the assertion follows. □

The essential point in the proof of Proposition 4.2.2 is that every $\mathscr{A}_{\mathfrak{P}}$-measurable unbiased estimator is the conditional expectation of any unbiased estimator. This idea occurs first in Padmanabhan (1970, p. 109, Theorem 1), under the redundant assumption that the $\mathscr{A}_{\mathfrak{P}}$-measurable estimator minimizes the quadratic risk. Without the redundant assumption this assertion occurs in Schmetterer (1974, p. 61, Satz 1). The idea that $\mathscr{A}_{\mathfrak{P}}$-measurable estimators are conditional expectations and *therefore* optimal for every convex loss function was not obvious from the beginning. This may be seen from Theorem 7 in Schmetterer (1960, p. 1161) which asserts the optimality of $\mathscr{A}_{\mathfrak{P}}$-measurable estimators for the loss functions $u \to |u|^s$, $s \geq 1$. This is proved by means of Rao's generalized criterion, and not by the conditional expectation-property of $\hat{\kappa}_0$, which would imply optimality with respect to *every* convex loss function.

Corollary 4.2.3 *If $\hat{\kappa}_0$ is $\mathscr{A}_{P_0}$-measurable, then $\int \hat{\kappa}_0 dP_0 = 0$ implies $\hat{\kappa}_0 = 0$ P_0-a.e., i.e. $\mathscr{A}_{\mathfrak{P}}$ is 2-complete.*

Proof Apply relation (4.2.4) with $\hat{\kappa} \equiv 0$. □

In the following we discuss a generalization of Padmanabhan's result. By Proposition 4.2.2 every quadratically optimal bounded estimator is convex optimal. The appropriate generalization would be: If a bounded estimator is optimal with respect to *some* convex loss function then it is optimal with respect to *every* convex loss function.

Various papers by Schmetterer and Strasser are devoted to such a generalization. (See Strasser 1972, p. 110, Theorem 5.6 or Schmetterer 1977, p. 499, Satz 4.1; see also Schmetterer 1974 p. 60.)

The best result, obtained by Schmetterer (1974, pp. 61/2, Satz 2), is convex optimality of estimators $\hat{\kappa}_0$ which are optimal with respect to some convex loss function ℓ_0 with increasing derivative ℓ_0' that is bounded and continuous and which fulfills

$$\sup_{|t|>t_0} \ell_0(t)/t < \infty \quad \text{for some } t_0.$$

For such loss functions, $\int \ell_0'(\hat{\kappa}_0) v dP = 0$ for $v \in \mathscr{V}_2$ and $P \in \mathfrak{P}$ (see relation 6, p. 62) which implies $\int \ell(\hat{\kappa}_0) dP \leq \int \ell(\hat{\kappa}) dP$ for every convex loss function and every $P \in \mathfrak{P}$, whence

$$\int \ell(\hat{\kappa}_0 - \mu) dP \leq \int \ell(\hat{\kappa} - \mu) dP$$

for every $\mu \in \mathbb{R}$. If $\int \kappa_0 dP = \int \kappa_0 dP = \kappa(P)$, this implies

$$\int \ell(\hat{\kappa}_0 - \kappa(P)) dP \leq \int \ell(\kappa_0 - \kappa(P)) dP \quad \text{for every } P \in \mathfrak{P}. \tag{4.2.5}$$

The restrictive condition on the loss function ℓ_0 is, perhaps, responsible for the fact that this result is neglected in the literature.

There is another point in the paper by Schmetterer and Strasser which might cause some irritation: Optimality with respect to a loss function ℓ is defined by (4.2.5). For the quadratic loss function,

$$\int (\hat{\kappa}_0 - \kappa(P))^2 dP \leq \int (\hat{\kappa} - \kappa(P))^2 dP$$

is, under the condition $\int \hat{\kappa}_0 dP = \kappa(P)$ and $\int \hat{\kappa} dP = \kappa(P)$, equivalent to

$$\int \hat{\kappa}_0^2 dP \leq \int \hat{\kappa}^2 dP.$$

Schmetterer and Strasser start from the condition (their relation (2), p. 60)

$$\int \ell(\hat{\kappa}_0) dP \leq \int \ell(\hat{\kappa}) dP \quad \text{for } \int \hat{\kappa} dP = \int \hat{\kappa}_0 dP = \kappa(P) \text{ and } P \in \mathfrak{P},$$

which is not the same as (4.2.5) unless $\ell(u) = u^2$. Yet, since the final result,

$$\int \ell(\hat{\kappa}_0) dP \leq \int \ell(\kappa_0) dP \quad \text{for every convex } \ell,$$

refers to each $P \in \mathfrak{P}$ separately, the proof by Schmetterer and Strasser can be carried through with P fixed, i.e., with $\mathfrak{P} = \{P_0\}$ in which case the loss function may be taken to be $u \to \ell_0(u - \kappa(P_0))$.

The propositions stated above have nothing to do with the sufficiency of $\mathscr{A}_{\mathfrak{P}}$. However, if a 2-complete sufficient sub-σ-field of $\mathfrak{P}$ does exist, then this is the σ-field $\mathscr{A}_0$ defined by (4.2.1). $\mathscr{A}_{\mathfrak{P}}$ "recovers" the σ-field underlying the Bahadur–Rao–Lehmann–Scheffé Theorem.

The following Proposition states Bahadur's converse: If for every unbiasedly estimable functional there is a quadratically optimal unbiased estimator, then there exists a sufficient sub-σ-field. With this result, Bahadur answers a question which no statistician had ever asked. What the statistician is interested in is an optimal unbiased estimator for a given functional. Whether *every* unbiasedly estimable functional admits an optimal unbiased estimator is of no relevance for his problem.

Proposition 4.2.4 (Bahadur's converse) *If for every $A \in \mathscr{A}$ there is a quadratically optimal estimator, say $\hat{\kappa}_A$, for the functional $P \to P(A)$, then $\mathscr{A}_{\mathfrak{P}}$ is sufficient (and complete), and $\hat{\kappa}_A$ is for every $P \in \mathfrak{P}$ a conditional expectation of 1_A, given $\mathscr{A}_{\mathfrak{P}}$.*

Proof If $\hat{\kappa}_A$ is an unbiased estimator of $P \to P(A)$, we have

$$\int \hat{\kappa}_A v \, dP = 0 \quad \text{for } v \in \mathscr{V}_{\mathfrak{P}} \text{ and } P \in \mathfrak{P}.$$

By Proposition 4.2.1(ii) this implies that $\hat{\kappa}_A$ is $\mathscr{A}_{\mathfrak{P}}$-measurable. By Proposition 4.2.2, $\hat{\kappa}_A$ is for every $P \in \mathfrak{P}$ a conditional expectation of 1_A, given $\mathscr{A}_{\mathfrak{P}}$. By Definition 2.1.2 this implies that $\mathscr{A}_{\mathfrak{P}}$ is sufficient.

Bahadur's proof (1957), followed by Strasser (1985, pp. 171/2), uses the existence of optimal estimators for bounded densities.

Sufficiency, which was the main point in Bahadur's paper, is neglected by Schmetterer: "It is more difficult to prove under some more conditions that $\mathscr{A}_{\mathfrak{P}}$ is also sufficient" (see 1966, p. 252 and 1974, p. 289).

4.3 Unbiased Estimation of Probabilities and Densities

Probabilities are a typical example of a functional for which unbiasedness of estimators is a natural requirement. To start with: The existence of unbiased estimators for every sample size is guaranteed. For every $A \in \mathscr{A}$, $(x_1, \ldots, x_n) \to n^{-1} \sum_{\nu=1}^{n} 1_A(x_\nu)$ is an unbiased estimator of the functional $P \to P(A)$.

Under the conditions indicated in Sect. 2.2, for every statistic $S_n : X^n \to Y$ there exists a Markov kernel $M_n \mid Y \times \mathscr{A}$ such that

$$\int M_n(y, A) P^n \circ S_n(dy) = P(A) \quad \text{for } P \in \mathfrak{P},\, A \in \mathscr{A}.$$

That means: For every $A \in \mathscr{A}$, the function $(x_1, \ldots, x_n) \to M_n(S_n(x_1, \ldots, x_n), A)$ is an unbiased estimator of the functional $P \to P(A)$. Recall that this estimator is of minimal convex risk among all unbiased estimators if $\{P^n \circ S_n : P \in \mathfrak{P}\}$ is complete. In this case, $(x_1, \ldots, x_n) \to \int k(x) M_n(S_n(x_1, \ldots, x_n), dx)$ is of minimal convex risk among all unbiased estimators for $\kappa(P) = \int k(x) P(dx)$.

The practical question then becomes: How can an unbiased estimator M_n be obtained? If for every $P \in \mathfrak{P}$ there exists a μ-density q_P, then an unbiased estimator of $P \to P(A)$ can be obtained from an unbiased estimator of $q_P(\xi)$, $\xi \in X$. Assume that, more generally, there is a function $\hat{p}(\xi, y)$ and a probability measure $Q_P|\mathscr{B}$ such that, for every $P \in \mathfrak{P}$,

$$\int \hat{p}(\xi, y) Q_P(dy) = q_P(\xi) \quad \text{for } \xi \in X.$$

Then $\hat{M}_n(y, A) := \int 1_A(\xi) \hat{p}(\xi, y) \mu(d\xi)$ fulfills for every $A \in \mathscr{A}$ the relation

$$\int \hat{M}_n(y, A) Q_P(dy) = P(A) \quad \text{for } P \in \mathfrak{P}.$$

Applied with Q_P replaced by $P^n \circ S_n$ (and $\hat{p}(\xi, y)$ replaced by $p_n(\xi, y)$) this leads to $M_n(S_n(\cdot), A)$ as an unbiased estimator of $P(A)$.

The relation between unbiased density estimators and unbiased estimators of probabilities was first used by Kolmogorov in connection with the normal distribution. He attributes this idea to Linnik. (See Kolmogorov 1950, p. 388, footnote 9.) The relation between unbiased density estimators and unbiased estimators of probabilities was later stated as a formal theorem by Seheult and Quesenberry (1971, p. 1435, Theorem 1).

If $\{P^n \circ S_n : P \in \mathfrak{P}\}$ is complete, an unbiased density estimator $p_n(\xi; S_n(\cdot))$ minimizes the convex risk for every $\xi \in X$, i.e.

$$\int \ell(p_n(\xi; S_n(x_1, \ldots, x_n)) - q_P(\xi))P^n(d(x_1, \ldots, x_n))$$
$$\leq \int \ell(\hat{p}_n(\xi; x_1, \ldots, x_n) - q_P(\xi))P^n(d(x_1, \ldots, x_n))$$

for every $\xi \in X$, $P \in \mathfrak{P}$, and convex ℓ if $\hat{p}_n(\xi; \cdot)$ is unbiased for $q_P(\xi)$. Integrating over ξ with respect to μ, this implies that $p_n(\xi; S_n(\cdot))$ minimizes, in the class of all unbiased density estimators, the risk with respect to any of the loss functions

$$\Delta(p, q) = \int \ell(p(\xi) - q(\xi))\mu(d\xi).$$

Needless to say that here the natural choice is $\ell(u) = |u|$. Since

$$\frac{1}{2}\int |p(\xi) - q(\xi)|\mu(d\xi) = \sup_{A \in \mathscr{A}} |P(A) - Q(A)|,$$

this implies that $M_n(S_n(\cdot), \cdot)|\mathscr{A}$, evaluated as a probability measure, minimizes the sup-distance in the class of all unbiased estimators of $P|\mathscr{A}$. Yet, even more is true: The Rao–Blackwell–Lehmann–Scheffé Theorem implies that $M_n(S_n(\cdot), A)$ minimizes—for every $A \in \mathscr{A}$—the convex risk in the class of all unbiased estimators of $P \to P(A)$.

The question remains how an unbiased estimator of $q_P(\xi)$ can be obtained. Generalizing the ideas applied by Kolmogorov (1950, Sect. 9, pp. 389–392) for the normal distribution, Lumel'skii and Sapozhnikov (1969, p. 357, Theorem 1) suggest the following general procedure:

Let $\mathfrak{P}$ be dominated by μ. Assume that for every $P \in \mathfrak{P}$, $P^n \circ S_n|\mathscr{B}$ has a ν-density, say $h_P^{(n)}$, and that the joint distribution of $(x_1, S_n(x_1, \ldots, x_n))$ under P^n has $\mu \times \nu$-density $(\xi, y) \to h_P^{(n)}(\xi, y)$. Then $p_n(\xi, y) := h_P^{(n)}(\xi, y)/h_P^{(n)}(y)$ is, thanks to the sufficiency of S_n, independent of P, and $p_n(\xi, S_n(\cdot))$ is unbiased for $h_P(\xi)$. The computation of $h_P^{(n)}$ becomes simple if $S_n(x_1, \ldots, x_n) = \sum_{\nu=1}^n x_\nu$. (See Pfanzagl 1994, p. 118.)

The literature provides numerous examples of unbiased estimators of probabilities and the pertaining estimators of densities. A somewhat disturbing phenomenon, to be found in all these examples: If $\mathfrak{P}$ is a parametric family, the optimal unbiased estimator is not a member of this family. As an example we mention that the optimal unbiased estimator of $N(\mu, 1)(A)$ in the family $\{N(\mu, 1) : \mu \in \mathbb{R}\}$ is

$$(x_1, \ldots, x_n) \to N(\overline{x}_n, (n-1)/n)(A).$$

The optimal unbiased estimator of $N(\mu, \sigma^2)(A)$ in the family $\{N(\mu, \sigma^2) : \mu \in \mathbb{R}, \sigma^2 > 0\}$ is

$$(x_1, \ldots, x_n) \to \int 1_A(\xi)p^{(n)}(\xi; \overline{x}_n, s_n)d\xi$$

with $s_n = n^{-1} \sum_{\nu=1}^{n} (x_\nu - \overline{x}_n)^2$ and

$$p^{(n)}(\xi; \mu, \sigma) = c_n \frac{1}{\sigma} \left(1 - \frac{1}{n-1} \frac{(\xi - \mu)^2}{\sigma^2} \right)^{(n-4)/2}.$$

(Find out which of the versions of c_n offered in the literature comes closest to the truth: Kolmogorov 1950, pp. 391/2; Barton 1961, p. 228; Basu 1964, p. 219; Lumel'skii and Sapozhnikov (1969), p. 360, specialized for $p = 1$.)

4.4 The Cramér–Rao Bound

The first general result on mean unbiased estimators is the so-called Cramér–Rao bound, a classical example of multiple discoveries. With a straightforward proof, this result has an unusual number of fathers: Aitken and Silverstone (1942), Fréchet (1943, p. 185), Darmois (1945, p. 9, with a reference to Fréchet 1943), Rao (1945, p. 83 and 1947, p. 281) and Cramér (1946, p. 480, relation 23.3.3a). Following "Stigler's law of eponymy" it was named after the last two of these. Savage (1954, p. 238) therefore suggested the now widely used name "information bound".

The straightforward argument: Let $\{P_\vartheta : \vartheta \in \Theta\}$, $\Theta \subset \mathbb{R}$, be a parametric family with $p(\cdot, \vartheta) \in dP_\vartheta/d\mu$. Write $p^\bullet(x, \vartheta) = \partial_\vartheta p(x, \vartheta)$ and $\ell^\bullet(x, \vartheta) = \partial_\vartheta \log p(x, \vartheta) = p^\bullet(x, \vartheta)/p(x, \vartheta)$. If the estimator $\hat{\vartheta} : X \to \mathbb{R}$ is unbiased for ϑ, then

$$\int (\hat{\vartheta}(x) - \vartheta) p(x, \vartheta) \mu(dx) = 0 \quad \text{for } \vartheta \in \Theta. \tag{4.4.1}$$

Differentiation with respect to ϑ leads to

$$\int (\hat{\vartheta}(x) - \vartheta) \ell^\bullet(x, \vartheta) \; P_\vartheta(dx) = 1. \tag{4.4.2}$$

Using Schwarz' inequality, this implies

$$\int (\hat{\vartheta}(x) - \vartheta)^2 P_\vartheta(dx) \geq 1 \Big/ \int \ell^\bullet(x, \vartheta)^2 P_\vartheta(dx). \tag{4.4.3}$$

The step from (4.4.1) to (4.4.2) depends on the equality

$$\partial_\vartheta \int \hat{\vartheta}(x) p(x, \vartheta) \mu(dx) = \int \hat{\vartheta}(x) \partial_\vartheta p(x, \vartheta) \mu(dx), \tag{4.4.4}$$

i.e., on the interchange of differentiation and integration. This relation was neither taken for granted, nor considered as a regularity condition in the early papers. It was neglected in Rao (1945). Rao (1949, p. 216, Theorem 3) gives conditions on the family that imply relation (4.4.4).

A weak condition that ensures the validity of (4.4.4) at ϑ_0 (even for k-parameter families) can be found in Witting (1985, p. 319, Satz 2.136): L_2-differentiability of the family $\{P_\vartheta : \vartheta \in \Theta\}$ at ϑ_0, and the existence of a function $\hat{\vartheta}$ such that $\int \hat{\vartheta} dP_\vartheta = \vartheta$ and $\vartheta \to \int \hat{\vartheta}^2 dP_\vartheta$ is locally bounded at ϑ_0. Simons and Woodroofe (1983, p. 76, Corollary 2) show, under slightly weaker conditions, that (4.4.3) holds μ-a.e. (see also Witting 1985, p. 327, Aufgabe 2.33).

In "Kendall's Advanced Theory of Statistics", Stuart and Ord (1991) pass over such points. They confine themselves to the statement $\int p^\bullet(x, \vartheta)dx = 0$ is the only [!] condition for the validity of the Cramér–Rao bound (see p. 616). This slip subsists in all editions even though it was pointed out already in Polfeldt (1970, p. 23).

The Cramér–Rao bound is attainable in one special case only: If the family $\{P_\vartheta : \vartheta \in \Theta\}$ is exponential with density

$$p(x, \vartheta) = C(\vartheta) \exp[\vartheta T(x)] \quad \text{and} \quad \int T(x) P_\vartheta(dx) = \vartheta.$$

This is implicitly already contained in Aitken and Silverstone (1942, p. 188), who show that an unbiased estimator $\vartheta^{(n)}$ achieves the minimal variance iff

$$\sum_{\nu=1}^{n} \ell^\bullet(x_\nu, \vartheta) = C(\vartheta)(\vartheta^{(n)}(x) - \vartheta).$$

A corresponding relation occurs in Cramér (1946, p. 475).

Even for exponential families, unbiased estimators attaining the variance bound exist certain functions of ϑ only. Rao (1949, pp. 214/5) gives the following example. For the Γ-distribution with scale parameter a and known shape parameter b, the estimator $\kappa^{(n)}(x_1, \ldots, x_n) = (nb - 1)/\overline{x}_n$ is unbiased for $\kappa(\Gamma_{a,b}) = a^{-1}$ and has variance $1/a^2(nb - 2)$, whereas the Cramér–Rao bound is $1/a^2 nb$. He then shows that no continuous function of $\overline{x}_n$ other than $\kappa^{(n)}$ can be unbiased. Hence there is no continuous function of $\overline{x}_n$ attaining the bound. One year later, Rao could have argued that $\kappa^{(n)}$ is the function of a complete sufficient statistic, hence $1/a^2(nb-2)$ the minimal variance for unbiased estimators. Rao's example establishes only the existence of some family for which the bound is not attainable. He obviously had missed what was implicitly already contained in the paper by Aitken and Silverstone (cited by Rao), namely that the bound is attainable for special exponential families only. (See also Witting 1985, Sect. 3.7.1.)

A paper that represents the present state of the art is Müller-Funk et al. (1989). (See also Witting 1985, Sect. 3.7.1.)

Looking back on the hundreds of papers devoted to the Cramér–Rao bound, it is hard to understand the attention paid to the conditions needed in the proof of a bound if this bound is not attainable anyway. Moreover, the variance as a measure for the quality of an estimator is more than doubtful. Hence, we will abstain from discussing similar bounds put forward by Hammersley (1950) or Chapman and Robbins (1951).

Some scholars hold the opinion that bounds are meaningful even if they are not attainable, but they withhold their arguments supporting this opinion. Andersen (1970, p. 85, with respect to another bound which is unattainable, too) claimed: "... in situations where the lower bound is not attained, [it] provides us with a denominator for an efficiency measure". Similarly, Barnett (1975, p. 126): "[The Cramér–Rao bound] provides an absolute standard against which to measure estimators in a wide range of situations." Fisz (1963, p. 470, Definition 13.5.1) *defines* an unbiased estimator as "most efficient" if it attains the Cramér–Rao bound, neglecting the presence of models with an unbiased estimator of minimal convex risk, larger than the Cramér–Rao bound.

Some scholars think that the Cramér–Rao bound is at least valid asymptotically. This opinion results from the fact that in highly regular parametric families the Cramér–Rao bound happens to coincide with the asymptotic bound provided by the Convolution Theorem for regular estimator sequences. Even respected authors like Witting and Müller-Funk (1995, p. 198) cannot resist the temptation to pretend a connection which does not exist: The nature of these bounds is totally different, and so are the proofs. This can be seen from examples where the bound for the asymptotic variance of regular estimator sequences is attained, whereas the best sequences of unbiased estimators have a larger asymptotic variance (see Portnoy 1977, for a somewhat artificial, and Pfanzagl 1993, pp. 74–76, for a more natural example). This refutes a widely held opinion that the Cramér–Rao bound is always asymptotically attainable.

4.5 Optimal Median Unbiased Estimators

Recall that two estimators of a functional $\kappa : \mathfrak{P} \to \mathbb{R}$ may be comparable with respect to their concentration on arbitrary intervals $(\kappa(P) - t', \kappa(P) + t'')$ only if $\kappa(P)$ is a common quantile. In natural applications, this common quantile $\kappa(P)$ is the median, and one can expect that for certain models the class of all median unbiased estimators of $\kappa(P)$ contains an element which is optimal in the sense of being maximally concentrated in every interval containing $\kappa(P)$.

In contrast, optimality for mean unbiased estimators just means "minimal convex risk". Since mean unbiasedness and median unbiasedness are two fundamentally different conditions on the location of an estimator, it makes no sense to compare the quality of a mean unbiased estimator with the quality of a median unbiased one. Moreover, the optimality of a median unbiased estimator, expressed in terms of concentration intervals, is of a clear operational significance, whereas the optimality of a mean unbiased estimator is based on the concept of convex risk, the interpretation of which is open to question.

Yet, it might be worthwhile to have a look at estimators that are both mean as well as median unbiased, say $\overline{x}_n$ as an estimator for μ in the family $\{N(\mu, \sigma^2)^n :$

$\mu \in \mathbb{R}, \sigma^2 > 0\}$. Most textbooks restrict themselves to stating that $\overline{x}_n$ minimizes the quadratic risk in the class of all mean unbiased estimators. They do not consider worth mentioning that no median unbiased estimator can be more concentrated than $\overline{x}_n$ in any of the intervals $(\mu - t', \mu + t'')$.

In the following we discuss conditions under which (optimal) median unbiased estimators exist. To outline the basic ideas, we consider a family $\{P_\vartheta : \vartheta \in \Theta\}$ in which $\Theta \subset \mathbb{R}$ is an interval. Let $S : (X, \mathscr{A}) \to (\mathbb{R}, \mathbb{B})$ be a statistic such that the family $\{P_\vartheta \circ S : \vartheta \in \Theta\}$ is stochastically isotone, say increasing. Then

$$F_\vartheta(u) := P_\vartheta\{x \in X : S(x) \leq u\}$$

is, for fixed u, decreasing in ϑ. To avoid technicalities, we assume for the moment that $u \to F_\vartheta(u)$ is increasing and continuous. If $u(\vartheta)$ is such that $F_\vartheta(u(\vartheta)) = 1/2$, then $x \to u^{-1}(S(x))$ is a median unbiased estimator of ϑ.

The median unbiased estimators thus obtained will be maximally concentrated on arbitrary intervals containing ϑ: if S is sufficient, and if the densities $p(x, \vartheta) = h(x)g(S(x), \vartheta)$ have monotone likelihood ratios in S, i.e., if $y \to g(y, \vartheta_2)/g(y, \vartheta_1)$ is increasing if $\vartheta_1 < \vartheta_2$. Monotonicity of likelihood ratios implies that the family $\{P_\vartheta \circ S : \vartheta \in \Theta\}$ is stochastically increasing, so that median unbiased estimators can be obtained as indicated above.

Stochastic monotonicity suffices for the existence of a median unbiased estimator. It does, however, not guarantee that this estimator is reasonable (see Pfanzagl 1972, p. 160, Example 3.16). If S is sufficient for $\{P_\vartheta : \vartheta \in \Theta\}$ and if $\{P_\vartheta \circ S : \vartheta \in \Theta\}$ has monotone likelihood ratios, then a median unbiased estimator is maximally concentrated on all intervals $(\vartheta - t', \vartheta + t'')$ if it is a monotone function of S. Because of the m.l.r. property, any set $\{x \in X : m(S(x)) \geq \vartheta'\}$, with m increasing, is, according to the Neyman–Pearson Lemma, most powerful for every testing problem $P_{\vartheta'} : P_\vartheta$ with $\vartheta > \vartheta'$. Since

$$P_{\vartheta'}\{m \circ S \geq \vartheta'\} = \frac{1}{2} = P_{\vartheta'}\{\hat{\vartheta} \geq \vartheta'\}$$

if $\hat{\vartheta}$ is median unbiased, this implies

$$P_\vartheta\{m \circ S \geq \vartheta'\} \geq P_\vartheta\{\hat{\vartheta} \geq \vartheta'\},$$

hence

$$P_\vartheta\{\vartheta' \leq m \circ S < \vartheta\} \geq P_\vartheta\{\vartheta' \leq \hat{\vartheta} < \vartheta\}.$$

The same argument yields

$$P_\vartheta\{\vartheta \leq m \circ S < \vartheta''\} \geq P_\vartheta\{\vartheta \leq \hat{\vartheta} < \vartheta''\} \text{ for } \vartheta'' > \vartheta.$$

What has been said so far is essentially presented in Lehmann (1959, Sect. 5, pp. 78–83). Lehmann also indicates how randomization can be used to obtain median unbiased estimators in the more general case where $P_\vartheta \circ S$ may contain atoms.

Lehmann is aware of the fact that critical regions $\{x \in X : S(x) \geq c\}$ do not only maximize the power for testing ϑ'' against $\vartheta'' > \vartheta$; they also minimize the power for alternatives $\vartheta' < \vartheta$ (see his pp. 68/9, Theorem 2). Yet he arrives at the optimality assertion (see p. 83) by means of a different argument: He considers median unbiased estimators as a boundary case of two-sided confidence intervals $(\underline{\vartheta}, \bar{\vartheta})$ with $P_\vartheta\{\vartheta < \underline{\vartheta}\} = \alpha_1$, and $P_\vartheta\{\vartheta > \bar{\vartheta}\} = \alpha_2$ for $\alpha_1 = \alpha_2 = 1/2$, using a loss function like

$$\ell(\vartheta; \underline{\vartheta}, \bar{\vartheta}) = \begin{cases} \vartheta - \underline{\vartheta} & \\ \bar{\vartheta} - \underline{\vartheta} & \text{if} \\ \bar{\vartheta} - \vartheta & \end{cases} \begin{array}{c} \vartheta < \underline{\vartheta}, \\ \underline{\vartheta} \leq \vartheta \leq \bar{\vartheta}, \\ \bar{\vartheta} < \vartheta. \end{array}$$

Birnbaum (Birnbaum 1964, p. 27) attributes the optimality result for median unbiased estimators to Birnbaum (1961), where it is contained implicitly in Lemma 2, p. 121. Pfanzagl (1970, p. 33, Theorem 1.12) treats the general case of randomized estimators.

According to Brown et al. (1976, p. 719, Corollary 4.1), the following is true under the m.l.r. conditions indicated above: For any median unbiased estimator which is not a contraction of S (almost everywhere), there exists a median unbiased contraction of S which is strictly better. This result is an analogue to the Rao–Blackwell–Lehmann–Scheffé Theorem for mean unbiased estimators.

A straightforward application of these results is to exponential families with densities of the form

$$C(\vartheta)h(x)\exp[a(\vartheta)S(x)] \tag{4.5.1}$$

with a monotone function a. These families have monotone likelihood ratios for every sample size. In fact, this is almost the only application. According to a result obtained by Borges and Pfanzagl (1963, p. 112, Theorem 1) a one-parameter family of mutually absolutely continuous probability measures with monotone likelihood ratios for every sample size is necessarily of the type (4.5.1).

Relevant for applications are results on the existence of optimal median unbiased estimators for families with nuisance parameters. Such a result may be found in Pfanzagl (1979, p. 188, Theorem). Here is an important special case.

Assume that $P_{\vartheta,\eta}$, $\vartheta \in \Theta$ (an interval in $\mathbb{R}$) and $\eta \in H$ (and abstract parameter space) has densities

$$C(\vartheta, \eta)h(x)\exp\left[a(\vartheta)S(x) + \sum_{i=1}^{k} a_i(\vartheta, \eta)T_i(x)\right]$$

with $S : (X, \mathscr{A}) \to (\mathbb{R}, \mathbb{B})$ and $T_i : (X, \mathscr{A}) \to (\mathbb{R}, \mathbb{B})$.

If the function $\vartheta \to a(\vartheta, \eta)$ is increasing and continuous for every $\eta \in H$, and if *$\{(a_1(\vartheta, \eta), \ldots, a_k(\vartheta, \eta)) : \eta \in H\}$ has for every $\vartheta \in \Theta$ a nonempty interior in*

$\mathbb{R}^k$, *then there exists a maximally concentrated median unbiased estimator of* ϑ *for every sample size.*

Taken at its surface, the Rao–Blackwell–Lehmann–Scheffé Theorem seems to provide a much more general assertion about the existence of optimal mean unbiased estimators (as compared to the above theorem on median unbiased estimators). However, this impression is misleading. The typical case in which the existence of a complete sufficient statistic is guaranteed for every sample size is the exponential family, again. Moreover, the Rao–Blackwell–Lehmann–Scheffé Theorem presumes that mean unbiased estimators exist, and that an "initial" mean unbiased estimator is known. As against that, *median* unbiased estimators always exist under the conditions indicated above.

4.6 Confidence Procedures

Let $\mathfrak{P}|\mathscr{A}$ be a family of probability measures, and let $\kappa : \mathfrak{P} \to \mathbb{R}^k$ be a functional. If $\hat{\kappa} : X \to \mathbb{R}^k$ is an estimator for κ, and a realization x from P is available, then one can compute the estimate $\hat{\kappa}(x)$. What does the observed value $\hat{\kappa}(x)$ tell us about $\kappa(P)$? One might surmise that $\kappa(P)$ will be somewhere close to $\kappa(x)$. In which sense can this be made precise?

The appropriate answer is given by a *confidence procedure*, i.e. a function K mapping a point $x \in X$ into a subset $K(x)$ of the family $\mathfrak{P}$ such that

$$\{x \in X : t \in K(x)\} \in \mathscr{A} \quad \text{for } t \in \mathbb{R}^k.$$

If $\mathfrak{P} = \{P_\vartheta : \vartheta \in \Theta\}$ with $\Theta \subset \mathbb{R}^k$ is a parametric family of probability measures, we write the confidence procedure as a map from X into the (Lebesgue measurable) subsets of $\mathbb{R}^k$. For one-dimensional parametric families, the confidence sets $K(x)$ are usually intervals $[\underline{\vartheta}(x), \overline{\vartheta}(x)]$ or half rays $(-\infty, \overline{\vartheta}(x)]$ or $[\underline{\vartheta}(x), \infty]$.

Let $\Theta \subset \mathbb{R}^k$, and for each ϑ let $H_\vartheta \subset \Theta$. For $\Theta \subset \mathbb{R}$, we will usually take $H_\vartheta = \{\vartheta\}$ or $H_\vartheta = (-\infty, \vartheta]$ or $H_\vartheta = [\vartheta, \infty)$. A confidence procedure K has *significance level* $1 - \alpha$ for H if

$$P_\tau\{\vartheta \in K\} \geq 1 - \alpha \quad \text{for } \tau \in H_\vartheta,\ \vartheta \in \Theta.$$

If $1 - \alpha$ is close to 1 we will be confident that the unknown value $\kappa(P)$ is in the observed set $K(x)$.

The notion of a confidence interval as a set of parameter values containing the true parameter value with "high probability" is used in a vague form by several writers in the 19th century, the first being perhaps Laplace (1812), who gives an asymptotic confidence interval for the parameter of the binomial distribution (2nd Book, Sect. 16). It also appears in Gauss (1816). Other examples can be found in Fourier (1826), Cournot (1843, pp. 185/6) and Lexis (1875).

The first formally correct statement of confidence sets as random objects containing the fixed parameter with prescribed probability is given by Wilson (1927). Other examples of conceptually precise confidence statements prior to Neyman's general theory are Working and Hotelling (1929), Hotelling (1931) and Clopper and Pearson (1934).

When Neyman entered the scene (1934, Appendix, Note 1, and 1937, 1938) he was obviously not aware of Wilson's paper. His intention was to make clear that the concept of confidence, being based on the usual concept of probability, was something else than Fisher's concept of "fiducial distributions" (1930), a concept based on principles that cannot be deduced from the rules of ordinary logic. (See in particular Neyman 1941.)

Lehmann (1959) was the first textbook dealing with confidence sets. As a book with emphasis on test theory, it treats confidence intervals more or less as an appendix to test theory. It obtains confidence sets by inverting critical regions (Lemma 5.5.1, p. 179). This accounts for the restriction to confidence intervals for the parameter of a univariate family. As an appendix to test theory, the author borrows the concept for describing properties of confidence intervals (such as unbiasedness) from the corresponding properties of tests. This prevents more or less the development of a genuine theory of confidence sets. For the case of univariate families, something like optimality was obtained for exponential (more generally: m.l.r.-) families.

The theory of confidence sets has virtually disappeared in the more advanced textbooks on mathematical statistics—Bickel et al. (1993) has two references to confidence sets; Strasser (1985) has none. The fact that the theory of confidence sets had lost its adequate treatment is, perhaps due to the inadequate starting point, in Lehmann (1959), to consider where confidence procedures are considered as an appendix to test theory. Even the most recent book which contains a longer chapter on confidence sets ("The theory of confidence sets", Chap. IV, pp. 254–267 in Schmetterer 1974), still deals with confidence sets as an appendix to test theory. The number of references to confidence procedures is two in Lehmann and Casella (1998) and zero in Bickel et al. (1993) and in Strasser. See however Shao (2003), Chap. 7.

A confidence procedure K^* with confidence level $1-\alpha$ is called *uniformly most accurate* (for H) if for every confidence procedure K^* with confidence level $1-\alpha$ we have

$$P_\tau(\vartheta \in K^*) \leq P_\tau(\vartheta \in K) \quad \text{for } \tau \notin H_\vartheta,\ \vartheta \in \Theta.$$

Confidence sets can be obtained by inverting acceptance regions of tests. Let $\Theta \subset \mathbb{R}^k$, and for each ϑ let H_ϑ be a hypothesis and $A(\vartheta)$ an acceptance region with level α. For $x \in X$ set

$$K(x) = \{\vartheta \in \Theta : x \in A(\vartheta)\}.$$

Then K is a confidence procedure with significance level $1-\alpha$.

If $\Theta \subset \mathbb{R}$ and the acceptance region $A(\vartheta)$ is uniformly most powerful for H_ϑ for each $\vartheta \in \Theta$, then the confidence procedure obtained by inversion is uniformly most accurate for H.

The idea to obtain confidence sets by inversion of acceptance regions meets the following problem: Whereas the shape of a acceptance region is of no relevance for the power of a test, the shape of a confidence set is crucial for its interpretation. Think of examples where confidence sets obtained by inversion of (optimal) tests are not connected.

Pratt (1961) and Ghosh (1961) relate the false covering probabilities of confidence sets $K(x)$ to their expected size. With $\Theta \subset \mathbb{R}^k$ and λ denoting Lebesgue measure on $\mathbb{R}^k$, Fubini's Theorem implies

$$\int \lambda(K(x))\, P(dx) = \int P\{x \in X : \tau \in K(x)\}\lambda(d\tau)$$

for any probability measure P, in particular the Ghosh–Pratt identity

$$\int \lambda(K(x))\, P_\vartheta(dx) = \int_{\tau \neq \vartheta} P_\vartheta\{x \in X : \tau \in K(x)\}\lambda(d\tau).$$

We now consider the problem of an upper confidence bound with covering probability $\beta \in (0, 1)$ for the parameter ϑ in a univariate family $P_\vartheta|\mathscr{A}$, $\vartheta \in \Theta \subset \mathbb{R}$, i.e. a function $\overline{\vartheta}_\beta : X \to \mathbb{R}$ such that

$$P_\vartheta\{x \in X : \vartheta \leq \overline{\vartheta}_\beta(x)\} = \beta \quad \text{for every } \vartheta \in \Theta. \tag{4.6.1}$$

For a given ϑ the ideal answer would be $q_\beta(\vartheta)$, the β-quantile of P_ϑ. A function $x \to \overline{\vartheta}_\beta(x)$ which meets this requirement for every $\vartheta \in \Theta$, should be close to $q_\beta(\vartheta)$ for every $\vartheta \in \Theta$. If we consider $\overline{\vartheta}_\beta$ as an estimator of $q_\beta(\vartheta)$, the $\overline{\vartheta}_\beta$ should be concentrated about $q_\beta(\vartheta)$ as closely as possible. Under special conditions there is a precise answer to this vague question. Let us say that a functional f_0 is *more concentrated* about $q(\vartheta)$ than the functional f_1 if

$$P_\vartheta\{x \in X : f_0(x) \in I\} \geq P_\vartheta\{x \in X : f_1(x) \in I\}$$

for every interval I containing $q(\vartheta)$. With this terminology, *no confidence bound* ϑ_β *fulfilling condition* (4.6.1) *can be more concentrated about* $q_\beta(\vartheta)$ *than a confidence bound depending on X through* $T(x)$ *only*. An earlier version of this result occurs in Pfanzagl (1994, p. 173, Theorem 5.4.3).

References

Aitken, A. C., & Silverstone, H. (1942). On the estimation of statistical parameters. *Proceedings of the Royal Society of Edinburgh Section A*, *61*, 186–194.

Andersen, E. B. (1970). Asymptotic properties of conditional maximum-likelihood estimators. Journal of the Royal Statistical Society. *Series B*, *32*, 283–301.

Bahadur, R. R. (1957). On unbiased estimates of uniformly minimum variance. *Sankhyā, 18*, 211–224.
Barankin, E. W. (1949). Locally best unbiased estimates. *The Annals of Mathematical Statistics, 20*, 477–501.
Barankin, E. W. (1950). Extension of a theorem of Blackwell. *The Annals of Mathematical Statistics, 21*, 280–284.
Barankin, E. W. (1951). Conditional expectation and convex functions. In *Proceedings of the Second Berkeley Symposium on Mathematical Statistics and Probability* (pp. 167–169). Berkeley: University of California Press.
Barnett, V. (1975). *Comparative statistical inference*. New York: Wiley.
Barton, D. E. (1961). Unbiased estimation of a set of probabilities. *Biometrika, 48*, 227–229.
Basu, A. P. (1964). Estimators of reliability for some distributions useful in life testing. *Technometrics, 6*, 215–219.
Basu, D. (1955). A note on the theory of unbiased estimation. *The Annals of Mathematical Statistics, 26*, 345–348.
Bednarek-Kozek, B., & Kozek, A. (1978). *Two examples of strictly convex non-universal loss functions* (p. 133). Preprint: Institute of Mathematics Polish Academy of Sciences.
Bickel, P. J., Klaassen, C. A. J., Ritov, Y., & Wellner, J. A. (1993). *Efficient and adaptive estimation for semiparametric models*. Baltimore: Johns Hopkins University Press (1998 Springer Paperback).
Birnbaum, A. (1961). A unified theory of estimation, I. *The Annals of Mathematical Statistics, 32*, 112–135.
Birnbaum, A. (1964). Median-unbiased estimators. *Bulletin of Mathematics and Statistics, 11*, 25–34.
Blackwell, D. (1947). Conditional expectation and unbiased sequential estimation. *The Annals of Mathematical Statistics, 18*, 105–110.
Bomze, I. M. (1986). Measurable supports, reducible spaces and the structure of the optimal σ-field in unbiased estimation. *Monatshefte für Mathematik, 101*, 27–38.
Bomze, I. M. (1990). *A functional analytic approach to statistical experiments* (Vol. 237). Pitman research notes in mathematics series. Harlow: Longman.
Borges, R., & Pfanzagl, J. (1963). A characterization of the one-parameter exponential family of distributions by monotonicity of likelihood ratios. *Z. Wahrscheinlichkeitstheorie verw. Gebiete, 2*, 111–117.
Brown, L. D., Cohen, A., & Strawderman, W. E. (1976). A complete class theorem for strict monotone likelihood ratio with applications. *The Annals of Statistics, 4*, 712–722.
Chapman, D. G., & Robbins, H. (1951). Minimum variance estimation without regularity assumptions. *The Annals of Mathematical Statistics, 22*, 581–586.
Clopper, C. J., & Pearson, E. S. (1934). The use of confidence or fiducial limits illustrated in the case of the binomial. *Biometrika, 26*, 404–413.
Cournot, A. A. (1843). *Exposition de la théorie des chances et des probabilités*. Paris: Hachette.
Cramér, H. (1946). *Mathematical methods of statistics*. Princeton: Princeton University Press.
Darmois, G. (1945). Sur les lois limites de la dispersion de certain estimations. *Revue de l'Institut International de Statistique, 13*, 9–15.
Eberl, W. (1984). On unbiased estimation with convex loss functions. In E. J. Dudewitz, D. Plachky, & P. K. Sen (Eds.), *Recent results in estimation theory and related topics* (pp. 177–192). Statistics and Decisions, Supplement, Issue 1.
Eberl, W., & Moeschlin, O. (1982). *Mathematische Statistik*. Berlin: De Gruyter.
Fisher, R. A. (1930). Inverse probability. *Mathematical Proceedings of the Cambridge Philosophical Society, 26*, 528–535.
Fisz, M. (1963). *Probability theory and mathematical statistics* (3rd ed.). (1st ed. 1954 in Polish, 2nd ed. 1958 in Polish and German). New York: Wiley.

Fourier, J. B. J. (1826). Mémoire sur les résultats moens déduits d'un grand nombre d'observations. Recherches statistiques sur la ville de Paris et le département de la Seine. *Reprinted in Œuvres, 2*, 525–545.

Fréchet, M. (1943). Sur l'extension de certaines évaluations statistiques de petits échantillons. *Revue de l'Institut International de Statistique, 11*, 182–205.

Gauss, C. F. (1816). Bestimmung der Genauigkeit der Beobachtungen. Carl Friedrich Gauss Werke *4*, 109–117, Königliche Gesellschaft der Wissenschaften, Göttingen.

Ghosh, J. K. (1961). On the relation among shortest confidence intervals of different types. *Calcutta Statistical Association Bulletin, 10*, 147–152.

Halmos, P. R. (1946). The theory of unbiased estimation. *The Annals of Mathematical Statistics, 17*, 34–43.

Hammersley, J. M. (1950). On estimating restricted parameters. *Journal of the Royal Statistical Society. Series B, 12*, 192–229; Discussion, 230–240.

Heizmann, H.-H. (1989). UMVU-Schätzer und ihre Struktur. *Systems in Economics, 112*. Athenäum-Verlag.

Heyer, H. (1973). *Mathematische Theorie statistischer Experimente. Hochschultext*. Berlin: Springer.

Heyer, H. (1982). *Theory of statistical experiments. Springer series in statistics*. Berlin: Springer.

Hodges, J. L, Jr., & Lehmann, E. L. (1950). Some problems in minimax point estimation. *The Annals of Mathematical Statistics, 21*, 182–197.

Hotelling, H. (1931). The generalization of Student's ratio. *The Annals of Mathematical Statistics, 2*, 360–378.

Kolmogorov, A. N. (1950). Unbiased estimators. *Izvestiya Akademii Nauk S.S.S.R. Seriya Matematicheskaya, 14*, 303–326. Translation: pp. 369–394 in: Selected works of A.N. Kolmogorov (Vol. II). Probability and mathematical statistics. A. N. Shiryaev (ed.), Mathematics and its applications (Soviet series) (Vol. 26). Dordrecht: Kluwer Academic (1992).

Kozek, A. (1977). On the theory of estimation with convex loss functions. In R. Bartoszynski, E. Fidelis, & W. Klonecki (Eds.), *Proceedings of the Symposium to Honour Jerzy Neyman* (pp. 177–202). Warszawa: PWN-Polish Scientific Publishers.

Kozek, A. (1980). On two necessary and sufficient σ-fields and on universal loss functions. *Probability and Mathematical Statistics, 1*, 29–47.

Krámli, A. (1967). A remark to a paper of L. *Schmetterer. Studia Scientiarum Mathematicarum Hungarica, 2*, 159–161.

Laplace, P. S. (1812). *Théorie analytique des probabilités*. Paris: Courcier.

Lehmann, E. L. (1959). *Testing statistical hypotheses*. New York: Wiley.

Lehmann, E. L., & Casella, G. (1998). *Theory of point estimation* (2nd ed.). Berlin: Springer.

Lehmann, E. L., & Scheffé, H. (1950, 1955, 1956). Completeness, similar regions and unbiased estimation. *Sankhyā, 10*, 305–340; *15*, 219–236; *Correction, 17* 250.

Lexis, W. (1875). *Einleitung in die Theorie der Bevölkerungsstatistik*. Straßburg: Trübner.

Linnik, Yu V, & Rukhin, A. L. (1971). Convex loss functions in the theory of unbiased estimation. *Soviet Mathematics Doklady, 12*, 839–842.

Lumel'skii, Ya. P., & Sapozhnikov, P. N., (1969). Unbiased estimates of density functions. *Theory of Probability and Its Applications, 14*, 357–364.

Müller-Funk, U., Pukelsheim, F., & Witting, H. (1989). On the attainment of the Cramér-Rao bound in L_r-differentiable families of distributions. *The Annals of Statistics, 17*, 1742–1748.

Neyman, J. (1934). On the two different aspects of the representative method: The method of stratified sampling and the method of purposive selection. *The Journal of the Royal Statistical Society, 97*, 558–625.

Neyman, J. (1937). Outline of a theory of statistical estimation based on the classical theory of probability. Philosophical Transactions of the Royal Society of London. *Series A, 236*, 333–380.

Neyman, J. (1938). Léstimation statistique traitée comme un probléme classique de probabilité. *Actualités Sci. Indust., 739*, 25–57.

Neyman, J. (1941). Fiducial argument and the theory of confidence intervals. *Biometrika, 32*, 128–150.
Padmanabhan, A. R. (1970). Some results on minimum variance unbiased estimation. *Sankhyā Ser. A, 32*, 107–114.
Pfanzagl, J. (1970). Median-unbiased estimators for M.L.R.-families. *Metrika, 15*, 30–39.
Pfanzagl, J. (1972). On median-unbiased estimates. *Metrika, 18*, 154–173.
Pfanzagl, J. (1979). On optimal median-unbiased estimators in the presence of nuisance parameters. *The Annals of Statistics, 7*, 187–193.
Pfanzagl, J. (1993). Sequences of optimal unbiased estimators need not be asymptotically optimal. *Scandinavian Journal of Statistics, 20*, 73–76.
Pfanzagl, J. (1994). *Parametric statistical theory*. Berlin: De Gruyter.
Polfeldt, T. (1970). Asymptotic results in non-regular estimation. Skand. Aktuarietidskr. Supplement 1–2.
Portnoy, S. (1977). Asymptotic efficiency of minimum variance unbiased estimators. *The Annals of Statistics, 5*, 522–529.
Pratt, J. W. (1961). Length of confidence intervals. *Journal of the American Statistical Association, 56*, 549–567.
Rao, C. R. (1945). Information and accuracy attainable in the estimation of statistical parameters. *Bulletin of Calcutta Mathematical Society, 37*, 81–91.
Rao, C. R. (1947). Minimum variance estimation of several parameters. *Proceedings of the Cambridge Philosophical Society, 43*, 280–283.
Rao, C. R. (1949). Sufficient statistics and minimum variance estimates. *Proceedings of the Cambridge Philosophical Society, 45*, 213–218.
Rao, C. R. (1952). Some theorems on minimum variance estimation. *Sankhyā, 12*, 27–42.
Savage, L. J. (1954). *The foundations of statistics*. New York: Wiley.
Schmetterer, L. (1960). On unbiased estimation. *The Annals of Mathematical Statistics, 31*, 1154–1163.
Schmetterer, L. (1966). On the asymptotic efficiency of estimates. In Research Papers in Statistics. Festschrift for J. Neyman & F. N. David (Eds.) (pp. 301–317). New York: Wiley.
Schmetterer, L. (1974). *Introduction to mathematical statistics*. Translation of the 2nd German edition 1966. Berlin: Springer.
Schmetterer, L. (1977). Einige Resultate aus der Theorie erwartungstreuer Schätzungen. In *Transactions of the Seventh Prague Conference on Information Theory, Statistical Decision Functions, Random processes* (Vol. B, pp. 489–503).
Schmetterer, L., & Strasser, H. (1974). *Zur Theorie der erwartungstreuen Schätzungen*. Anzeiger, Österreichische Akadademie der Wissenschaften, Mathematisch-Naturwissenschaftliche. Klasse, *6*, 59–66.
Seheult, A. H., & Quesenberry, C. P. (1971). On unbiased estimation of density functions. *The Annals of Mathematical Statistics, 42*, 1434–1438.
Shao, J. (2003). *Mathematical statistics* (2nd ed.). Berlin: Springer.
Simons, G., & Woodroofe, M. (1983). The Cramér-Rao inequality holds almost everywhere. In M. H. Rizvi, J. S. Rustagi, & D. Siegmund (Eds.), *Papers in Honor of Herman Chernoff* (pp. 69–83). New York: Academic Press.
Stein, Ch. (1950). Unbiased estimates of minimum variance. *The Annals of Mathematical Statistics, 21*, 406–415.
Strasser, H. (1972). Sufficiency and unbiased estimation. *Metrika, 19*, 98–114.
Strasser, H. (1985). *Mathematical theory of statistics*. Berlin: De Gruyter.
Stuart, A., & Ord, K. (1991). *Kendall's advanced theory of statistics. Classical inference and relationship* (5th ed., Vol. 2). London: Edward Arnold.
Wilson, E. B. (1927). Probable inference, the law of succession, and statistical inference. *Journal of the American Statistical Association, 22*, 209–212.
Witting, H. (1985). *Mathematische Statistik I*. Parametrische Verfahren bei festem Stichprobenumfang: Teubner.

Witting, H., & Müller-Funk, U. (1995). *Mathematische Statistik II*. Asymptotische Statistik: Parametrische Modelle und nichtparametrische Funktionale. Teubner.

Working, H., & Hotelling, H. (1929). Applications of the theory of error to the interpretation of trends. *Journal of the American Statistical Association*, *24*(165A), 73–85.

Chapter 5
Asymptotic Optimality of Estimators

5.1 Introduction

Nonasymptotic theory, which flourished between 1940 and 1960, ended up as a collection of results that does not amount to a consistent theory.

Important parts of nonasymptotic theory are, in fact, confined to exponential families. Even in this restricted domain, the applicability depends on accidental features of the model.

Example (i) If $\mathfrak{P} := \{N(\mu_1, \sigma^2)^n \times N(\mu_2, \sigma^2)^n : \mu_i \in \mathbb{R},\ \sigma^2 > 0\}$, then $\overline{x}_{1n} - \overline{x}_{2n}$ is a mean unbiased estimator of $\mu_1 - \mu_2$ which minimizes the convex risk. (Recall that $(\overline{x}_{1n}, \overline{x}_{2n}, s_{1n}^2 + s_{2n}^2)$ is a complete sufficient statistic.)

(ii) For $\mathfrak{P} := \{N(\mu, \sigma_1^2)^n \times N(\mu, \sigma_2^2)^n : \mu \in \mathbb{R},\ \sigma_i^2 > 0\}$, the statistic s_{1n}^2/s_{2n}^2 is a plausible estimator of σ_1^2/σ_2^2, but the Rao–Blackwell–Lehmann–Scheffé Theorem does not apply: $(\overline{x}_{1n}, \overline{x}_{2n}, s_{1n}^2, s_{2n}^2)$ is minimal sufficient, but not complete (since $\overline{x}_{1n} - \bar{x}_{2n}$ has expectation 0). □

The limitations of nonasymptotic theory motivated statisticians to accept unreasonable optimality concepts (like minimal quadratic risk) whenever reasonable optimality concepts (like maximal concentration in intervals) are not compatible with nonasymptotic techniques. In some cases estimators fulfilling certain side conditions (like unbiasedness, equivariance etc.) do not exist (as in the case of the Behrens–Fisher problem), whereas estimators fulfilling the side conditions asymptotically are available. In other cases, relaxing side conditions from "exact" to "approximate" (say unbiasedness versus asymptotic mean unbiasedness) may lead to estimator sequences that are asymptotically better.

By relaxing "strict" conditions (like unbiasedness or similarity) to "approximate" conditions, and "optimality" to "approximate optimality", one obtains a theory of almost universal applicability. A problem which can be successfully addressed using asymptotic techniques is the choice of good confidence intervals, say $(\underline{\vartheta}^{(n)}, \bar{\vartheta}^{(n)})$. For intervals with a prescribed covering probability (i.e., $P_\vartheta^n\{\vartheta \in (\underline{\vartheta}^{(n)}, \bar{\vartheta}^{(n)})\} \geq \alpha$) one

J. Pfanzagl, *Mathematical Statistics*, Springer Series in Statistics,
DOI 10.1007/978-3-642-31084-3_5

may require that the expected length, $\int(\bar{\vartheta}^{(n)}(\mathbf{x}_n) - \underline{\vartheta}^{(n)}(\mathbf{x}_n))P_\vartheta^n(d\mathbf{x}_n)$, is minimal, or, alternatively, that $(\underline{\vartheta}^{(n)}, \bar{\vartheta}^{(n)})$ is most accurate in the sense that it covers the parameter values $\vartheta - n^{-1/2}r$ and $\vartheta + n^{-1/2}r$ with minimal probability. Strictly speaking, these two conditions are incompatible, but with increasing sample size, the contradiction fades away.

The preponderance of fixed-sample-size theory in many textbooks (such as Lehmann and Casella 1998 with nearly 400 pages on fixed sample-size procedures and less than 100 pages on asymptotic theory) indicates how uneasy some statisticians felt about results which are only asymptotically true. However, fixed sample-size results are exact within the model only. In most applications it is impossible to assess the model error; it is comparatively easy to assess the error of asymptotic assertions within the model. While the error of an asymptotic assertion decreases with increasing sample size, the model error remains constant. Hence asymptotic assertions based on more general models will usually be more accurate than "exact" assertions referring to a wrong special model.

The shift of the theory from "exact" to "asymptotic" is illustrated by two survey papers on "Nonparametric Statistics": Schmetterer (1959), covering the period until the mid-fifties, hardly mentions asymptotic results; Witting (1998), who covers the following 40 years, focuses exclusively on asymptotics.

Even though asymptotic methods are necessary to obtain a generally applicable theory, the problem remains how asymptotic results can be interpreted.

Consistency

For convenience of notation we write $\mathbf{x} = (x_1, x_2, \ldots)$ and $\mathbf{x}_n = (x_1, \ldots, x_n)$. We sometimes consider estimators $\kappa^{(n)}$ as functions of $\mathbf{x} \in X^{\mathbb{N}}$ rather than $\mathbf{x}_n \in X^n$. The estimator sequence $\kappa^{(n)}$, $n \in \mathbb{N}$, for a real-valued functional $\kappa(P)$ is *consistent* at P if, for every $\varepsilon > 0$,

$$P^n\{\mathbf{x}_n \in X^n : |\kappa^{(n)}(\mathbf{x}_n) - \kappa(P)| < \varepsilon\} \to 1.$$

We write $\kappa^{(n)} - \kappa(P) = o(n^0, P^n)$ or $\kappa^{(n)} \to \kappa(P)\ (P^n)$.

The estimator sequence is *strongly consistent* at P if, for every $\varepsilon > 0$,

$$P^n\{\mathbf{x}_n \in X^n : |\kappa^{(n)}(\mathbf{x}_n) - \kappa(P)| < \varepsilon \text{ for almost all } n \in \mathbb{N}\} \to 1.$$

Strong consistency of $\kappa^{(n)}$ is equivalent to the convergence of $\kappa^{(n)}(\mathbf{x}))$, $n \in \mathbb{N}$, to $\kappa(P)$ for $P^{\mathbb{N}}$–a.a. $\mathbf{x} \in X^{\mathbb{N}}$. To see this, use the equivalence between

$$\mathbf{x} \in A_n \text{ for } n \geq n(\mathbf{x}) \quad \text{and} \quad \mathbf{x} \in \bigcup_{m=1}^{\infty} \bigcap_{n=m}^{\infty} A_n$$

with $A_n = \{\mathbf{x} \in X^{\mathbb{N}} : |\kappa^{(n)}(\mathbf{x}) - \kappa(P)| < \varepsilon\}$.

The concept of "strong consistency" was introduced into statistical theory by Doob (1934) and has been used by Dugué (1937, p. 327), Wald (1949) and many others since. It reinforces the assertion of consistency from a mathematical point of view, but it adds nothing to its operational significance for finite n, even if one rewrites $(\kappa^{(n)}(\mathbf{x}))_{n\in\mathbb{N}} \to \kappa(P)$ for $P^{\mathbb{N}}$-a.a. $\mathbf{x} \in X^{\mathbb{N}}$ as

$$P^{\mathbb{N}} \bigcup_{n=m}^{\infty} \{\mathbf{x} \in X^{\mathbb{N}} : |\kappa^{(n)} - \kappa(P)| > \varepsilon\} \leq \varepsilon \text{ for } m \geq m_\varepsilon.$$

The assertions of statistical theory refer to the probability of

$$P^{n_0}\{\mathbf{x}_{n_0} \in X^{n_0} : |\kappa^{(n_0)}(\mathbf{x}_{n_0}) - \kappa(P)| < \varepsilon\}$$

for a fixed sample size n_0. Assertions about "what might happen" if the observations were continued from $x_1, \ldots, x_{n_0}$ to $x_{n_0+1}, \ldots$ are not of primary interest.

The concept of consistency (as opposed to strong consistency) was easy to deal with using n-fold products of Lebesgue densities. Dealing with convergence $P^{\mathbb{N}}$-a.e. in a mathematically precise way was impossible prior to Kolmogorov's trailblazing book (1933). It appears that R.A. Fisher's unwillingness to handle a mathematically more complex concept like "strong consistency" lies at the root of his deviating concept of consistency.

> A statistic ...is a consistent estimate of any parameter, if when calculated from an indefinitely large sample it tends to be accurately equal to that parameter.

This definition appeared in Fisher (1925, p. 702), and still held firm in Fisher (1959, pp. 143–146).

Consistent estimator sequences can be used as a starting point for the construction of optimal estimator sequences. Consistency of the ML-sequences is needed to obtain their limit distribution. Some authors require to choose among several solutions of the likelihood equations a consistent one. Yet, consistency is a property of an estimator sequence, not of a particular estimator.

Rules that guarantee the consistency of an estimator sequence are of questionable significance if they refer to sample sizes that will never occur. From the methodological point of view, the situation is even worse in the case of Cramér's Consistency Theorem (see Sect. 5.2). Cramér's Consistency Theorem is valid only if the likelihood equation has for every sample size and every (!) sample $(x_1, \ldots, x_n)$ one solution only. There is no theorem based on properties of the basic family $\{P_\vartheta : \vartheta \in \Theta\}$ that guarantees that these conditions are fulfilled. (The same mistake occurs in Schmetterer 1966, p. 373, Satz 3.5.)

How could a property (like consistency) that lacks operational significance, contribute to the proof of an operationally significant assertion (like the asymptotic distribution of the ML-sequence)? An asymptotic assertion about the estimator $\kappa^{(n)}$ should be based on properties for the sample size n only. That $N(0, \sigma^2(\vartheta))$ is the limit distribution of the estimator sequence $\vartheta^{(n)}$, $n \in \mathbb{N}$, can be turned into the operational assertion that the true distribution of $P_\vartheta^{n_0} \circ n_0^{1/2}(\vartheta^{(n_0)} - \vartheta)$ can be approximated by $N0, \sigma^2(\vartheta))$, provided n_0 is large. In principle, this assertion can be verified by computations. As far as *consistency* of $\vartheta^{(n)}$, $n \in \mathbb{N}$, there is no assertion about $\vartheta^{(n_0)}$ which could be verified by computations. Assertions about $P_\vartheta^{(n_0)} \circ n_0^{1/2}(\vartheta^{(n_0)} - \vartheta)$ should be based on the observations $(x_1, \ldots, x_{n_0})$ and not on properties of $\vartheta^{(n)}$ for samples $(x_1, \ldots, x_n)$ that will never occur.

Limit Distributions

Consistency is just an ancillary concept. What makes asymptotic theory useful are assertions about the (weak) convergence of the standardized estimator sequences to a limit distribution, for example

$$Q_P^{(n)} := P^n \circ n^{1/2}(\kappa^{(n)} - \kappa(P)) \Rightarrow Q_P.$$

Comparing the asymptotic performance of estimator sequences according to their limit distribution goes back at least to Laplace and Gauss. As an example we mention a result of Laplace (1812) and Gauss (1816), who showed that the estimator sequence $(n^{-1}\sum_{\nu=1}^n (x_\nu - \overline{x}_n)^2)^{1/2}$ for σ in the model $N(0, \sigma^2)$ is asymptotically better than $\sqrt{\pi/2}n^{-1}\sum_{\nu=1}^n |x_\nu - \overline{x}_n|$.

The concept of a limit distribution requires thinking of an estimator $\kappa_0^{(n_0)}$ as an element of an infinite sequence $\kappa^{(n)}$, $n \in \mathbb{N}$. If the definition of $\kappa^{(n)}$ follows the same rule for every $n \in \mathbb{N}$, there is no question as to which limit distribution should be used for such an approximation. In principle, an estimator for a given sample size can be considered as an element of an infinite number of sequences, and the question is open as to which limit distribution should be used for such approximations. For example, if $\vartheta^{(25)}(x_1, \ldots, x_{25}) := (1/15)\sum_{\nu=6}^{20} x_{\nu:n}$, what is $\vartheta^{(n)}$ for arbitrary n? Is it $(1/(n - 10))\sum_{\nu=6}^{n-5} x_{\nu:n}$? Or perhaps $(1/(n - 2\sqrt{n}))\sum_{\nu=\sqrt{n}+1}^{n-\sqrt{n}} x_{\nu:n}$? The appropriate answer is: neither. The problem is not to guess which estimator the experimenter would have chosen for sample sizes n other than $n = 25$, but to create an estimator sequence the limit distribution of which renders a good approximation to $P_\vartheta^{25} \circ 25^{1/2}(\vartheta^{(25)} - \vartheta)$.

To make this point clearer, consider the example of estimating the difference between a_1 and a_2, based on a sample of size (n_1, n_2) from $\Gamma(a_1, b_1)$ and $\Gamma(a_2, b_2)$. What is of interest is the distribution of $a_1^{(n_1)} - a_2^{(n_2)}$ under $\Gamma(a_1, b_1)^{n_1} \times \Gamma(a_2, b_2)^{n_2}$. The application of asymptotic theory requires determining the standardized limit distribution of $a_1^{(n_1)} - a_2^{(n_2)}$, as (n_1, n_2) tends to infinity. The natural idea is to choose the sequence $(n_1/(n_1 + n_2), n_2/(n_1 + n_2))$, $n \in \mathbb{N}$, but it is not clear whether this is the best choice for the approximation of $c(n_1, n_2)(a_1^{(n_1)} - a_2^{(n_2)})$ by means of a limit distribution. Further, it would make no sense to consider an infinite sequence of experiments where, say, the size n of the combined sample is divided into n_1 and n_2 such that the costs are minimized. This may be relevant for the experimenter, but it has nothing to do with the question of how the distribution of $c(n_1, n_2)(a_1^{(n_1)} - a_2^{(n_2)})$, with n_1, n_2 fixed, can be approximated.

Examples of this kind illustrate what the role of a limit distribution really is: If $\vartheta^{(n)}$ is an estimator for which $P_\vartheta^n \circ n^{1/2}(\vartheta^{(n)} - \vartheta)$ can be approximated with reasonable accuracy, nobody will care for the limit (if any) to which $P_\vartheta^n \circ n^{1/2}(\vartheta^{(n)} - \vartheta)$, $n \in \mathbb{N}$, converges. The limit distribution is only relevant if no other approximations to $P_\vartheta^n \circ n^{1/2}(\vartheta^{(n)} - \vartheta)$ are available.

For parametric families, the main result on the asymptotic performance of ML-sequences is usually stated as follows[1] (see Lehmann and Casella 1998, p. 447, Theorem 6.3.7, Witting and Müller-Funk 1995, p. 202, Satz 6.3.5): Every consistent sequence of ML estimators converges weakly to $N(0, \Lambda(\vartheta))$.

Even supplied by conditions which guarantee the consistency of (every? some?) ML-sequence, this assertion is different from what the statistician would like to know: that $N(0, \Lambda(\vartheta))$ is a good approximation to the distribution of the ML estimator under P_ϑ^n when n is large. The latter property has nothing to do with the performance of the ML-sequence as the sample size tends to infinity.

The condition that $\vartheta^{(n)}$ should belong to a consistent sequence is meaningless. To give operational significance to such a mathematical assertion, one must, again, keep error bounds in one's mind. Weak convergence to a limit distribution should always include the connotation of error bounds. In ideal situations, there exists $K > 0$ such that

$$|Q_P^{(n)}(I) - Q_P(I)| \leq K n^{-1/2} \text{ for every } P \in \mathfrak{P}, \text{ every interval } I, \text{ and every } n \in \mathbb{N}. \tag{5.1.1}$$

Even if such an assertion can be proved for a certain class of estimator sequences (say ML), it will, in general, not be useful for practical purposes, unless K is given explicitly and is not very large. The proof of an assertion like (5.1.1) for a general class of estimators, say ML estimators, requires several intermediate steps which, taken together, lead to a constant K which is far too large. In general, practically useful information about the accuracy of $Q_P(I)$ as an estimate of $P^n\{n^{1/2}(\kappa^{(n)} - \kappa(P)) \in I\}$ can only be obtained using simulations. Refinements of the normal approximation by means of asymptotic expansions are more useful than bounds for the error of the normal approximation.

Basing the comparison of estimator sequences on the standardized version of the limit distribution is generally accepted. One might question, of course, whether the comparison of standardized multivariate estimators according to their concentration on symmetric convex sets is more informative than the component-wise comparison on intervals containing 0. But one might have reservations about using joint (limit) distributions as the appropriate concept for judging the quality of estimator sequences if the standardizing factors c_n are different for different components.

Uniformity on $\mathfrak{P}$

For two different probability measures P_0 and P_1, the sup-distance between P_0^n and P_1^n tends to 1 as n tends to infinity. Hence the asymptotic performance of $P_0^n \circ n^{1/2}(\kappa^{(n)} - \kappa(P_0))$, $n \in \mathbb{N}$, is unrelated to the asymptotic performance of $P_1^n \circ n^{1/2}(\kappa^{(n)} - \kappa(P_1))$, $n \in \mathbb{N}$. One might change the asymptotic performance of a given estimator sequence at P_0 without affecting its asymptotic performance for any $P \neq P_0$. This makes it impossible to speak of a limit distribution which is optimal for every $P \in \mathfrak{P}$, unless one restricts the consideration to estimator sequences which

[1] A formulation of this result giving full attention to all details as in Schmetterer (1966, p. 388, Satz 3.10) may leave the reader in a state of desperation.

behave, in some sense, uniformly in P. Uniformity on $\mathfrak{P}$, which was required by Rao (1963) [p. 194], is meaningful from an operational point of view. Wolfowitz (1965) [p. 250] makes a strong case for requiring uniformity on $\mathfrak{P}$ as the only operationally meaningful condition. What is technically useful for asymptotic theory is: for each P_0, uniformity on a neighborhood shrinking to P_0 at the appropriate rate.

Section 5.6 deals with the question which kind of uniform convergence is operationally significant and—at the same time—strong enough to guarantee that the limit distribution provides for large samples a good approximation to the true distribution of the estimator sequence.

Asymptotic Optimality

To avoid the "evil of superefficiency" (see Ibragimov and Has'minskii 1981, p. 100, and Stoica and Ottersten 1996), a condition weaker than "uniformity on $P \in \mathfrak{P}$" suffices: "Regularity" of estimator sequences $\kappa^{(n)}$, $n \in \mathbb{N}$, defined as convergence of $P_n^n \circ n^{1/2}(\kappa^{(n)} - \kappa(P_n))$, $n \in \mathbb{N}$, to the same limit distribution for certain sequences $P_n \to P_0$. The relation between this concept of "regular" and "uniform" convergence will be discussed in Sect. 5.10.

Proving the asymptotic optimality of ML-sequences was in the focus of statistical theory for decades. First solid results had been attained around 1930; the final result, obtained by Kaufman (1966), comes close to what is now known as "Convolution Theorem", see Sect. 5.13. For parametric families, this bound, $N(0, \Lambda(\vartheta))$, is attained by ML-sequences.

The attempts at proving the asymptotic optimality of ML-sequences started without a clear concept of what "optimality" means for multivariate estimator sequences. The Convolution Theorem established the concentration on convex sets, symmetric about the origin, as the appropriate expression of asymptotic optimality.

In its present form, the Convolution Theorem asserts that the bound for the concentration of regular estimator sequences is determined by the canonical gradient of the functional $P \to \kappa(P)$. It is necessary to restrict attention to regular estimator sequences in order to obtain limit distributions which are approximations to the true distribution of the estimator.

"Nice" limit distributions are operationally useful in asymptotic theory. An indispensable property of a limit distribution is its continuity as a function of $P \in \mathfrak{P}$. Hence it seems natural to replace conditions on the estimator sequence (like regularity) by conditions on the limit distribution; in other words: To present theorems on bounds for estimator sequences with "nice" limit distributions instead of Theorems on bounds for the concentration of regular estimator sequences. In Sect. 5.14 it will be shown that the usual bound for regular estimator sequences is also valid for estimator sequences with "nice" limit distributions, and that such estimator sequences are automatically regular.

Initially the purpose of asymptotic theory was to obtain approximate assertions about the distribution of estimators. This can be achieved by assertions about the limit distribution of *standardized* estimators. For parametric families P_ϑ this means an approximation to $P_\vartheta^n \circ c_n(\vartheta^{(n)} - \vartheta)$. For general families of probability measures, it means an approximation, say Q_P, to

$$Q_P^{(n)} = P^n \circ n^{1/2}(\kappa^{(n)} - \kappa(P)),$$

where $\kappa^{(n)} : X^n \to \mathbb{R}^p$ is an estimator of the functional $\kappa : \mathfrak{P} \to \mathbb{R}^p$. The usual answer to this problem is the assertion that $Q_P^{(n)}$, $n \in \mathbb{N}$, converges *weakly* to Q_P (for every (?) $P \in \mathfrak{P}$). The desired answer is something different: For which purposes can Q_P be used as a surrogate for $Q_P^{(n)}$; for instance: Can Q_P be used for computing a confidence set for $\kappa(P)$?

For such purposes, conditions on the relation between $Q_P^{(n)}$, $n \in \mathbb{N}$, and Q_P stronger than just convergence for each P separately are required. Such conditions are "regular convergence", "locally uniform convergence", "continuous convergence", etc. None of these conditions is distinguished by being appropriate for every purpose. However, *regular* convergence suffices for the purpose of developing the concept of an "optimal limit distribution" (maximal concentration on convex subsets about the origin in the Convolution Theorem).

Yet limit distributions of regularly convergent sequences are not necessarily continuous. The optimal limit distributions (as distinguished by the Convolution Theorem). Estimator sequences converging to the optimal limit distributions are automatically regular (but not necessarily locally uniformly convergent).

5.2 Maximum Likelihood

The Prehistory

In 1912 (Fisher 1912), a self-confident undergraduate, R.A. Fisher, believed to have discovered a new, "absolute" method of estimation, Maximum Likelihood (ML). Surprisingly, his paper was published at a time when this "new" method had been known under various names and with varying justifications for 150 years. That the basic idea already occurs in Lambert (1760), was only discovered by Sheynin in (1966). But it was well known that Bernoulli (1778) had suggested the parameter value as an estimator—the location in his case—"... qui maxima gaudet probabilitate" (Todhunter 1865, pp. 236/7). In the 19th century, the method was well known to astronomers (Encke 1832–34), could be found in various textbooks (Czuber 1891) and was used by Pearson (1896) [pp. 262–265] to determine the sample correlation coefficient as an estimator of the population correlation coefficient. The basic idea of maximum likelihood is so obvious that it was "discovered" once again by Zermelo (1929), a mathematician without any ties to statistics. Though some of the authors came to the ML idea via a Bayesian approach using a uniform prior (see Hald 1999, on this particular point), the opinion was widespread that it was the performance of the estimators that counted, not the "metaphysical principle" on which the estimator was grounded. As an example we can cite Hald (1998) [p. 499] :

> Helmert (1876) borrowed the idea from inverse probability that the "best" estimate is obtained by maximizing the probability density of the sample and having found the estimate, he evaluated its properties by studying its sampling distribution.

Fisher (1912) [p. 155] claims that his new method leads to optimal estimators. It appears that Fisher was at this time convinced of optimum properties even for finite samples. The further development of his thought over the next decade is carefully examined by Stigler (2005).

At the beginning of the 20th century, the time was ripe for a general theorem on the asymptotic distribution of ML-sequences. The result (for an arbitrary number of parameters) is basically contained in the paper by Pearson and Filon (1898), written in the spirit of "inverse probability", and with the (misguided) interpretation that their result refers to moment estimators. Fisher (1922) [p. 329, footnote] heavily criticizes these points. If he had been familiar with the literature of his time he could have cited the papers by Edgeworth (1908/1909), which appeared not in some obscure journal, but in the Journal of the Royal Statistical Society. In these papers, Edgeworth puts forward a correct interpretation of the results of Pearson and Filon, and he shows that, in a variety of examples, the asymptotic variance of the ML-sequence is not larger than that of other estimator sequences. 1. For symmetric densities, the asymptotic variance of the ML-sequence is at least as good as the sample mean and the sample median. 2. For Pearsons's Type III distributions, the ML-sequence is at least as good as the moment estimators. 3. Among the estimating equations $\sum_{\nu=1}^{n} g(x_\nu - \vartheta) = 0$, the one with $g(x) = p'(x)/p(x)$ (corresponding to the likelihood equation for a shift parameter) yields the estimator sequence with the smallest asymptotic variance.

Instead of welcoming the results of Edgeworth (after they had become known to him) as supporting his claim of the superiority of the ML method, Fisher continued to quarrel about the distinction between "maximum likelihood" and "maximal inverse probability", though Edgeworth had stated explicitly (1908/1909, p. 82) that his results are "free from the speculative character which attaches to inverse probability". Is it legitimate to refuse giving mathematical results proper recognition with the argument that they are based on the wrong "philosophy"?

The relations between the results of Pearson and Filon (1898), Edgeworth (1908/1909) and Fisher are discussed in detail by a number of competent scholars, like Edwards (1974, 1992), Pratt (1976), Savage (1976), Hald (1998, Chap. 28) and (most interesting) Stigler (2007).

Even though the ML principle (in one or the other version) was familiar, general results about the asymptotic distribution of the ML-sequence were late to arrive. It might be a promising project to find out how scientists determined the accuracy of their ML estimators before such general results were available. As an example, we can mention a paper by Schrödinger (1915) on measurements with particles subject to Brownian motion. On p. 290, Schrödinger writes:

> Es gilt nun, aus einer Beobachtungsreihe ... diejenigen Werte [der beiden Parameter] zu berechnen, welche durch diese Beobachtungsreihe zu den wahrscheinlichsten gemacht werden; ferner die Fehler zu berechnen, die beiden Werten wahrscheinlich oder im Mittel anhaften.

What Schrödinger (Sect. 6, pp. 294/5) does to determine the accuracy of his ML estimators was, probably, the common practice among physicists at this time: With an explicit formula for the ML estimator $\vartheta^{(n)}$ at hand, he computes $\sigma_n^2(\vartheta) := \int (\vartheta^{(n)} -$

$\vartheta)^2 dP_\vartheta^n$, and bases a measure of accuracy on $\sigma_n(\vartheta^{(n)}(x))$. In the particular case dealt with by Schrödinger, $\sigma_n(\vartheta)$ is asymptotically equivalent to $n^{-1/2}\sigma(\vartheta)$.

In the following sections we deal with various proofs of consistency and asymptotic normality of the ML-sequences. At the technical level of Fisher's writings, consistency had no relevance. (Fisher's concept of "consistency" meant something else.) His derivation of the asymptotic variance of the ML-sequence (1922, pp. 328/9) is intelligible only for readers familiar with Fisher's style. The emphasis of Fisher's papers (1922, pp. 330–32, 1925, p. 707) is on the asymptotic optimality of the ML-sequence. His intention is to show that the asymptotic variance of asymptotically normal estimator sequences cannot be smaller than $\left(\int \ell^\bullet(x,\vartheta)^2 P_\vartheta(dx)\right)^{-1}$. Fisher's efforts to support his claim to the superiority of the ML method by means of mathematical arguments motivated mathematicians and statisticians with a background in probability theory to turn conjectures and assertions about ML-sequences into mathematical theorems. The results will be discussed in the following sections.

What, in the end, is Fisher's contribution to the theory of ML estimators? It is the concept of "maximum likelihood", which he introduced in (1922, p. 326).

When one dealt with the asymptotic theory of ML-sequences in a mathematically precise way, it soon became clear that the consistency is more difficult to establish than the asymptotic distribution, given consistency. Occasionally, the consistency can be proved directly, if the estimators are given by explicit formulas. Hence it is preferable to present the asymptotic theory of ML-sequences in two parts: (a) a consistency theorem, (b) a theorem about the asymptotic distribution of consistent ML-sequences.

Remark Assertions about the distribution of $n^{1/2}(\vartheta^{(n)} - \vartheta)$ under P_ϑ^n are possible only if $\vartheta^{(n)}(\mathbf{x}_n)$ is a measurable function of $\mathbf{x}_n$. If $\vartheta^{(n)}(\mathbf{x}_n)$ is defined by

$$\prod_{\nu=1}^n p(x_\nu, \vartheta^{(n)}(\mathbf{x}_n)) = \sup_{\vartheta\in\Theta} \prod_{\nu=1}^n p(x_\nu, \vartheta),$$

say, the measurability of $\vartheta^{(n)}$ can be guaranteed by some kind of "Measurable Selection Theorem". An early example of such a theorem occurs in Schmetterer (1966) [p. 375, Lemma 3.3], who seems to have been unaware of selection theorems in various connections. Now measurable selection theorems occur occasionally in advanced textbooks on mathematical statistics (Strasser 1985, p. 34, Theorem 6.10.1; Pfanzagl 1994, p. 21, Theorem 6.7.22; Witting and Müller-Funk 1995, p. 173, Hilfssatz 6.7). A paper on the measurability of ML estimators far off the mainstream is Reiss (1973).

Consistency of ML-Sequences

The consistency of an estimator sequence is just an intermediate step on the way to its asymptotic distribution. Nevertheless, the literature on consistency is immense. The following considerations are restricted to weak consistency of ML-sequences for one-parameter families. The generalizations to strong consistency, perhaps locally uniform, and to multi-parameter families are straightforward if the arguments for the simplest case are understood.

The prototype of a consistency theorem is based on 1. conditions on the parameter space Θ and the family $\{P_\vartheta : \vartheta \in \Theta\}$ and 2. a prescription as to how the estimator is defined, for every $n \in \mathbb{N}$. Not all theorems declared as consistency theorems are of this type: Some of them include ad hoc assumptions about the ML-sequence, such as "The likelihood equation has a unique root for every $n \in \mathbb{N}$". Such an assumption makes the whole theorem meaningless, unless it can be verified within the given framework.

Conditions on estimator sequences concerning "all sample sizes" remind us of the problems with the operational significance of asymptotic assertions in general. One might be acquiesced if for the sample size n_0 in question there were only one solution. Can our confidence in the quality of the ML estimator for the sample size n_0 really depend on conditions concerning sample sizes which will never occur? Yet the consistency theorem is valid only if this follows from the regularity conditions. Consistency as an extra-condition is with no operational significance.

Basically, the ML estimator $\vartheta^{(n)}(\mathbf{x}_n)$ is defined by

$$\prod_{\nu=1}^{n} p(x_\nu, \vartheta^{(n)}(\mathbf{x}_n)) \geq \prod_{\nu=1}^{n} p(x_\nu, \vartheta) \quad \text{for } \vartheta \in \Theta.$$

When $\Theta \subset \mathbb{R}^k$, we have the necessary condition that $\vartheta^{(n)}(\mathbf{x}_n)$ is a solution of the likelihood equation

$$\sum_{\nu=1}^{n} \ell^{(i)}(x_\nu, \vartheta) = 0, \quad i = 1, \ldots, k.$$

It is possible that solutions of the estimating equation give reasonable results although the maximizer of the likelihood function does not; see Kraft and Le Cam (1956).

Notice the basic distinction: To prove consistency based on the original definition, only topological conditions (on Θ and the function $\vartheta \to p(x, \vartheta)$) are required. Obtaining consistency for solutions of the likelihood equation not only requires $\vartheta \to p(x, \vartheta)$ to be differentiable; it leads to problems if the likelihood equation has several solutions. (Recall the result of Reeds 1985, that for the location family of Cauchy distributions, $k_n(\mathbf{x}_n) - 1$ is asymptotically distributed according to the Poisson distribution with parameter $1/\pi$, where $k_n(\mathbf{x}_n)$ is the number of solutions of the likelihood equation.)

It does not make sense to state that among the roots of the estimating question for the sample size n_0 there is (at least) one which is an element of a consistent estimator sequence. To say that $\vartheta^{(n_0)}$ is an element of a consistent estimator sequence is meaningless, anyway. What is needed is a guide which root to choose.

Some Technical Questions

As soon as mathematically oriented statisticians entered the scene, they discovered a number of problems which nobody had thought about before. The most elementary one: Assertions about consistency and asymptotic normality require the estimators to be measurable. Even if the ML estimator is not uniquely determined, there is always a measurable version if the density $p(x, \cdot)$ is continuous for every $x \in X$.

To prepare for the following sections on "consistency" and "asymptotic normality", we will discuss some other mathematical problems typical for such asymptotic considerations at a more general level.

Let $(X, \mathscr{A})$ be a measurable space, $(\Theta, \mathscr{U})$ a topological space endowed with the Borel algebra $\mathbb{B}$ generated by $\mathscr{U}$, and $f : X \times \Theta \to \mathbb{R}$ a function which is for each y measurable in x, and for each x continuous in y. We might mention in passing that f is then $\mathscr{A} \times \mathbb{B}$-measurable. This allows the application of Fubini's Theorem.

Before discussing some of the consistency theorems from a historical point of view, it is advisable to point to one of the typical technical problems: To strengthen

$$n^{-1} \sum_{\nu=1}^{n} f(x_\nu, \vartheta_0) \to \int f(\cdot, \vartheta_0) dP_{\vartheta_0} \quad (P^n_{\vartheta_0})$$

to

$$n^{-1} \sum_{\nu=1}^{n} f(x_\nu, \vartheta^{(n)}(\mathbf{x}_n)) \to \int f(\cdot, \vartheta_0) dP_{\vartheta_0} \quad (P^n_{\vartheta_0}) \tag{5.2.1}$$

if $\vartheta^{(n)} \to \vartheta_0$ $(P^n_{\vartheta_0})$. Not all authors had the right idea about which conditions on f going beyond continuity of $\vartheta \to f(x, \vartheta)$ for every $x \in X$, are needed for this step: Continuity of $\vartheta \to f(x, \vartheta)$ uniformly in x, as required by Dugué (1936, 1937), is much too strong. Let $\Theta \subset \mathbb{R}$ and $f^\bullet(x, \vartheta) = \partial_\vartheta f(x, \vartheta)$. In their "fundamental" Lemma 2.1, Gong and Sameniego (1981, pp. 862/3) require, in addition to $\int |f(\cdot, \vartheta_0)| dP_{\vartheta_0}) < \infty$, a condition corresponding to $\int \sup_{\vartheta \in U} |f^{(1)}(\cdot, \vartheta)| dP_{\vartheta_0} < \infty$ for an open subset U containing ϑ_0. In fact, the condition

$$\int \sup_{\vartheta \in U} |f(\cdot, \vartheta)| dP_{\vartheta_0} < \infty \quad \text{for some } U \ni \vartheta_0 \tag{5.2.2}$$

suffices. Since $\vartheta \to f(x, \vartheta)$ is continuous, this implies for every $\varepsilon > 0$ the existence of $U_\varepsilon \ni \vartheta_0$ such that

$$\int \sup_{\vartheta \in U_\varepsilon} |f(\cdot, \vartheta) - f(\cdot, \vartheta_0)| dP_{\vartheta_0} < \varepsilon. \tag{5.2.3}$$

By the same argument, condition (5.2.2) implies that

$$n^{-1} \sum_{\nu=1}^{n} \int_0^1 f(x_\nu, (1-u)\vartheta_0 + u\vartheta^{(n)}(\mathbf{x}_n)) du \to \int f(\cdot, \vartheta_0) dP_{\vartheta_0} \quad (P^n_{\vartheta_0})$$

if $\vartheta^{(n)} \to \vartheta_0$ $(P^n_{\vartheta_0})$, a relation useful in connection with Taylor expansions.

Some proofs concerning the asymptotic distribution of ML-sequences are unnecessarily complicated because they use a Taylor expansion with a remainder term which is not to the purpose. With $f^\bullet$ and $f^{\bullet\bullet}$ denoting the first and second derivative of $f(x,\vartheta)$ with respect to ϑ, most authors use an expansion like

$$n^{-1}\sum_{\nu=1}^{n} f(x_\nu,\vartheta) = n^{-1}\sum_{\nu=1}^{n} f(x_\nu,\vartheta_0 + (\vartheta-\vartheta_0)\, n^{-1}\sum_{\nu=1}^{n} f^\bullet(x_\nu,\vartheta)$$
$$+\frac{1}{2}(\vartheta-\vartheta_0)^2 n^{-1}\sum_{\nu=1}^{n} f^{\bullet\bullet}(x_\nu,\bar\vartheta_n(\mathbf{x}_n)) \tag{5.2.4}$$

with $\bar\vartheta(\mathbf{x}_n)$ between ϑ_0 and ϑ. However, to know the latter is not enough. Since $\bar\vartheta^{(n)}(\mathbf{x}_n)$ depends on ϑ_0 as well as on ϑ, one runs into trouble if (5.2.4) is applied with ϑ replaced by some $\vartheta^{(n)}(bx_n)$. For this purpose, the expansion

$$n^{-1}\sum_{\nu=1}^{n} f(x_\nu,\vartheta) = n^{-1}\sum_{\nu=1}^{n} f(x_\nu,\vartheta_0)$$
$$+(\vartheta-\vartheta_0)\, n^{-1}\sum_{\nu=1}^{n}\int_0^1 f^{(1)}(x_\nu,(1-u)\vartheta_0+u\vartheta)du \tag{5.2.5}$$

is technically easier to handle (and avoids, in addition, the use of $f^{\bullet\bullet}$).

There is no need to list all papers using Taylor expansions with an unsuitable remainder term. It suffices to mention that this tradition goes back to Cramér (1946a) [p. 501] for $f=\ell^\bullet$.

Finally, one can criticize the fact that many authors (including Dugué, Cramér, and M.G. Kendall) simply use the symbol P (or even "prob") instead of P_ϑ, P_ϑ^n, etc., so that one has to study the details of the proof to find out what the regularity conditions really are.

Wald's Consistency Proof

For any set $V\subset\Theta\subset\mathbb{R}^k$, let

$$\overline{\ell}(x,V) := \sup\{\ell(x,\vartheta):\vartheta\in V\}.$$

If $\vartheta\to\ell(x,\vartheta)$ is continuous for every $x\in X$, the function $\overline{\ell}(\cdot,V)$ is measurable. Since $\int(\ell(x,\vartheta)-\ell(x,\vartheta_0))P_{\vartheta_0}(dx)<0$ for $\vartheta\neq\vartheta_0$, there exists an open set $V\ni\vartheta$ not containing ϑ_0 which fulfills

$$\int(\overline{\ell}(x,V)-\ell(x,\vartheta_0))P_{\vartheta_0}(dx)<0, \tag{5.2.6}$$

provided $\int\overline{\ell}(x,V)P_0(dx)<\infty$ for some V. Let $X_V\subset X^{\mathbb{N}}$ denote the set of all $\mathbf{x}\in X^{\mathbb{N}}$ such that

$$\frac{1}{n}\sum_{\nu=1}^{n}(\overline{\ell}(x_\nu, V) - \ell(x_\nu, \vartheta_0)) \geq 0 \quad \text{for infinitely many } n \in \mathbb{N}.$$

By the Strong Law of Large Numbers, relation (5.2.6) implies that $P_{\vartheta_0}^{\mathbb{N}}(X_V) = 0$. For $n \in \mathbb{N}$, let $\vartheta^{(n)}(\mathbf{x})$ be such that $\sum_{\nu=1}^{n} \ell(x_\nu, \vartheta^{(n)}(\mathbf{x})) \geq \sum_{\nu=1}^{n} \ell(x_\nu, \vartheta)$ for $\vartheta \in \Theta$. Then $\vartheta^{(n)}(\mathbf{x}) \in V$ implies

$$n^{-1}\sum_{\nu=1}^{n} \ell(x_\nu, \vartheta_0) \leq n^{-1}\sum_{\nu=1}^{n} \ell(x_\nu, \vartheta^{(n)}(\mathbf{x})) \leq n^{-1}\sum_{\nu=1}^{n} \overline{\ell}(x_\nu, V).$$

Hence the set of all $\mathbf{x} \in X^{\mathbb{N}}$ such that $\vartheta^{(n)}(\mathbf{x}) \in V$ for infinitely many $n \in \mathbb{N}$ is a subset of the $P_{\vartheta_0}^{\mathbb{N}}$-null set X_V. The immediate consequence: The ML estimator $\vartheta_{(\mathbf{x})}^{(n)}$ cannot be for infinitely many n in a set V fulfilling (5.2.6) (except for $\mathbf{x}$ in a set of $P_{\vartheta_0}^{\mathbb{N}}$-measure zero). In particular: $(\vartheta^{(n)}(\mathbf{x}))_{n\in\mathbb{N}}$ *cannot converge to some* $\vartheta \neq \vartheta_0$ except for $\mathbf{x}$ in a set of $P_{\vartheta_0}^{\mathbb{N}}$-measure zero.

If Θ is compact, the set $\{\vartheta \in \Theta : |\vartheta - \vartheta_0| \geq \varepsilon\}$ is compact and can, therefore, be covered by a finite number of sets $V_{\vartheta_i} : i = 1, \ldots, m$, fulfilling (5.2.6). Hence for $P_{\vartheta_0}^{\mathbb{N}}$-a.a. $\mathbf{x} \in X^{\mathbb{N}}$, the relation $|\vartheta^{(n)}(\mathbf{x}) - \vartheta_0| \geq \varepsilon$ holds for finitely many n only. That means: $\vartheta_{(\mathbf{x})}^{(n)}$, $n \in \mathbb{N}$, converges to ϑ_0 except for $\mathbf{x}$ in a set of $P_{\vartheta_0}^{\mathbb{N}}$-measure 0.

This is Wald's *Consistency Theorem,* the earliest consistency theorem which is mathematically correct. It is widely ignored because it requires Θ to be compact. Various examples show that the compactness condition cannot be omitted without further ado. The compactness condition, however, may be replaced by the condition that for every neighbourhood U_0 of ϑ_0, the set U_0^c can be covered by a finite number of sets V_i fulfilling condition (5.2.6). (See e.g. the Covering Condition 6.3.8 in Pfanzagl 1994, p. 194, or condition 6.1.57 in Witting and Müller-Funk 1995, p. 201, Satz 6.34.) Schmetterer (1966) [Satz 3.8, p. 384] is the first textbook presenting a (somewhat confused) version of Wald's strong consistency theorem. Schmetterer does not mention the compactness condition explicitly, but all his considerations (including the definition of the ML estimator) are restricted to some compact subset of Θ (called Γ_0) which restricts the assertion of consistency to compact subsets of Θ.

Kendall and Stuart (1961) [pp. 39–41] think they can do better than Wald:

> This direct proof of consistency is a simplified form of Wald's proof. Its generality is clear from the absence of any regularity conditions on the distribution $[p(x, \vartheta)]$.

This is supplemented by their Exercise 18.35, p. 74: "... show that many *inconsistent* ML estimators, as well as consistent ML estimators, exist". The proof in Kendall and Stuart (see p. 40) is based on their relations (18.20),

$$\lim_{n\to\infty} P_{\vartheta_0}^{n}\left\{\sum_{\nu=1}^{n} \ell(x_\nu, \vartheta) < \sum_{\nu=1}^{n} \ell(x_\nu, \vartheta_0)\right\} = 1 \quad \text{for every } \vartheta \neq \vartheta_0,$$

and (18.21),

$$\sum_{\nu=1}^{n} \ell(x_\nu, \vartheta^{(n)}(\mathbf{x}_n)) \geq \sum_{\nu=1}^{n} \ell(x_\nu, \vartheta_0) \quad \text{for } n \in \mathbb{N} \text{ and } \mathbf{x}_n \in \mathbb{R}^n,$$

"since, by (18.20), (18.21) only holds with probability zero for any $\vartheta \neq \vartheta_0$ [?] it follows that $\lim_{n\to\infty} P_{\vartheta_0}^{\mathbb{N}}\{\vartheta^{(n)} = \vartheta_0\} = 1$."

The idea that, because of (18.20) and (18.21), the sequence $\vartheta^{(n)}$, $n \in \mathbb{N}$, cannot stay away from ϑ_0 with positive probability seems to be obvious, but is not really true. To make it precise one has to consider the performance of the estimator sequence $\vartheta^{(n)}(\mathbf{x})$, $n \in \mathbb{N}$, on $\mathbf{x} \in X^{\mathbb{N}}$, as done by Wald, an idea which was not understood by the authors.

If the argument of Kendall and Stuart might have been acceptable in the thirties, its weakness must have been clear to anybody who had understood Wald's paper from (1949). Thirty years later (in "Kendall's Advanced Theory of Statistics" by Stuart and Ord 1991) not much had been changed. From (18.20) and (18.21) the authors now argue as follows: "as $n\to\infty$, $\sum_{\nu=1}^{n} \ell(x_\nu, \hat{\vartheta}^{(n)}(\mathbf{x}_n))$ cannot take any other value than $\sum_{\nu=1}^{n} \ell(x_\nu, \vartheta_0)$. If $\sum_{\nu=1}^{n} \ell(x_\nu, \vartheta)$ is identifiable [?] this implies that $P_{\vartheta_0}^{\mathbb{N}}\{\lim_{n\to\infty} \vartheta^{(n)} = \vartheta_0\} = 1$".

A second approach to consistency, using the existence of $\ell^{\bullet\bullet}$ (see Kendall and Stuart, 1961, p. 41 and 1991, p. 660), follows Cramér's lead, described below, both in the proof as in its misinterpretation.

Cramér's Consistency Proof

Even now, most textbooks refer to Cramér (1946a) as the author of the first consistency proof. In fact, the basic idea is of tempting simplicity. A concise version of the proof starts with the expansion

$$n^{-1}\sum_{\nu=1}^{n} \ell^{\bullet}(x_\nu, \vartheta) = n^{-1}\sum_{\nu=1}^{n} \ell^{\bullet}(x_\nu, \vartheta_0) + (\vartheta - \vartheta_0)n^{-1}\sum_{\nu=1}^{n} k(x_\nu, \vartheta_0, \vartheta) \quad (5.2.7)$$

with

$$k(x, \vartheta_0, \vartheta) = \int_0^1 \ell^{\bullet\bullet}(x, (1+u)\vartheta_0 + u\vartheta)du.$$

If $\vartheta \to \int \ell^{\bullet\bullet}(x, \vartheta)P_{\vartheta_0}(dx)$ is continuous at $\vartheta = \vartheta_0$, then $\vartheta \to \int k(x, \vartheta_0, \vartheta)P_{\vartheta_0}(dx)$ is continuous at $\vartheta = \vartheta_0$. Since $\int k(x, \vartheta_0, \vartheta_0)P_{\vartheta_0}(dx) = \int \ell^{\bullet\bullet}(x, \vartheta_0)P_{\vartheta_0}(dx) < 0$, it follows that $n^{-1}\sum_{\nu=1}^{n} k(x_\nu, \vartheta_0, \vartheta)$ converges in $P_{\vartheta_0}^{\mathbb{N}}$-measure to a negative value if ϑ is sufficiently close to ϑ_0. Together with $n^{-1}\sum_{\nu=1}^{n} \ell^{\bullet}(x_\nu, \vartheta_0) \to 0$ $(P_{\vartheta_0}^n)$, relation (5.2.7) implies that for $\delta > 0$ and sufficiently small,

$$P_{\vartheta_0}^n\left\{n^{-1}\sum_{\nu=1}^{n} \ell^{\bullet}(x_\nu, \vartheta_0 + \delta) < 0 < n^{-1}\sum_{\nu=1}^{n} \ell^{\bullet}(x_\nu, \vartheta_0 - \delta)\right\} \to 1. \quad (5.2.8)$$

Since $\vartheta \to n^{-1} \sum_{\nu=1}^{n} \ell^\bullet(x_\nu, \vartheta)$ is continuous in a neighbourhood of ϑ_0, there is a solution of $\sum_{\nu=1}^{n} \ell^\bullet(x_\nu, \vartheta) = 0$ in $(\vartheta_0 - \delta, \vartheta_0 + \delta)$. In other words: For every sufficiently small δ, the $P_{\vartheta_0}^{\mathbb{N}}$-probability that the likelihood equation has a solution in the interval $(\vartheta_0 - \delta, \vartheta_0 + \delta)$ converges to 1.

It is surprising to see how a mathematician with Cramér's stature deals with this approach (see p. 501). For the proof of (5.2.8) he requires the existence of functions H_i, $i = 1, 2, 3$, such that $|\ell^{(i)}(x, \vartheta)| \leq H_i(x)$ for $x \in X$ and $\vartheta \in \Theta$. If Cramér says that H_i, $i = 1, 2$, are "integrable over $(-\infty, +\infty)$", he means P_ϑ-integrable (for all $\vartheta \in \Theta$, or for $\vartheta = \vartheta_0$?). Local versions of these conditions would suffice, and even for Cramér's crude proof, based on the expansion

$$\ell^\bullet(x, \vartheta) = \ell^\bullet(x, \vartheta_0) + (\vartheta - \vartheta_0)\ell^{\bullet\bullet}(x, \vartheta_0) + \frac{1}{2}\delta(\vartheta - \vartheta_0)^2 H_3(x) \text{ with } |\delta| < 1,$$

his condition $\sup_{\vartheta \in \Theta} \int H_3(x) P_\vartheta(dx) < \infty$ could be replaced by $\int H_3(x) P_{\vartheta_0}(dx) < \infty$.

Cramér did not take much advantage of what other authors had done before him. Wald (1941a, b) paper was, perhaps, not accessible to Cramér when he wrote the first version of his book, but he retained his unsatisfactory global regularity conditions in all subsequent editions. More surprising than the technical infelicities is the misleading interpretation Cramér gives to his result, namely: That there is a sequence of solutions of the likelihood equation which is consistent for every $\vartheta \in \Theta$. To recall: He just proved that for every $\vartheta_0 \in \Theta$ and every neighbourhood of ϑ_0, the $P_{\vartheta_0}^n$-probability for a solution in this neighbourhood converges to 1 as n tends to infinity. This does not deliver what Cramér had promised on p. 500: "It will be shown that, under general conditions, the likelihood equation ... has a solution which converges in probability to the true value of [ϑ] as $n \to \infty$". Cramér's result implies consistency under an additional condition: That the likelihood equation has one solution only for every $n \in \mathbb{N}$ and all $(x_1, \ldots, x_n)$. It would certainly be reassuring if this holds true for the actual sample size n_0. But what is required is uniqueness for *every* sample size, a condition which casts some doubts on the interpretation of asymptotic results.

Though Cramér's paralogism became known soon (Wald 1949, p. 595, footnote 3), the message did not spread quickly. Cramér's result was presented as a consistency proof in a number of textbooks. As an example we might mention Schmetterer (1956), at this time the mathematically most advanced textbook. In Satz 8, pp. 223/4, Schmetterer asserts the existence of a sequence of solutions of the likelihood equation which is consistent for every $\vartheta \in \Theta$, an open rectangle in $\mathbb{R}^k$. His argument is dubious. After having proved a k-dimensional version of Cramér's result (following Chanda 1954), he makes a vault to the interpretation as a consistency theorem (p. 231). It is, by the way, not straightforward to transfer Cramér's argument "the interval $(\vartheta_0 - \delta, \vartheta_0 + \delta)$ contains a solution of the likelihood equation" from $\mathbb{R}^1$ to $\mathbb{R}^k$. An error in Chanda's proof was discovered by Tarone and Gruenhage (1975) [p. 903].

Foutz (1977) tries to simplify the technical details of this proof by means of the Inverse Function Theorem, but he, too, follows Cramér in the misinterpretation of his result.

In fact, it requires additional steps to transform Cramér's result into an assertion about the existence of a consistent sequence. Introducing

$$A_n(\mathbf{x}_n) := \Big\{\vartheta \in \Theta : \sum_{\nu=1}^{n} \ell^{\bullet}(x_\nu, \vartheta) = 0\Big\},$$

Cramér's result can be written as

$$P^n_{\vartheta_0}\{A_n \cap (\vartheta_0 - \delta, \vartheta_0 + \delta) \neq \emptyset\} \to 1 \quad \text{for every } \delta > 0.$$

For $k \in \mathbb{N}$ let N_k be such that

$$P^n_{\vartheta_0}\{A_{N_k} \cap (\vartheta_0 - 1/k, \vartheta_0 + 1/k) \neq \emptyset\} > 1 - 1/k.$$

W.l.g.: $N_{k+1} > N_k$. With k_n denoting the largest integer k such that $N_k \leq n$, this implies

$$P^n_{\vartheta_0}\{A_n \cap (\vartheta_0 - 1/k_n, \vartheta_0 + 1/k_n) \neq \emptyset\} \to 1.$$

For every $\mathbf{x}_n \in X^n$ let

$$\vartheta^{(n)}(\mathbf{x}_n) \in A_n(\mathbf{x}_n) \cap (\vartheta_0 - 1/k_n, \vartheta_0 + 1/k_n)$$

if $A_n(\mathbf{x}_n) \cap (\vartheta_0 - 1/k_n, \vartheta_0 + 1/k_n) \neq \emptyset$, and $\vartheta^{(n)}(\mathbf{x}_n) = \vartheta_0$ otherwise. The estimator sequence $(\vartheta^{(n)}(\mathbf{x}_n))_{n\in\mathbb{N}}$ thus defined converges to ϑ_0 for every $\mathbf{x}_n \in X^n$, and it is a solution of the likelihood equation with $P^n_{\vartheta_0}$-probability converging to 1.

Why do some authors go to the trouble of constructing such an estimator sequence? The answer is easy in the case of Serfling (1980) [p. 145, Theorem (i)]: He had missed the dependence of this sequence on the arbitrary ϑ_0. The motivation is less clear in the case of Schmetterer (1974) [pp. 296/7, Theorem 3.1], who explicitly mentions the dependence on ϑ_0 (p. 303). This estimator sequence occurs again in Schmetterer's Theorem 3.9, p. 316, now without a caveat.

There is no need for a discussion of the papers by Huzurbazar (1948), Gurland (1954), Kulldorff (1956), Chan (1967) etc., which treat the problem of consistency and asymptotic normality of ML-sequences under slightly modified regularity conditions, as they offer nothing new from a methodological point of view.

Remark If there is some estimator sequence, say $(\tilde{\vartheta}^{(n)})_{n\in\mathbb{N}}$, which is consistent for $\vartheta \in \Theta$, then there is also an estimator sequence consistent for $\vartheta \in \Theta$ which is, with P^n_ϑ-probability tending to 1, a solution of the likelihood equation. (Hint: choose for $\vartheta^{(n)}(\mathbf{x}_n)$ the element in $A_n(\mathbf{x}_n)$ closest to $\tilde{\vartheta}^{(n)}(\mathbf{x}_n)$. If $\tilde{\vartheta}^{(n)}(\mathbf{x}_n) \in (\vartheta_0 - \delta, \vartheta_0 + \delta)$ and $A_n(\mathbf{x}_n) \cap (\vartheta_0 - \delta, \vartheta_0 + \delta) \neq \emptyset$, then there is $\vartheta^{(n)}(\mathbf{x}_n) \in A_n(\mathbf{x}_n)$ such that $|\tilde{\vartheta}^{(n)}(\mathbf{x}_n) - \vartheta^{(n)}(\mathbf{x}_n)| < 2\delta$, whence $|\vartheta^{(n)}(\mathbf{x}_n) - \vartheta_0| < 3\delta$. Since $\tilde{\vartheta}^{(n)}(\mathbf{x}_n) \in (\vartheta_0 -$

$\delta, \vartheta_0 + \delta)$ and $A_n(\mathbf{x}_n) \cap (\vartheta_0 - \delta, \vartheta_0 + \delta) \neq \emptyset$ are events with $P^n_{\vartheta_0}$-probability converging to 1, the same is true of $|\vartheta^{(n)}(\mathbf{x}_n) - \vartheta_0| < 3\delta$. (See Weiss and Wolfowitz 1966, p. 77.)

Examples of Consistency Theorems Prior to Cramér

The first attempt at a consistency proof is due to Hotelling (1930). His idea to obtain a general theorem by approximating an arbitrary family with Lebesgue densities by a family with Lebesgue densities constant on intervals is not very transparent. It could, perhaps, lead to a consistency proof for a family of discrete distributions but certainly not to a consistency proof in the generality strived for by Hotelling.

Doob (1934, p. 759) promises "... for the first time a complete proof of the validity of the method of maximum likelihood of R.A. Fisher ...", but his Theorem 5, p. 767, does not live up to this promise. The assumptions of this theorem include a condition on the asymptotic behaviour of the ML-sequence $(\vartheta^{(n)}(\mathbf{x}))_{n\in\mathbb{N}}$, namely: There is a set $X_0 \subset X^{\mathbb{N}}$ of $P^{\mathbb{N}}$-measure 1 such that for every $\mathbf{x} \in X_0$ and every subsequence $\mathbb{N}_0 \subset \mathbb{N}$

$$\overline{p}_n(x, \mathbf{x}) := \sup\{p(x, \vartheta^{(m)}(\mathbf{x}))/p(x) : m \in \mathbb{N}_0, m \geq n\}$$

is a continuous function of x, and $\log^+ \overline{p}_n(\cdot, \mathbf{x})$ is P-integrable. Among his conclusions: If $\int \limsup_{n\in\mathbb{N}_0} p(x, \vartheta^{(n)}(\mathbf{x}))dx \leq 1$, then $\lim_{n\in\mathbb{N}_0} p(x, \vartheta^{(n)}(\mathbf{x})) = p(x)$ for $\mathbf{x} \in X_0$ and P-a.a. $x \in X$.

This result can, at best, be considered as a lemma from which a consistency theorem could be derived. Notice that it contains no assumptions on the parameter (like continuity of $\vartheta \to p(x, \vartheta)$), and that the distinguished probability measure P is not necessarily a member of the family. Hence it needs conditions on the family which imply the conditions of Doob's Theorem 5, and a condition which leads from $p(x, \vartheta^{(n)}(\mathbf{x})) \to p(x, \vartheta_0)$ to $\vartheta^{(n)}(\mathbf{x}) \to \vartheta_0$. Doob confines himself to demonstrating the usefulness of his theorem by showing (see p. 768) that it establishes the consistency of $\overline{x}_n$ as an estimator for ϑ in the family $\{N(\vartheta, 1) : \vartheta \in \mathbb{R}\}$.

Since Doob's theorem needs further specifications in order to arrive at a consistency theorem for ML-sequences, the question appears to be of minor importance whether it is true in its own framework. Doubts are in order since in the proof of Theorem 5, Doob makes use of Theorem 4, p. 766, which is obviously wrong. Doob claims: "Its proof is simple and will be omitted." He seems to have overlooked the fact that the points of discontinuity may depend on the functions g. (See in this connection also Wald 1949, p. 595, footnote 2.)

Starting in 1936 and 1937, Dugué made several attempts to deliver precise proofs of the consistency and asymptotic normality of the ML-sequence (Dugué 1936, 1937). According to Le Cam (1953) [p. 279] "... the proofs are not rigorous and the mistakes are apparent". In a more transparent version, Dugué's arguments are presented (almost unchanged) in 1958, and here the weak points are easier to spot (Dugué 1958).

From $\sum_{\nu=1}^n \ell^\bullet(x_\nu, \vartheta^{(n)}(\mathbf{x})) = 0$ for $\mathbf{x} \in X^{\mathbb{N}}$, $n \in \mathbb{N}$, and $n^{-1}\sum_{\nu=1}^n \ell^\bullet(x_\nu, \vartheta_0) \to 0$ $(P^{\mathbb{N}}_{\vartheta_0})$, Dugué concludes (Theorem I, p. 140/1) that

$$n^{-1}\sum_{\nu=1}^{n}(\ell^{\bullet}(x_\nu,\vartheta_0)-\ell^{\bullet}(x_\nu,\vartheta^{(n)}(\mathbf{x})))\to 0\quad (P_{\vartheta_0}^{\mathbb{N}}).$$

Now Dugué makes an assumption which excludes most applications: That $\vartheta \to \ell^{\bullet}(x,\vartheta)$ is continuous for $\vartheta = \vartheta_0$, uniformly in $x \in X$, which implies: For every $\varepsilon > 0$ there exists δ_ε such that

$$\left|n^{-1}\sum_{\nu=1}^{n}(\ell^{\bullet}(x_\nu,\vartheta)-\ell^{\bullet}(x_\nu,\vartheta_0))\right| < \varepsilon \quad \text{for } n\in\mathbb{N} \text{ and } \mathbf{x}_n \in X^{\mathbb{N}},$$

provided $|\vartheta - \vartheta_0| < \delta_\varepsilon$. Without further arguments, Dugué then states that, conversely, for every $\varepsilon > 0$, there is δ_ε such that

$$\left|n^{-1}\sum_{\nu=1}^{n}(\ell^{\bullet}(x_\nu,\vartheta)-\ell^{\bullet}(x_\nu,\vartheta_0))\right| < \delta_\varepsilon \text{ implies } |\vartheta-\vartheta_0| < \varepsilon, \tag{5.2.9}$$

provided the function $\vartheta \to n^{-1}\sum_{\nu=1}^{n}\ell^{\bullet}(x_\nu,\vartheta)$ is one-one, so that (5.2.9) implies that $\vartheta^{(n)} \to \vartheta_0\ (P_{\vartheta_0}^{\mathbb{N}})$.

Wilks (1962) seeks to provide more elegant consistency proofs (Theorem 12.3.2, p. 360 for $\Theta \subset \mathbb{R}$ and 12.7.2, pp. 379/380 for $\Theta \subset \mathbb{R}^k$) by using the following auxiliary result 4.3.8, p. 105: If $f_n(\cdot,\vartheta) \to g(\vartheta)\ (P^{\mathbb{N}})$ uniformly for ϑ in a neighbourhood of ϑ_0, with g continuous at $\vartheta = \vartheta_0$, then $\vartheta^{(n)} \to \vartheta_0\ (P^{\mathbb{N}})$ implies

$$f_n(\cdot,\vartheta^{(n)}(\cdot)) \to g(\vartheta_0)\quad (P^{\mathbb{N}}).$$

This is what one needs for $f_n(\mathbf{x},\vartheta) = n^{-1}\sum_{\nu=1}^{n} f(x_\nu,\vartheta)$ with $P = P_{\vartheta_0}$, in which case

$$g(\vartheta) = \int f(x,\vartheta)P_{\vartheta_0}(dx).$$

Wilks' proof contains the usual mistake: Presented in a more transparent way, he concludes from

$$P_{\vartheta_0}^n\{|f_n(\cdot,\vartheta)-g(\vartheta)|<\varepsilon\} > 1-\varepsilon \quad \text{for } \vartheta \in U_\varepsilon,\ n \geq n_\varepsilon$$

and

$$P_{\vartheta_0}^n\{\vartheta^{(n)} \in U_\varepsilon\} > 1-\varepsilon \quad \text{for } n \geq n'_\varepsilon,$$

together with the continuity of g at ϑ_0, that

$$P_{\vartheta_0}^n\{|f_n(\cdot,\vartheta^{(n)})-g(\vartheta_0)|<\varepsilon\} > 1-\varepsilon \quad \text{for } n \geq n''_\varepsilon,$$

a mistake already noted in the review by Hoeffding (1962) [p. 1469]: The argument requires $P^n_{\vartheta_0}\{\sup_{\vartheta\in U_\varepsilon}|f_n(\cdot,\vartheta)-g(\vartheta)|<\varepsilon\}>1-\varepsilon$ for $n\geq n_\varepsilon$. This is, by far, not the only weak point in Wilks' book. The 7-page review by Hoeffding contains an impressive list of shortcomings.

A consistency theorem for a general family of probability measures (with conditions for consistency which are necessary and sufficient) can be found in Pfanzagl 1969a, p. 258, Theorem 2.6. See also Landers (1972).

Examples of Inconsistent ML-Sequences

The intuitive appeal of the ML method is so strong that one might be inclined to consider all conditions used for the consistency theorem (in particular the compactness of Θ) as being artifacts of the technique of the proof. It is, therefore, important to point to the numerous examples of nice-looking families where one or the other condition is violated, and where the ML-sequence fails to be consistent. First, we will concentrate our attention on counterexamples for parametric families and the i.i.d. case.

The first counterexamples, provided by Kraft and Le Cam (1956), present families where, with probability tending to unity, the ML estimator exists for every sample size and yet fails to be consistent, though consistent estimator sequences exist. A slight shortcoming of these examples: The parameter space Θ is an open set, but not an interval, and the densities are not continuous functions. Bahadur (1958) has various clever examples, but they are all nonparametric. Ferguson (1982) [Sects. 2 and 3] presents examples of one-parameter families with $\Theta=[0,1]$ satisfying Cramér's conditions such that every ML-sequence converges to 1 if $\vartheta\in[1/2,1]$. Observe that these are more than examples of inconsistent ML-sequences: They are also counterexamples to Cramér's interpretation of his "consistency" theorem. Another example of an inconsistent ML-sequence can be found in Pfanzagl (1994) [Example 6.6.2, pp. 209–210].

Many textbooks cite Neyman and Scott (1948) [p. 7] for an example of an inconsistent ML-sequence. Here is the simplest version of this example: Let (x_ν,y_ν), $\nu=1,\ldots,n$, be distributed as $\times_{\nu=1}^n N(\vartheta_\nu,\sigma^2)^2$ with $\sigma^2>0$ and $\vartheta_\nu\in\mathbb{R}$ for $\nu=1,\ldots,n$. The ML estimators are

$$\vartheta_\nu^{(n)}((x_1,y_1),\ldots,(x_n,y_n))=\frac{1}{2}(x_\nu+y_\nu)\quad\text{for }\nu=1,\ldots,n$$

and

$$\sigma_n^2((x_1,y_1),\ldots,(x_n,y_n))=\frac{1}{4n}\sum_{\nu=1}^n(x_\nu-y_\nu)^2.$$

The clue is that $(\sigma_n^2)_{n\in\mathbb{N}}$ converges to $\sigma^2/2$. This is not really a "counterexample", since the usual consistency theorems refer to the i.i.d. case. In cases with varying unknown nuisance parameters, inconsistency is not unexpected. Another example of this kind can be found in M. Ghosh (1995) [p. 166].

Asymptotic Normality of ML-Sequences

As soon as the consistency of an ML-sequence is established, its asymptotic distribution is easy to obtain. Let $\Theta \subset \mathbb{R}$ be open, and $P_\vartheta|\mathscr{A}$ a probability measure with μ-density $p(\cdot,\vartheta)$. Under suitable conditions on $\ell^\bullet$ and $\ell^{\bullet\bullet}$, a consistent ML-sequence $(\vartheta^{(n)})_{n\in\mathbb{N}}$ fulfills

$$P_{\vartheta_0}^{\mathbb{N}} \circ n^{1/2}(\vartheta^{(n)} - \vartheta_0) \Rightarrow N(0, \sigma^2(\vartheta_0)),$$

where $\sigma^2(\vartheta_0) = I(\vartheta_0)^{-1}$ with $I(\vartheta) = \int \ell^\bullet(x,\vartheta)^2 P_\vartheta(dx)$ the *Fisher information.*

The proof starts from the expansion of $\vartheta \to n^{-1/2}\sum_{\nu=1}^n \ell^\bullet(x_\nu,\vartheta)$ which can conveniently be written as (see (5.2.5))

$$n^{-1/2}\sum_{\nu=1}^n \ell^\bullet(x_\nu, \vartheta^{(n)}(\mathbf{x}_n)) = n^{-1/2}\sum_{\nu=1}^n \ell^\bullet(x_\nu,\vartheta_0)$$
$$+n^{1/2}(\vartheta^{(n)}(\mathbf{x}_n) - \vartheta_0)n^{-1}\sum_{\nu=1}^n k(x_\nu,\vartheta_0,\vartheta^{(n)}(\mathbf{x}_n)) \qquad (5.2.10)$$

with

$$k(x,\vartheta_0,\vartheta) := \int_0^1 \ell^{\bullet\bullet}(x,(1-u)\vartheta_0 + u\vartheta)du.$$

This expansion is valid if $\vartheta \to \ell^{\bullet\bullet}(x,\vartheta)$ is continuous in a neighbourhood of ϑ_0.

If $\sum_{\nu=1}^n \ell^\bullet(x_\nu,\vartheta^{(n)}(\mathbf{x}_n)) = 0$ for $\mathbf{x}_n \in X^n$, relation (5.2.10) implies that

$$n^{1/2}(\vartheta^{(n)}(\mathbf{x}_n) - \vartheta_0) = \frac{-n^{-1/2}\sum_{\nu=1}^n \ell^\bullet(x_\nu,\vartheta_0)}{n^{-1}\sum_{\nu=1}^n k(x_\nu,\vartheta_0,\vartheta^{(n)}(\mathbf{x}_n))} + o(n^0, P_{\vartheta_0}^n). \qquad (5.2.11)$$

If $\vartheta \to \ell^{\bullet\bullet}(x,\vartheta)$ is continuous and fulfills condition (5.2.2), then $\vartheta^{(n)} \to \vartheta_0$ $(P_{\vartheta_0}^n)$ implies (see (5.2.1))

$$n^{-1}\sum_{\nu=1}^n k(x_\nu,\vartheta_0,\vartheta^{(n)}(\mathbf{x}_n)) \to \int \ell^{\bullet\bullet}(x,\vartheta_0)P_{\vartheta_0}(dx) \quad (P_{\vartheta_0}^n). \qquad (5.2.12)$$

Hence $P_{\vartheta_0}^n \circ n^{1/2}(\vartheta^{(n)} - \vartheta_0)$, $n \in \mathbb{N}$, is asymptotically normal with variance

$$\frac{-\int \ell^\bullet(x,\vartheta_0)^2 P_{\vartheta_0}(dx)}{\left(\int \ell^{\bullet\bullet}(x,\vartheta_0)P_{\vartheta_0}(dx)\right)^2},$$

which reduces to $\sigma^2(\vartheta_0)$ under the condition

$$\int \ell^\bullet(x,\vartheta_0)^2 P_{\vartheta_0}(dx) + \int \ell^{\bullet\bullet}(x,\vartheta_0)P_{\vartheta_0}(dx) = 0. \qquad (5.2.13)$$

Simple though this proof may be, various authors were unable to find reasonable conditions which guarantee (5.2.12) for the sequence of their remainder terms. Recall that Dugué (1937, pp. 328/331) requires that $\vartheta \to \ell^{\bullet\bullet}(x, \vartheta)$ is continuous at ϑ_0, uniformly in $x \in X$ (a condition which excludes, for instance, the application to $\{N(0, \sigma^2) : \sigma > 0\}$). The fact that Dugué still uses this condition in 1958, Theorem II, p. 143, more than 17 years after Wald (1941b), is hard to explain.

Dugué's paper (1937) is just a simplified version of what Doob (1934) [Theorem 6, pp. 770/1], who presented the first valid proof, had already written. Yet Doob's proof is also unsatisfactory, since he starts from a Taylor expansion of $\vartheta \to \ell(x, \vartheta)$, which he uses to derive the expansion of $\vartheta \to \ell^\bullet(x, \vartheta)$ by means of differentiation, a procedure which requires a more complex condition on the remainder term.

Remark Under suitable regularity conditions on the family $\{P_\vartheta : \vartheta \in \Theta\}$, the relations (5.2.11)–(5.2.13) imply that

$$n^{1/2}(\vartheta^{(n)}(\mathbf{x}_n) - \vartheta_0) - \sigma^2(\vartheta_0)n^{-1/2}\sum_{\nu=1}^{n} \ell^\bullet(x_\nu, \vartheta_0) \to 0 \quad (P^n_{\vartheta_0})_{n\in\mathbb{N}}.$$

This implies that the ML-sequences are regular, a usually neglected fact which is needed to give operational significance to the assertion that the limit distribution $N(0, \sigma^2(\vartheta_0))$ is optimal. (For more see Sect. 5.13.)

Maximum Probability Estimators

The MP estimator (= maximum probability estimator) is based on a smoothed version of the density $p^{(n)}(\cdot, \vartheta)$ of the observations, defined as

$$p_r^{(n)}(\mathbf{x}_n, \vartheta) := \frac{c_n}{2r}\int 1_{(\vartheta - c_n^{-1}r, \vartheta + c_n^{-1}r)}(\tau)p^{(n)}(\mathbf{x}_n, \tau)d\tau.$$

The MP estimator $\vartheta_r^{(n)}$ is the value of ϑ which maximizes $\vartheta \to p_r^{(n)}(\mathbf{x}_n, \vartheta)$ (approximately). In an elementary but poorly structured proof, Weiss and Wolfowitz (1967b, Theorem, pp. 196/7 and 1974, Theorem 3.1, p. 17 and pp. 20/21) obtain a result which, in a simplified version, reads as follows: If $\vartheta_r^{(n)}$, $n \in \mathbb{N}$, standardized by c_n, $n \in \mathbb{N}$, converges regularly to some limit distribution, say Q_r, then $Q(-r, r) \le Q_r(-r, r)$ for any regularly attainable limit distribution Q.

In spite of the analogy between MP sequences and the ML-sequence in the definition, the result is of a different nature:

(i) There are no general conditions under which $\vartheta_r^{(n)}$, $n \in \mathbb{N}$, is a regular estimator sequence in the sense discussed in Sect. 5.12.

(ii) There is no general result expressing Q_r in terms of inherent properties of the model. To obtain the "optimal" limit distribution Q_r one needs to determine the MP estimator $\vartheta_r^{(n)}$ for every $n \in \mathbb{N}$ and to find its asymptotic distribution.

(iii) The optimum property of Q_r refers to the interval $(-r, r)$ only.

In (1967) and (1974), Weiss and Wolfowitz present various examples for the successful applications of the MP theory. Grossmann (1979, 1981) shows that the shortcomings indicated above do not arise in regular models. His main result is Theorem 3.2 in (1981, pp. 99/100): If for $s, t \in \mathbb{R}$,

$$\lim_{n\to\infty} nH^2(P_{\vartheta+c_n^{-1}s}, P_{\vartheta+c_n^{-1}t}) = (s-t)^2 L(\vartheta)^2$$

with $L(\vartheta) = \int \ell(\cdot, \vartheta)^2 dP_\vartheta$, and

$$\lim_{n\to\infty} nP_{\vartheta+c_n^{-1}s}\{p(x, \vartheta + c_n^{-1}t)/p(x, \vartheta + c_n^{-1}s) < 1 - \varepsilon\} = 0 \quad \text{for every } \varepsilon > 0,$$

then $c_n(\vartheta_r^{(n)} - \vartheta)$, $n \in \mathbb{N}$, converges for every $r > 0$ regularly to the optimal limit distribution given by the underlying LAN condition.

The following example, put forward by Akahira and Takeuchi (1979), demonstrates some shortcomings of the MP theory.

Example Let $\{P_\vartheta : \vartheta \in \mathbb{R}\}$ be the shift parameter family, generated by the truncated distribution

$$p(x) = K \exp[-x^2/2] 1_{(-1,1)}(x) \quad \text{with } K = 2\phi(1) - 1.$$

The authors determine the MP estimator $\vartheta_r^{(n)}$ and they show that

$$\lim_{n\to\infty} P_\vartheta^{(n)}\{n|\vartheta_r^{(n)} - \vartheta| \le t\} = 1 - \exp[-k(t + \min\{t, r\})] \tag{5.2.14}$$

with $k = c/\sqrt{e}$. Applied with $r = t$, relation (5.2.14) yields

$$\lim_{n\to\infty} P_\vartheta^n\{n|\vartheta_t^{(n)} - \vartheta| \le t\} = 1 - \exp[-2kt] \quad \text{for } t > 0;$$

hence the optimum property of $(Q_t^{(n)})_{n\in\mathbb{N}}$ implies

$$\lim_{n\to\infty} P_\vartheta^n\{n|\vartheta^{(n)} - \vartheta| \le t\} \le 1 - \exp[-2kt] \quad \text{for } t > 0 \tag{5.2.15}$$

for any regular estimator sequence $(\vartheta^{(n)})_{n\in\mathbb{N}}$.

The right-hand side of (5.2.15) is just a bound. For $t > 0$ it is attained by $\vartheta_t^{(n)}$, $n \in \mathbb{N}$, but there is no general result which guarantees the existence of an estimator sequence $\vartheta^{(n)}$, $n \in \mathbb{N}$, such that

$$\lim_{n\to\infty} P_\vartheta^n\{n|\vartheta^{(n)} - \vartheta| \le t\} = 1 - \exp[-2kt] \quad \text{for every } t > 0. \tag{5.2.16}$$

Among the MP sequences there is none: We have

$$\lim_{n\to\infty} P_\vartheta^n\{n|\vartheta_r^{(n)} - \vartheta| \le t\} = 1 - \exp[-2kt] \quad \text{for } t \in (0, r],$$

but

$$\lim_{n\to\infty} P_\vartheta^n\{n|\vartheta_r^{(n)} - \vartheta| \le t\} = 1 - \exp[-k(t+r)] < 1 - \exp[-2kt] \quad \text{for } t > r.$$

It is a particular property of this model (not the outgrowth of a general theorem) that an estimator sequence attaining the bound (5.5.1) does exist, namely

$$\hat{\vartheta}^{(n)}(\mathbf{x}_n) := \frac{1}{2}(x_{1:n} + x_{n:n}).$$

We might mention in passing that the asymptotic efficiency of the ML-sequence is 1/2. According to Akahira and Takeuchi (1979) [p. 137] the ML estimator is

$$\vartheta^{(n)}(x_1,\ldots,x_n) = \begin{cases} x_{1:n}+1 & \overline{x}_n > x_{1:n}+1, \\ \overline{x}_n & \text{if} \quad x_{n:n}-1 \le \overline{x}_n \le x_{1:n}+1, \\ x_{n:n}-1 & \overline{x}_n < x_{n:n}-1, \end{cases}$$

hence

$$\lim_{n\to\infty} P_\vartheta^n\{n|\vartheta^{(n)} - \vartheta| \le t\} = 1 - \exp[-kt].$$

5.3 Convergence of Distributions

For $n \in \mathbb{N}$ let $P_n|(X, \mathscr{A})$ be a probability measure. The following concepts for the convergence of P_n, $n \in \mathbb{N}$, to P_0, *setwise* and *uniform*, are in use.

$$P_n(A) \to P_0(A) \quad \text{for every } A \in \mathscr{A} \tag{5.3.1}$$

and

$$\sup_{A\in\mathscr{A}} |P_n(A) - P_0(A)| \to 0. \tag{5.3.2}$$

Scheffé's Lemma. For $n \in \mathbb{N}$, let p_n be a P_0-density of P_n. Then

$$p_n(x) \to p_0(x) \quad \text{for } x \in X \quad \text{implies (4.3.2).}$$

(See Scheffé 1947, pp. 434–438.)

Lehmann (1959) [pp. 351/2, Lemma 4] was the first to recognize the importance of what is now generally known as Scheffé's Lemma. It is not always reproduced correctly (see Tong 1990, p. 211).

Remark In the early days of Mathematical Statistics, it was easy for a scholar to get his name attached to a result that was generally known. In fact, Scheffé's Lemma is a special case of Vitali's Theorem. It goes back to Riesz (1928/9) [I, p. 350 and II, p. 182.]

Let X be a metric space, endowed with its Borel-algebra $\mathscr{A}$. Let P and $P_n, n \in \mathbb{N}$, be probability measures on $\mathscr{A}$. We define *weak convergence* of P_n to P as follows:

$$P_n \Rightarrow P \quad \text{if} \quad P_n(A) \to P(A) \tag{5.3.3}$$

for every set $A \in \mathscr{A}$ the boundary of which is of P-measure 0.

Remark It is well known that the bound of every convex set in $\mathbb{B}^p$ is of λ^p-measure zero. This result extends from λ^p to all measures that are a product of non-atomic components. (Hájek, 1971, p. 234. See also Gänssler and Stute, 1976, p. 54, Lemma 2.20.)

Various equivalent definitions of weak convergence are often cited as *Portmanteau Theorem* (see e.g. Billingsley 1968 pp. 11/12, Theorem 2.1, Witting and Müller-Funk 1995 p. 48, Satz 5.40, Elstrodt 2000 p. 385, Theorem 4.10, Bickel et al. 1993 p. 478, Lemma 1.8.3. Some of these equivalences were already known to Alexandrov (1940–1943). A full list of these equivalences can be found in Ash (2000) and Parthasarathy (1967).

The story behind the enigmatic name "Portmanteau Theorem": Billingsley had discovered Alexandrov's equivalences independently. Following Stigler's law of eponymy he refused to speak of "Alexandrov's Theorem". Instead, he invented the name "Portmanteau Theorem", a name which calls for a detailed explanation (see Witting and Müller-Funk, p. 48, and Elstrodt, p. 384).

Among the relations equivalent to (5.3.3), the following is, perhaps, the most useful:

$$P_n \Rightarrow P \quad \text{if} \quad \int f(x) P_n(dx) \to \int f(x) P(dx) \tag{5.3.4}$$

for every bounded and uniformly continuous function f.

For $X = \mathbb{R}^k$ this is equivalent to

$$P_n(B_r) \to P(B_r)$$

for every rectangle $B_r = \{x \in \mathbb{R}^k : x_i \leq r_i, i = 1, \dots, k\}$ whose boundary has P-measure 0. See also Billingsley (1968) [p. 17]. In applications, P has a Lebesgue density, and all rectangles have boundaries with P-measure 0.

C.R. Rao, Wolfowitz, Kaufman and Inagaki require weak convergence uniformly in r. (See e.g. Rao 1963, p. 194, Definition 3c, Wolfowitz 1965, p. 25, Kaufman 1966, p. 155.) It appears that these authors have missed that $P_n(B_r) - P(B_r) \to 0$ for every $r \in \mathbb{R}^k$ already implies uniformity in r, i.e.

$$\sup_{r \in \mathbb{R}} |P_n(B_r) - P(B_r)| \to 0.$$

In particular, weak convergence to a limit distribution P with $P \ll \lambda^k$ implies uniform convergence on all rectangles. In this case every convex set has a boundary of P-measure 0. Hence $P_n \Rightarrow P$ implies

$$P_n(C) \to P(C) \quad \text{for every } C \in \mathscr{C}, \tag{5.3.5}$$

with $\mathscr{C}$ denoting the family of convex sets $C \in \mathbb{B}^k$. According to a theorem of Ranga Rao (1962) [p. 665, Theorem 4.2] (see Fabian 1970, p. 142, Theorem 4.1 for an alternative proof), the convergence in (5.3.5) is uniform with respect to $C \in \mathscr{C}$, i.e.

$$\sup_{C \in \mathscr{C}} |P_n(C) - P(C)| \to 0.$$

Let $\mathscr{L}$ denote the set of all subconvex functions $\ell : \mathbb{R}^k \to [0, 1]$, and $\mathscr{L}_u$ the subset of uniformly continuous functions. These sets are weak convergence determining classes. We will often express weak convergence of P_n to P by

$$\int \ell d P_n \to \int \ell d P \quad \text{for } \ell \in \mathscr{L}_u. \tag{5.3.6}$$

Proposition 5.3.1 *Let $P \ll \lambda^k$. Then $P_n \Rightarrow P$ implies*

$$\sup_{\ell \in \mathscr{L}_u} \left| \int \ell d P_n - \int \ell d P \right| \to 0.$$

Proof Since $\{x \in \mathbb{R}^k : \ell(x) \le t\}$ is convex for every $t \ge 0$, for every $\varepsilon > 0$, there is n_ε such that for $n \ge n_\varepsilon$, $t > 0$ and $\ell \in \mathscr{L}_u$,

$$\left| P_n\{x \in \mathbb{R}^k : \ell(x) \le t\} - P\{x \in \mathbb{R}^k : \ell(x) \le t\} \right| < \varepsilon.$$

This implies that

$$\left| P_n\{x \in \mathbb{R}^k : \ell(x) > t\} - P\{x \in \mathbb{R}^k : \ell(x) > t\} \right| < \varepsilon \quad \text{for } n \ge n_\varepsilon.$$

Hence the relation

$$\int \ell(x) P_n(dx) = \int_0^1 P_n\{x \in \mathbb{R}^k : \ell(x) > t\} dt$$

implies

$$\left| \int \ell d P_n - \int \ell d P \right| < \varepsilon \quad \text{for } n \ge n_\varepsilon.$$

□

Remark If $\ell : \mathbb{R}^k \to [0, 1]$ is continuous and $\{x \in \mathbb{R}^k : \ell(x) \le t\}$ is bounded for every $t \in [0, 1)$, then ℓ is uniformly continuous.

Proof Since ℓ is continuous and $|\ell(u') - \ell(u'')| < \varepsilon$ on the complement of the compact set $\{u \in \mathbb{R}^k : \ell(u) \le 1 - \varepsilon\}$, ℓ is uniformly continuous on $\mathbb{R}^k$ (see Hewitt and Stromberg, 1965, pp. 87/88, Theorem 7.17). □

Now let $\mathfrak{P}$ be a family of probability measures on a measurable space $(X, \mathscr{A})$, and let κ be a real-valued functional on $\mathfrak{P}$. Let $\kappa^{(n)}$ be an estimator for $\kappa(P)$. Asymptotic statistical theory deals with the weak convergence of probability measures $Q_P^{(n)} := P^n \circ n^{1/2}(\kappa^{(n)} - \kappa(P))$. In the case of a parametric family $\mathfrak{P} = \{P_\vartheta : \vartheta \in \Theta\}$ we will write $Q_\vartheta^{(n)}$ in place of $Q_{P_\vartheta}^{(n)}$. The phenomenon of superefficiency demonstrates that it does not always suffice to consider the performance of $Q_P^{(n)}$, $n \in \mathbb{N}$, for P fixed. Significant assertions about the asymptotic performance of $Q_P^{(n)}$, $n \in \mathbb{N}$, under $P \in \mathfrak{P}$ depend on the way in which $Q_P^{(n)}$, $n \in \mathbb{N}$, depend on P.

In the following we use the *sup-distance* between probability measures $P, Q \in \mathfrak{P}$,

$$d(P, Q) = \sup_{A \in \mathscr{A}} |P(A) - Q(A)|.$$

Proposition 5.3.2 *Let $n \in \mathbb{N}$ be fixed. If $P \to \kappa(P)$ is continuous, then $P \to \int \ell dQ_P^{(n)}$ is continuous for $\ell \in \mathscr{L}_u$.*

Proof We have

$$\left| \int \ell(n^{1/2}(\kappa^{(n)} - \kappa(P))dP^n - \int \ell(n^{1/2}(\kappa^{(n)} - \kappa(P_0))dP_0^n \right|$$
$$\le \int \left|\ell(n^{1/2}(\kappa^{(n)} - \kappa(P)) - \ell(n^{1/2}(\kappa^{(n)} - \kappa(P_0))\right| dP_0 + d(P^n, P_0^n),$$

and $d(P^n, P_0^n) \le nd(P, P_0)$ by Hoeffding and Wolfowitz (1958). Since ℓ is uniformly continuous, there exists for every $\varepsilon > 0$ a $\delta_\varepsilon > 0$ such that $n^{1/2}|\kappa(P) - \kappa(P_0)| < \delta_\varepsilon$ implies

$$|\ell(n^{1/2}(\kappa^{(n)} - \kappa(P)) - \ell(n^{1/2}(\kappa^{(n)} - \kappa(P_0))| < \varepsilon.$$

The assertion follows since κ is continuous. □

Proposition 5.3.3 *If $\ell \in \mathscr{L}_u$, κ is continuous at P_0 and $\int \ell dQ_P^{(n)} \to \int \ell dQ_P$ locally uniformly at P_0, then $P \to \int \ell dQ_P$ is continuous at P_0. For short: The limit distribution of a sequence converging locally uniformly everywhere is continuous.*

The proof follows as in Proposition 5.3.2.

If $Q_P^{(n)}$, $n \in \mathbb{N}$, converges to Q_P everywhere on $\mathfrak{P}$, then it converges locally uniformly on a dense subset of $\mathfrak{P}$. Hence the convergence is locally uniform on a dense subset of $\mathfrak{P}$. Therefore, $P \to Q_P$ is continuous on a dense subset of $\mathfrak{P}$.

Corollary 5.3.4 *Every locally uniformly attainable limit distribution is continuous.*

Remark A weaker type of convergence like "regular" convergence (i.e. $Q_{P_n}^{(n)} \to Q_{P_0}$ for $P_n = P_0(1 + n^{-1/2} g)$) is not strong enough to imply the continuity of $P \to Q_P$ at $P = P_0$.

We abstain from a definition of weak convergence of $Q^{(n)}$ to Q by some metric on the set of probability measures on $\mathbb{B}^k$. Different metrics for a weak convergence express different aspects of this concept and give different rates. In this connection we mention the paper by Kersting (1978). One could even question whether the concept of weak convergence as defined by (5.3.6) is too strong. If the true loss function, say ℓ, is known, it would be sufficient that $\int \ell(x) Q_\vartheta^{(n)}(dx) \to \int \ell(x) Q_\vartheta(dx)$ for the true loss function ℓ.

Let Θ be a metric space with metric ρ. *Locally uniform weak convergence* of $Q_\vartheta^{(n)}$ to Q_ϑ at ϑ_0 is defined by

$$\inf_{U \ni \vartheta_0} \limsup_{n \to \infty} \sup_{\vartheta \in U} \left| \int f(x) Q_\vartheta^{(n)}(dx) - \int f(x) Q_\vartheta(dx) \right| = 0$$

for every bounded and uniformly continuous function f, or by

$$\inf_{U \ni \vartheta_0} \limsup_{n \to \infty} \sup_{\vartheta \in U} |Q_\vartheta^{(n)}(A) - Q_\vartheta(A)| = 0$$

for every set A the boundary of which is of Q_ϑ-measure zero (e.g. for every convex set A if $Q_\vartheta \ll \lambda^k$) for every $\vartheta \in \Theta$. Uniformity in f or A is not required.

Proposition 5.3.5 *The following statements are equivalent.*

(i) $Q_\vartheta^{(n)}$ *converges to* Q_ϑ *locally uniformly weakly at* ϑ_0.
(ii) $\rho(\vartheta_n, \vartheta_0) \to 0$ *implies* $Q_{\vartheta_n}^{(n)} \Rightarrow Q_{\vartheta_0}$.

Either of these conditions implies that Q_ϑ *is continuous at* ϑ_0.

Proof Apply Proposition 5.3.12 below with $h_n(\vartheta) = \int f(x) Q_\vartheta^{(n)}(dx)$ and $h(\vartheta) = \int f(x) Q_\vartheta(dx)$. Recall that $\int f(x) Q_\vartheta(dx)$ is continuous at ϑ_0 as a consequence of Lemma 5.3.11. □

An application of Corollary 5.3.13 below leads to the following

Proposition 5.3.6 *If* $Q_\vartheta^{(n)}$ *converges to* Q_ϑ *uniformly on some set* $\Theta_0 \subset \Theta$, *then this convergence is continuous at every* $\vartheta_0 \in \Theta_0$. *If* Θ *is locally compact, then locally uniform weak convergence at every* $\vartheta \in \Theta$ *implies uniform weak convergence on every compact subset of* Θ.

Fundamental for the application of asymptotic techniques are appropriate versions of the Central Limit Theorem. A detailed presentation can be found in Bhattacharya and Ranga Rao (1976). The following Theorem 5.3.7 is a uniform version of the Central Limit Theorem that suffices for most purposes of the present text.

Theorem 5.3.7 *Let $h : X \times \Theta \to \mathbb{R}^k$ be such that $\int h(\cdot, \vartheta) dP = 0$ and*

$$\lim_{a\to\infty} \sup_{\vartheta\in\Theta} \int \|h(\cdot, \vartheta)\|^2 1_{(\|h(\cdot,\vartheta)\|>a)} dP_\vartheta = 0$$

and the smallest eigenvalue of $\Sigma(\vartheta) := \int h(\cdot, \vartheta)h(\cdot, \vartheta)^\top dP_\vartheta$ is bounded away from 0 on Θ. Then $\tilde{h}_n(\mathbf{x}_n) = n^{-1/2} \sum_{\nu=1}^n h(x_\nu, \vartheta)$ fulfills

$$\lim_{n\to\infty} \sup_{\vartheta\in\Theta} \sup_{C\in\mathscr{C}} \left| P_\vartheta^n\{\tilde{h}_n(\cdot, \vartheta) \in C\} - N(0, \Sigma(\vartheta))(C) \right| = 0.$$

Proof Pfanzagl (1994) [p. 259, Theorem 7.7.11]. □

Auxiliary Results on the Convergence of Functions on Metric Spaces

Let Θ be a metric space with metric ρ. A sequence of functions $h_n : \Theta \to \mathbb{R}, n \in \mathbb{N}$, converges to 0 *uniformly* on $\Theta_0 \subset \Theta$ if

$$\limsup_{n\to\infty} \sup_{\vartheta\in\Theta_0} |h_n(\vartheta)| = 0.$$

The sequence converges to 0 *locally uniformly* at ϑ_0 if

$$\inf_{U\ni\vartheta_0} \limsup_{n\to\infty} \sup_{\vartheta\in U} |h_n(\vartheta)| = 0.$$

If h_n converges to 0 uniformly on Θ_0, then it converges locally uniformly to 0 at every ϑ_0 in Θ_0.

Proposition 5.3.8 *The following statements are equivalent.*

(i) $h_n, n \in \mathbb{N}$, converges to 0, locally uniformly at ϑ_0.
(ii) $h_n(\vartheta_n) \to 0$ if $\rho(\vartheta_n, \vartheta_0) \to 0$.

Proof The relations (i) and (ii) can be rewritten as follows.

(i) $\inf_{U\ni\vartheta_0} \limsup_{n\to\infty} \sup_{\vartheta\in U} h_n(\vartheta) = 0$.
(ii) $\rho(\vartheta_n, \vartheta_0) \to 0$ implies $h_n(\vartheta_n) \to 0$.

The conclusion from (i) to (ii) is straightforward. It remains to show that "non (i)" implies "non (ii)". Now "non (i)" implies that

$$\alpha_0 := \inf_{U\ni\vartheta_0} \limsup_{n\to\infty} \sup_{\vartheta\in U} h_n(\vartheta) > 0. \tag{5.3.7}$$

Set $U_m := \{\vartheta \in \Theta : \rho(\vartheta, \vartheta_0) < 1/m\}$. For m sufficiently large,

$$\limsup_{n\to\infty} \sup_{\vartheta\in U_m} h_n(\vartheta) > \alpha_0/2.$$

Let

$$a_{n,m} := \sup_{\vartheta \in U_m} h_n(\vartheta).$$

Since (5.3.7) implies

$$n \to \infty a_{n,m} \geq \alpha_0/2 \quad \text{for } m \text{ sufficiently large,}$$

there exists a sequence $n(m)$, $m \in \mathbb{N}$, such that

$$\liminf_{m \in \mathbb{N}} a_{n(m),m} > 0.$$

Hint: If $n(m_0)$ is defined, choose $n > n(m_0)$ such that $a_{n,m_0+1} > \alpha/2$, i.e.

$$\sup_{\vartheta \in U_m} h_{n(m)}(\vartheta) \geq \alpha_0 \quad \text{for } m \in \mathbb{N}.$$

With $\vartheta_m \in U_m$ chosen such that $h_{n(m)}(\vartheta_m) > \sup_{\vartheta \in U_m} h_n(\vartheta) - 1/m$, we have

$$h_{n(m)}(\vartheta_m) \geq \alpha_0 > 0.$$

Since $\vartheta_m \in U_m$ for $m \in \mathbb{N}$, we have $\rho(\vartheta_m, \vartheta_0) \to 0$. Since $h_{n(m)}(\vartheta_m) \geq \alpha_0 > 0$, this implies "non (ii)". □

Lemma 5.3.9 *Assume that $\delta(h_n(\vartheta), h_0(\vartheta)) \to 0$ for every ϑ in a neighbourhood of ϑ_0, and that this convergence is locally uniform at ϑ_0. If every h_n is continuous at ϑ_0, then h_0 is continuous at ϑ_0.*

Proof We have

$$\delta(h_0(\vartheta), h_0(\vartheta_0)) \leq \delta(h_0(\vartheta), h_n(\vartheta)) + \delta(h_n(\vartheta), h_n(\vartheta_0)) + \delta(h_n(\vartheta_0), h_0(\vartheta_0)).$$

Now $\delta(h_n(\vartheta_0), h_0(\vartheta_0)) < \varepsilon$ for $n \geq n'_\varepsilon$ and $\delta(h_0(\vartheta), h_n(\vartheta)) < \varepsilon$ for $\vartheta \in U'_\varepsilon$ and $n \geq n'_\varepsilon$, and, with $n_\varepsilon := n'_\varepsilon \vee n''_\varepsilon$, $\delta(h_{n_\varepsilon}(\vartheta), h_{n_\varepsilon}(\vartheta_0)) < \varepsilon$ for $\vartheta \in U''_\varepsilon$. It follows that $\delta(h_0(\vartheta), h_0(\vartheta_0)) < 3\varepsilon$ for $\vartheta \in U'_\varepsilon \cap U''_\varepsilon$. □

A Remark on Continuous Convergence

Let Θ be a metric space with metric ρ, and let $h_n : \Theta \to \mathbb{R}$, $n \in \mathbb{N}$, be a sequence of functions.

Definition 5.3.10 The convergence of the sequence $(h_n)_{n \in \mathbb{N}}$ to h_0 is *continuous* at ϑ_0 if $\rho(\vartheta_n, \vartheta_0) \to 0$ implies $\delta(h_n(\vartheta_n), h_0(\vartheta_0)) \to 0$.

Continuous convergence was used in certain parts of mathematics (see e.g. Hahn 1932). It was introduced in asymptotic statistical theory by Schmetterer (1966). To motivate this step (see p. 301), he does not say more than

> it seems better to introduce the idea of continuous convergence. When the limit of a sequence of functions is continuous, the idea of continuous convergence is even more general [?] than the idea of uniform convergence.

Continuous convergence was used by Roussas (1972, Sects. 5.6 and 5.7) J.K. Ghosh (1985) mentions continuous convergence as "interesting and useful" (p. 315) without any further arguments.

At a surface inspection, continuous convergence looks more natural than regular convergence. Does it really make sense to restrict the condition of convergence to sequences $\vartheta_n = \vartheta_0 + n^{-1/2}a$? It does, from the technical point of view, since these are the sequences for which $(P^n_{\vartheta_n})_{n\in\mathbb{N}}$ is contiguous to $(P^n_{\vartheta_0})_{n\in\mathbb{N}}$. For sequences $(\vartheta_n)_{n\in\mathbb{N}}$ converging to ϑ_0 more slowy than $n^{-1/2}$ (i.e., sequences with $n^{1/2}|\vartheta_n - \vartheta_0| \to \infty$), convergence of $P^n_{\vartheta_n} \circ n^{1/2}(\vartheta^{(n)} - \vartheta_n)$, $n \in \mathbb{N}$, is no effective restriction on the asymptotic performance of the estimator sequences. Pay attention to examples where the asymptotic performance of $P^n_{\vartheta+n^{-1/4}} \circ n^{1/2}(\vartheta^{(n)} - \vartheta)$ is distinct from $P^n_{\vartheta+n^{-1/2}} \circ n^{1/2}(\vartheta^{(n)} - \vartheta)$. (See, e.g. Lehmann and Casella 1998, pp. 442/443, Example 2.7.)

Schmetterer's conclusion about the relation between "uniform convergence" and "continuous convergence" reads as follows (see 1966, p. 303):

> Continuous convergence and uniform convergence to a continuous function on a compact set are equivalent.

This equivalence remained unnoticed. Schmetterer's assertion seems to be much more complicated than our Corollary 5.3.14 below. The explanation: The relevance of this equivalence lies in its application to "continuous" resp. "uniform" convergence between probability measures. Since $\vartheta \to Q^{(n)}_\vartheta$ is continuous, one may use Corollary 5.3.14 which takes advantage of the continuity of the functions h_n. Schmetterer cites the following Lemma of Hahn (1932) [p. 222] showing that continuous convergence implies continuity of the limit distribution.

Lemma 5.3.11 *Let $\delta(h_n(\vartheta), h_0(\vartheta)) \to 0$ for every ϑ in a neighbourhood of ϑ_0. If the convergence of $(h_n)_{n\in\mathbb{N}}$, to h_0 is continuous at ϑ_0, then h_0 is continuous at ϑ_0. (No continuity of h_n is required!)*

Proof Let $(\vartheta_m)_{m\in\mathbb{N}} \to \vartheta_0$. By assumption, for every $m \in \mathbb{N}$ there is n_m such that

$$|h_n(\vartheta_m) - h_0(\vartheta_m)| < \frac{1}{m} \quad \text{for} \quad n \geq n_m.$$

W.l.g., $n_{m+1} > n_m$. For $N \in \mathbb{N}$, let $M(N)$ be the largest integer m such that $n_m \leq N$. We have $M(N) \to \infty$ and $|h_n(\vartheta_{M(N)}) - h_0(\vartheta_{M(N)})| < 1/M(N)$, hence

$$\lim_{N\to\infty} |h_N(\vartheta_{M(N)}) - h_0(\vartheta_{M(N)})| = 0. \tag{5.3.8}$$

Since $\vartheta_{M(N)} \to \vartheta_0$, the continuous convergence of h_n, $n \in \mathbb{N}$, to h_0 implies

$$\lim_{n\to\infty} h_N(\vartheta_{M(N)}) = h_0(\vartheta_0).$$

Together with (5.3.8) this implies

$$\lim_{N\to\infty} h_0(x_{M(N)}) = h_0(\vartheta_0). \tag{5.3.9}$$

Hence any sequence $(\vartheta_m)_{m\in\mathbb{N}}$ converging to x_0 contains a subsequence $(\vartheta_{M(N)})_{N\in\mathbb{N}}$ fulfilling (5.3.9), and this implies $\lim_{m\to\infty} h_0(\vartheta_m) = h_0(\vartheta_0)$. □

Proposition 5.3.12 *The following statements are equivalent.*

(i) h_n, $n \in \mathbb{N}$, *converges at* ϑ_0 *locally uniformly to* h_0, *and* h_0 *is continuous at* ϑ_0.
(ii) h_n, $n \in \mathbb{N}$, *converges to* h_0 *continuously at* ϑ_0, *i.e.* $\rho(\vartheta_n, \vartheta_0) \to 0$ *implies* $\delta(h_n(\vartheta_n), h(\vartheta_0)) \to 0$.

Proof (i) *implies* (ii). Let $f_n(\vartheta) := \delta(h_n(\vartheta), h_0(\vartheta))$. If $h_n \to h_0$ locally uniformly at ϑ_0, this implies $f_n(\vartheta) \to 0$ locally uniformly at ϑ_0. By Proposition 5.3.8 this implies $f_n(\vartheta_n) \to 0$ and $\delta(h_n(\vartheta_n), h_0(\vartheta_n) \to 0$ if $\rho(\vartheta_n, \vartheta_0) \to 0$. Since h_0 is, by assumption, continuous at ϑ_0, this implies $\delta(h_n(\vartheta_n), h_0(\vartheta_0) \to 0$, i.e. (ii).

(ii) *implies* (i). Let $f_n(\vartheta) := \delta(h_n(\vartheta), h_0(\vartheta_0))$. Relation (ii) implies $f_n(\vartheta_n) \to 0$ if $\rho(\vartheta_n, \vartheta_0) \to 0$. By Proposition 5.3.8 this implies $f_n(\vartheta) \to 0$ locally uniformly at ϑ_0, i.e. $\delta(h_n(\vartheta), h(\vartheta_0)) \to 0$ locally uniformly at ϑ_0. Since h is continuous at ϑ_0 by Proposition 5.3.8, this implies $\delta(h_n(\vartheta), h(\vartheta)) \to 0$ locally uniformly at ϑ_0, which is (i). □

Corollary 5.3.13 *If every* h_n *is continuous at* ϑ_0, *then locally uniform convergence of* h_n, $n \in \mathbb{N}$, *to* h_0 *is the same as continuous convergence (both at* ϑ_0*).*

Proof Follows from Proposition 5.3.8 since h_0 is continuous at ϑ_0 by Lemma 5.3.9. □

Corollary 5.3.14 *If a sequence of continuous functions* h_n *converges to the function* h *uniformly on some subset* $\Theta_0 \subset \Theta$, *then this convergence is continuous on* Θ_0.

Proof Follows from Corollary 5.3.13: Uniform convergence on Θ_0 implies locally uniform convergence at ϑ_0 for every $\vartheta_0 \in \Theta_0$. □

5.4 Consistency and $\sqrt{n}$-consistency of Estimator Sequences

In the theory of ML-sequences, consistency is the delicate point. Natural conditions suffice to prove the asymptotic optimality of a consistent ML-sequence. Conditions of a different kind are needed to establish consistency.

For different estimating procedures, consistency is proved under different types of regularity conditions. Hence it appears plausible that these conditions are contingent on the estimating procedure, and far from necessary. In the following we consider general theorems on the existence of consistent and $\sqrt{n}$-consistent estimator sequences.

Consistency

Let $\mathfrak{P}$ be a family of probability measures on a measurable space $(X, \mathscr{A})$, endowed with a metric Δ. Using ideas of Le Cam (1956) we present in the following a reasonably general consistency theorem for estimator sequences of $P \in \mathfrak{P}$, and we show how such estimator sequences can be transformed into estimator sequences of the parameter in case $\mathfrak{P}$ is a parametric family.

The first step consists in finding an estimator sequence of $P \in \mathfrak{P}$ assigning to $\mathbf{x}_n = (x_1, \ldots, x_n)$ a probability measure $P^{(n)}_{\mathbf{x}_n}|\mathscr{A}$. The estimator sequence $(P^{(n)})_{n\in\mathbb{N}}$ is *uniformly consistent* on $\mathfrak{P}_0 \subset \mathfrak{P}$ if

$$\sup_{P\in\mathfrak{P}_0} P^n\{\mathbf{x}_n \in X^n : \Delta(P^{(n)}_{\mathbf{x}_n}, P) > \varepsilon\} \to 0 \quad \text{for } \varepsilon > 0. \tag{5.4.1}$$

If Δ is bounded, an equivalent and perhaps more convenient version of (5.4.1) is

$$\sup_{P\in\mathfrak{P}_0} \int \Delta(P^{(n)}_{\mathbf{x}_n}, P)P^n(d\mathbf{x}_n) \to 0.$$

The usual starting point for consistency theorems is

$$\sup_{P\in\mathfrak{P}} P^n\{\mathbf{x}_n \in X^n : \Delta(Q^{(n)}_{\mathbf{x}_n}, P) > \varepsilon\} \to 0 \quad \text{for } \varepsilon > 0, \tag{5.4.2}$$

where $Q^{(n)}_{\mathbf{x}_n}$ is the *empirical distribution*, defined by

$$Q^{(n)}_{\mathbf{x}_n}(A) := n^{-1}\sum_{\nu=1}^{n} 1_A(x_\nu), \quad A \in \mathscr{A}.$$

Projecting $Q^{(n)}_{\mathbf{x}_n}$ approximately onto $\mathfrak{P}$, e.g., determining $P^{(n)}_{\mathbf{x}_n} \in \mathfrak{P}$ such that

$$\Delta(P^{(n)}_{\mathbf{x}_n}, Q^{(n)}_{\mathbf{x}_n}) < 2\inf\{\Delta(P, Q^{(n)}_{\mathbf{x}_n}) : P \in \mathfrak{P}\},$$

say, we obtain

$$\Delta(P^{(n)}_{\mathbf{x}_n}, P) \le \Delta(P^{(n)}_{\mathbf{x}_n}, Q^{(n)}_{\mathbf{x}_n}) + \Delta(Q^{(n)}_{\mathbf{x}_n}, P) \le 3\Delta(Q^{(n)}_{\mathbf{x}_n}, P),$$

which implies

$$\sup_{P\in\mathfrak{P}} P^n\{\mathbf{x}_n \in X^n : \Delta(P^{(n)}_{\mathbf{x}_n}, P) > \varepsilon\} \to 0 \quad \text{for } \varepsilon > 0. \tag{5.4.3}$$

This idea of a "minimum distance estimator" occurs first in Wolfowitz (1952) in connection with a particular problem. It is developed further in Wolfowitz (1953b, 1957). An example of a distance function Δ fulfilling (5.4.2) is

$$\Delta(P', P'') := \sum_{k=1}^{\infty} 2^{-k} \left| \int f_k dP' - \int f_k dP'' \right|, \tag{5.4.4}$$

where $f_k : X \to \mathbb{R}$, $k \in \mathbb{N}$, is a sequence of functions *identifying* P, i.e.,

$$\int f_k dP' = \int f_k dP'' \text{ for } k \in \mathbb{N} \text{ implies } P' = P''. \tag{5.4.5}$$

For probability measures on $\mathbb{B}^k$, a metric fulfilling (5.4.2) is the Kolmogorov metric, which will play a decisive role in the construction of $\sqrt{n}$-consistent estimator sequences, since it fulfills the stronger relation (5.4.12).

If the problem is to estimate P, it is essential to have a metric which admits a statistically meaningful interpretation. If $\mathfrak{P}$ is a parametric family, say $\{P_\vartheta, \vartheta \in \Theta\}$ with Θ a metric space, and the problem is to estimate ϑ, then (5.4.3) is just an intermediary result on the way to a consistent estimator sequence of ϑ. For this purpose, an artificial metric like (5.4.4) suffices.

For a parametric family, we often consider $P^{(n)}_{\mathbf{x}_n}$ of the form $P_{\vartheta^{(n)}(\mathbf{x}_n)}$, and relation (5.4.3) becomes

$$\sup_{\vartheta\in\Theta} P^n_\vartheta\big\{\mathbf{x}_n \in X^n : \Delta(P_{\vartheta^{(n)}(\mathbf{x}_n)}, P_\vartheta) > \varepsilon\big\} \to 0 \quad \text{for } \varepsilon > 0. \tag{5.4.6}$$

The conclusion from the stochastic convergence of $P_{\vartheta^{(n)}}$ to P_ϑ to the stochastic convergence of $\vartheta^{(n)}$ to ϑ is possible only if ϑ is identifiable, i.e., if $\vartheta' \neq \vartheta''$ implies $P_{\vartheta'} \neq P_{\vartheta''}$. Yet an estimator sequence $(\vartheta^{(n)})_{n\in\mathbb{N}}$ fulfilling (5.4.6) will not necessarily be consistent for ϑ: Even if $\vartheta \to P_\vartheta$ is continuous, $P_{\vartheta_n} \to P_\vartheta$ does not necessarily imply $\vartheta_n \to \vartheta$.

To be more specific, let now Θ be a metric space with metric ρ. The problem is to prove that

$$P^n_\vartheta\{\mathbf{x}_n \in X^n : \rho(\vartheta^{(n)}(\mathbf{x}_n), \vartheta) > \varepsilon\} \to 0 \quad \text{for } \varepsilon > 0$$

holds for every $\vartheta \in \Theta$, if possible even uniformly on compact subsets of Θ. To obtain such an estimator sequence, Le Cam uses in various places the idea that identifiability and continuity of $\vartheta \to P_\vartheta$ (with respect to ρ and Δ) imply that, on every compact subset of Θ, the map $\vartheta \to P_\vartheta$ is one-to-one, and the inverse map

$P_\vartheta \to \vartheta$ is continuous. This idea can be used to obtain from an estimator sequence $(\vartheta^{(n)})_{n\in\mathbb{N}}$ fulfilling (5.4.6) an estimator sequence converging stochastically to ϑ. To isolate the abstract core of this argument, we consider two arbitrary metrics ρ and ρ_* on Θ, where ρ is stronger than ρ_* (i.e., for every $\varepsilon > 0$ there exists δ_ε such that $\rho(\vartheta', \vartheta'') < \delta_\varepsilon$ implies $\rho_*(\vartheta', \vartheta'') < \varepsilon$).

Lemma 5.4.1 *Assume that $\Theta_m \subset \Theta$, $m \in \mathbb{N}$, is an increasing sequence of ρ-compact subsets. If there exists an estimator sequence $(\vartheta^{(n)})_{n\in\mathbb{N}}$, such that*

$$\sup_{\vartheta\in\Theta_m} P_\vartheta^n\{\mathbf{x}_n \in X^n : \rho_*(\vartheta^{(n)}(\mathbf{x}_n), \vartheta) > \varepsilon\} \to 0 \quad \text{for } m \in \mathbb{N} \text{ and } \varepsilon > 0, \tag{5.4.7}$$

then there exists an estimator sequence $(\hat{\vartheta}^{(n)})_{n\in\mathbb{N}}$ such that

$$\sup_{\vartheta\in\Theta_m} P_\vartheta^n\{\mathbf{x}_n \in X^n : \rho(\hat{\vartheta}^{(n)}(\mathbf{x}_n), \vartheta) > \varepsilon\} \to 0 \quad \text{for } m \in \mathbb{N} \text{ and } \varepsilon > 0. \tag{5.4.8}$$

For later use we remark that for $n \geq n_m$, say,

$$\rho_*(\hat{\vartheta}^{(n)}(\mathbf{x}_n), \vartheta) \leq 2\rho_*(\vartheta^{(n)}(\mathbf{x}_n), \vartheta) \quad \text{for } \mathbf{x}_n \in X^n \text{ and } \vartheta \in \Theta_m. \tag{5.4.9}$$

Remark If Θ is locally compact and σ-compact, the sequence Θ_m, $m \in \mathbb{N}$, with $\Theta = \bigcup_{m=1}^\infty \Theta_m$ can be chosen such that $\Theta_m \subset \Theta_{m+1}^0$. (See e.g. Ash 1972, p. 387, Theorem A5.15.) In this case, every compact subset of Θ is contained in some Θ_m, so that relation (5.4.8) holds with Θ_m replaced by an arbitrary compact subset of Θ. Observe that $\Theta = \bigcup_{m=1}^\infty \Theta_m$ with $\Theta_m \subset \Theta_{m+1}$ (instead of $\Theta_m \subset \Theta_{m+1}^0$) is not enough. This point is observed by Le Cam (1956, p. 136, or 1986, p. 605), but missed by other authors (Bickel et al. 1993, p. 43). For example, $\mathbb{R}$ is the union of the compact sets $\Theta_m = [-m, 1 - m^{-1}] \cup [1, m]$, $m \in \mathbb{N}$, but none of the Θ_m covers the compact set $[0, 1]$. □

Proof of Lemma 5.4.1. Let

$$\delta_m := \inf\{\rho_*(\vartheta', \vartheta'') : \vartheta', \vartheta'' \in \Theta_m,\ \rho(\vartheta', \vartheta'') > m^{-1}\}.$$

Since ρ is stronger than ρ_*, any ρ-compact subset Θ_m is ρ_*-compact, which implies $\delta_m > 0$. As a consequence of (5.5.1), for every $m \in \mathbb{N}$ there is N_m such that

$$\sup_{\vartheta\in\Theta_m} P_\vartheta^n\{\mathbf{x}_n \in X^n : \rho_*(\vartheta^{(n)}(\mathbf{x}_n), \vartheta) > \delta_m/2\} < m^{-1} \quad \text{for } n \geq N_m.$$

W.l.g. we assume that $N_{m+1} > N_m$. With m_n denoting the largest integer m such that $N_m \leq n$, we obtain

$$\sup_{\vartheta\in\Theta_{m_n}} P_\vartheta^n\{\mathbf{x}_n \in X^n : \rho_*(\vartheta^{(n)}(\mathbf{x}_n), \vartheta) > \delta_{m_n}/2\} < m_n^{-1} \quad \text{for } n \in \mathbb{N}. \tag{5.4.10}$$

Since $\vartheta' \to \rho_*(\vartheta', \vartheta)$ is continuous and Θ_m is ρ_*-compact, there exists $\hat{\vartheta}^{(n)}(\mathbf{x}_n) \in \Theta_m$ such that

$$\rho_*(\hat{\vartheta}^{(n)}(\mathbf{x}_n), \vartheta^{(n)}(\mathbf{x}_n)) = \inf\{\rho_*(\vartheta', \vartheta^{(n)}(\mathbf{x}_n)) : \vartheta' \in \Theta_{m_n}\}, \tag{5.4.11}$$

whence

$$\rho_*(\hat{\vartheta}^{(n)}(\mathbf{x}_n), \vartheta^{(n)}(\mathbf{x}_n)) \le \rho_*(\vartheta, \vartheta^{(n)}(\mathbf{x}_n)) \quad \text{if } \vartheta \in \Theta_{m_n}.$$

This, in turn, implies

$$\rho_*(\hat{\vartheta}^{(n)}(\mathbf{x}_n), \vartheta) \le 2\rho_*(\vartheta^{(n)}(\mathbf{x}_n), \vartheta) \quad \text{if } \vartheta \in \Theta_{m_n}.$$

Since $(m_n)_{n\in\mathbb{N}} \uparrow \infty$, relation (5.4.9) follows.

If $\vartheta, \hat{\vartheta}^{(n)}(\mathbf{x}_n) \in \Theta_{m_n}$, then $\rho(\hat{\vartheta}^{(n)}(\mathbf{x}_n), \vartheta) > m_n^{-1}$ implies $\rho_*(\hat{\vartheta}^{(n)}(\mathbf{x}_n), \vartheta) \ge \delta_{m_n}$, hence also $\rho_*(\vartheta^{(n)}(\mathbf{x}_n), \vartheta) \ge \delta_{m_n}/2$. Together with (5.4.10) this implies

$$\begin{aligned}&\sup_{\vartheta\in\Theta_{m_n}} P_\vartheta^n\{\mathbf{x}_n \in X^n : \rho(\hat{\vartheta}^{(n)}(\mathbf{x}_n), \vartheta) > m_n^{-1}\}\\ &\le \sup_{\vartheta\in\Theta_{m_n}} P_\vartheta^n\{\mathbf{x}_n \in X^n : \rho_*(\vartheta^{(n)}(\mathbf{x}_n), \vartheta) \ge \delta_{m_n}/2\} < m_n^{-1} \quad \text{for } n \in \mathbb{N}.\end{aligned}$$

Since $m_n \uparrow \infty$, this completes the proof of (5.4.8). □

This Lemma can be applied with $\rho_*(\vartheta', \vartheta'') := \Delta(P_{\vartheta'}, P_{\vartheta''})$. The metric ρ_* is weaker than ρ if $\vartheta \to P_\vartheta$ is continuous with respect to ρ and Δ. Hence the estimator sequence $(\hat{\vartheta}^{(n)})_{n\in\mathbb{N}}$ defined by (5.4.11) is uniformly consistent on compact subsets of Θ.

The metric Δ defined in (5.4.4) presumes the existence of a sequence of functions $f_k : \mathscr{X} \to \mathbb{R}$, $k \in \mathbb{N}$, identifying P (i.e., fulfilling (5.4.5)). If this condition is fulfilled, the map $\vartheta \to P_\vartheta$ is continuous with respect to ρ and Δ if the map $\vartheta \to \int f_k dP_\vartheta$ from Θ to $\mathbb{R}$ is ρ-continuous for every $k \in \mathbb{N}$. Though the metric Δ does not show up in the final result (5.4.8), it is still the continuity of $\vartheta \to \int f_k dP_\vartheta$ which needs an intuitive interpretation.

There are two sets of conditions leading to meaningful results:

(i) X is a complete separable metric space. According to Parthasarathy (1967) [p. 17, Theorem 6.6], there exists a sequence of bounded and continuous functions f_k, $k \in \mathbb{N}$, fulfilling (5.4.5). Hence $\vartheta \to \int f_k dP_\vartheta$ will be ρ-continuous if the map $\vartheta \to P_\vartheta$ is ρ-continuous with respect to the topology of weak convergence. This is a result slightly more general than Lemma 4 in Le Cam (1956) [p. 136] which is confined to Euclidean spaces X and Θ.

(ii) If $(X, \mathscr{A})$ is arbitrary, the natural condition is ρ-continuity of $\vartheta \to P_\vartheta(A)$ for $A \in \mathscr{A}$. Applying the metric Δ defined by (5.4.4) requires a countable subfamily $\{A_k : k \in \mathbb{N}\}$ identifying P (i.e.: (5.4.5) is fulfilled with $f_k = 1_{A_k}$). If $\mathscr{A}$ is countably generated, one may take for $\{A_k : k \in \mathbb{N}\}$ a countable field generating $\mathscr{A}$. Here is a less restrictive alternative: If there exists a countable dense subset

$\Theta_0 \subset \Theta$, then $\{P_\vartheta : \vartheta \in \Theta_0\}$ is dense in $\{P_\vartheta : \vartheta \in \Theta\}$ with respect to the topology generated by the neighbourhoods $\{P \in \mathfrak{P} : |P(A) - P_0(A)| < \varepsilon\}$, $A \in \mathscr{A}$, $\varepsilon > 0$. This implies (see Pfanzagl 1969b, p. 15, Lemma 5 or Strasser 1985, p. 17, Lemma 4.3) the existence of a countably generated sufficient σ-field $\mathscr{A}_0 \subset \mathscr{A}$, and one may define Δ with $\{A_k : k \in \mathbb{N}\}$ a countable field generating $\mathscr{A}_0$. Hence relation (5.4.5) is guaranteed if Θ is a separable metric space, and $\vartheta \to P_\vartheta(A)$ is continuous for every $A \in \mathscr{A}$.

Relevant for applications is $\sqrt{n}$-consistency rather than just consistency without a rate, which however requires more restrictive conditions concerning the dependence of P_ϑ on ϑ. Hence it is a matter of taste whether one considers the existence of just consistent estimator sequences as fundamental, or as l'art pour l'art.

$\sqrt{n}$-Consistency

Under more restrictive conditions on the distance function Δ and on the family $\{P_\vartheta : \vartheta \in \Theta\}$, the estimator sequence $(\hat{\vartheta}^{(n)})_{n\in\mathbb{N}}$ constructed in Lemma 5.4.1 is even $\sqrt{n}$-consistent. Observe that $\sqrt{n}$-consistency is necessary for the convergence of $P_\vartheta^n \circ n^{1/2}(\vartheta^{(n)} - \vartheta)$, $n \in \mathbb{N}$, to a limit distribution. Moreover, $\sqrt{n}$-consistent estimator sequences may, under suitable conditions, be improved to become asymptotically optimal.

Le Cam (1956) [p. 137, Lemma 5] uses the existence of a consistent estimator sequence to construct a $\sqrt{n}$-consistent sequence. In a slightly modified version, his definition (p. 137, relation 38) of a $\sqrt{n}$-consistent estimator sequence reads as follows:

An estimator sequence $(\vartheta^{(n)})_{n\in\mathbb{N}}$ is *uniformly consistent on compact subsets* if there exists a sequence $m_n \uparrow \infty$ such that

$$\sup_{\vartheta\in\Theta_{m_n}} P_\vartheta^n\{\mathbf{x}_n \in X^n : \rho(\vartheta^{(n)}(\mathbf{x}_n), \vartheta) > m_n^{-1}\} < m_n^{-1} \quad \text{for } n \in \mathbb{N}.$$

The *modified (ML) estimator* $\hat{\vartheta}^{(n)}$ is defined as an element of $U_n(\mathbf{x}_n) := \{\vartheta \in \Theta_{m_n} : \rho(\vartheta^{(n)}(\mathbf{x}_n), \vartheta) \le m_n^{-1}\}$ such that

$$\sum_{\nu=1}^{n} \ell(x_\nu, \hat{\vartheta}^{(n)}) \ge \sup_{\vartheta\in U_n(\mathbf{x}_n)} \sum_{\nu=1}^{n} \ell(x_\nu, \vartheta) - n^{-1}.$$

Le Cam's Lemma 5 asserts that the estimator sequence thus defined is $\sqrt{n}$-consistent (under suitable regularity conditions on the densities $p(\cdot, \vartheta)$). In this Lemma, X is Euclidean; the conditions on Θ remain vague (locally convex and without isolated points on p. 130, locally compact and σ-compact on p. 137). Le Cam makes no use of the fact that his $(\hat{\vartheta}^{(n)})_{n\in\mathbb{N}}$ is a consistent sequence of asymptotic ML estimators, and therefore asymptotically efficient. (To obtain an asymptotically efficient estimator sequence, he applies in Lemma 6, p. 138, the usual one-step improvement procedure to this $\sqrt{n}$-consistent estimator sequence.)

In 1966 (p. 183, Lemma 2), 1969 (pp. 103–107) and 1986 (p. 608, Proposition 1) Le Cam uses a different argument in the construction of $\sqrt{n}$-consistent estimator sequences. The starting point is $\sqrt{n}$-consistency of the empirical distribution, i.e., relation (5.4.2) is replaced by the stronger relation

$$\sup_{P \in \mathfrak{P}} P^n\{\mathbf{x}_n \in X^n : n^{1/2}\Delta(Q^{(n)}_{\mathbf{x}_n}, P) > t_n\} \to 0 \quad \text{for } (t_n)_{n\in\mathbb{N}} \uparrow \infty. \tag{5.4.12}$$

Proceeding as above, relation (5.4.12) implies that

$$\sup_{\vartheta \in \Theta} P^n\{\mathbf{x}_n \in X^n : n^{1/2}\rho_*(\vartheta^{(n)}(\mathbf{x}_n), \vartheta) > t_n\} \to 0 \quad \text{for } (t_n)_{n\in\mathbb{N}} \uparrow \infty. \tag{5.4.13}$$

Since (5.4.13) is stronger than (5.5.1), the same construction as in Lemma 5.4.1 applies, thus leading to an estimator sequence $(\hat{\vartheta}^{(n)})_{n\in\mathbb{N}}$ fulfilling (5.4.8) and (5.4.9). This implies that $(\hat{\vartheta}^{(n)})_{n\in\mathbb{N}}$ is ρ-consistent (by 5.4.8), and $\sqrt{n}$-consistent with respect to ρ_* (by (5.4.13) and (5.4.9), uniformly on every Θ_m. The following Lemma asserts that such estimator sequences are $\sqrt{n}$-consistent with respect to ρ, uniformly on every Θ_m, provided the following condition on the relation between ρ and ρ_* holds true. For some $\varepsilon > 0$,

$$c := \inf\{\rho_*(\vartheta', \vartheta'')/\rho(\vartheta', \vartheta'') : \vartheta', \vartheta'' \in \Theta,\ \rho(\vartheta', \vartheta'') < \varepsilon\} > 0. \tag{5.4.14}$$

This means: *If the construction of the consistent estimator sequence is based on the Kolmogorov metric, this estimator is automatically $\sqrt{n}$-consistent under the condition* (5.4.14).

Lemma 5.4.2 *Let ρ and ρ_* be arbitrary metrics fulfilling (5.4.14). If*

$$\sup_{\vartheta \in \Theta_m} P^n_\vartheta\{\mathbf{x}_n \in X^n : \rho(\hat{\vartheta}^{(n)}(\mathbf{x}_n), \vartheta) > \varepsilon\} \to 0 \quad \textit{for } \varepsilon > 0$$

and

$$\sup_{\vartheta \in \Theta_m} P^n_\vartheta\{\mathbf{x}_n \in X^n : n^{1/2}\rho_*(\hat{\vartheta}^{(n)}(\mathbf{x}_n), \vartheta) > t_n\} \to 0 \quad \textit{for } (t_n)_{n\in\mathbb{N}} \uparrow \infty,$$

then

$$\sup_{\vartheta \in \Theta_m} P^n_\vartheta\{\mathbf{x}_n \in X^n : n^{1/2}\rho(\hat{\vartheta}^{(n)}(\mathbf{x}_n), \vartheta) > t_n\} \to 0 \quad \textit{for } (t_n)_{n\in\mathbb{N}} \uparrow \infty.$$

Proof Follows immediately from

$$\begin{aligned}&\{n^{1/2}\rho(\hat{\vartheta}^{(n)}, \vartheta) > t_n\}\\ &\subset \{\rho(\hat{\vartheta}^{(n)}, \vartheta) > \varepsilon\} \cup \{\rho(\hat{\vartheta}^{(n)}, \vartheta) \le \varepsilon,\ n^{1/2}\rho(\hat{\vartheta}^{(n)}, \vartheta) > t_n\}\end{aligned}$$

$$\subset \{\rho(\hat{\vartheta}^{(n)}, \vartheta) > \varepsilon\} \cup \{n^{1/2}\rho_*(\hat{\vartheta}^{(n)}, \vartheta) > ct_n\}.$$

□

Lemma 5.4.2 yields the existence of $\sqrt{n}$-consistent estimator sequences if there is a metric Δ on $\mathfrak{P}$ fulfilling (5.4.12), and if the pertaining metric $\rho_*(\vartheta', \vartheta'') := \Delta(P_{\vartheta'}, P_{\vartheta''})$ fulfills condition (5.4.14).

For $P|\mathbb{B}^k$, relation (5.4.12) holds, in particular, for the Kolmogorov metric (but not, for instance, for the Prohorov-metric; see Kersting 1978).

Relation (5.4.14) for the Euclidean metric on Θ and the Kolmogorov metric for probability measures on the Borel field of [0, 1] was proved by Le Cam (1966, p. 184; 1969, pp. 106/107; 1986, pp. 606/607, Lemma 3) and, with a similar proof, for $P|\mathbb{B}^k$ in Bickel et al. (1993) [p. 42, relation (e)]. These proofs require that the densities are differentiable in some sense.

Le Cam suggests that his theorems, proved for probability measures on the Borel-field of [0, 1], hold in fact for probability measures on an arbitrary measure space $(X, \mathscr{A})$. "There is no loss of generality in assuming that the variables x_ν take their values in the interval [0, 1]" (see Le Cam 1966, p. 183 and, similarly, 1969, p. 104, and 1986, pp. 608/9). Le Cam's argument: Let $\{\vartheta_k : k \in \mathbb{N}\}$ be a dense subset of (the Euclidean) Θ, map X into $[0, \infty]^{\mathbb{N}}$ by assigning to x the point $(p(x, \vartheta_k))_{k\in\mathbb{N}} \in [0, \infty]^{\mathbb{N}}$, and map $[0, \infty]^{\mathbb{N}}$ in a one-to-one Borel way onto [0, 1]. "The sufficiency of $[x \to (p(x, \vartheta_k))_{k\in\mathbb{N}}]$ implies that no information is lost in the process. Thus, the general case can be reduced to the case where $(X, \mathscr{A})$ is the interval [0, 1] with its Borel sets" (Le Cam 1986, pp. 608/9).

This is an argument of almost Fisherian precision. One would wish that Le Cam had presented his idea in a precise manner, somewhere, instead of repeating these vague hints three times. It is hard to understand how Le Cam's idea should work. After all, one has not more than a measurable map $\chi|X$ into a subset of [0, 1]. Being an isomorphism, this map is sufficient in a trivial sense, and it remains unclear what the role of the sufficiency of $x \to (p(x, \vartheta_k))_{k\in\mathbb{N}}$ could be. What comes out in the end is that $n^{1/2}\Delta(P_{\vartheta^{(n)}(\mathbf{x}_n)} \circ \chi, P_\vartheta \circ \chi)$, $n \in \mathbb{N}$, is under P_ϑ^n asymptotically bounded if Δ is the Kolmogorov metric. To infer from this $\sqrt{n}$-consistency of $(\vartheta^{(n)})_{n\in\mathbb{N}}$ one would need a relation like (5.4.14) between ρ and

$$\rho_*(\vartheta', \vartheta'') := \Delta(P_{\vartheta'} \circ \chi, P_{\vartheta''} \circ \chi).$$

Apart from Le Cam (1986) there are not more than three text-books which contain general theorems on the existence of consistent estimator sequences. Ibragimov and Has'minskii (1981, p. 31, Theorem 4.1) prove the existence of $\sqrt{n}$-consistent estimator sequences for Euclidean Θ under the condition that $\vartheta \to P_\vartheta$ is continuous with respect to the sup-distance, and that $\inf\{d(P_\vartheta, P_{\vartheta_0}) : \|\vartheta - \vartheta_0\| > \delta\} > 0$ if $\delta > 0$. Rüschendorf (1988) [p. 68, Proposition 3.7] proves the existence of $\sqrt{n}$-consistent estimator sequences assuming Fréchet-differentiability of $\vartheta \to \Delta_k(P_\vartheta, P_{\vartheta_0})$. Bickel et al. (1993) [p. 42, Theorem 1] asserts for regular parametric models $P_\vartheta|\mathbb{B}^p$, $\vartheta \in \Theta$, the existence of uniformly $\sqrt{n}$-consistent estimator sequences, provided ϑ is identifiable. "Regular parametric model" means, roughly speaking, that the density has a continuous derivative. (See their Definition 2, p. 12, for a precise definition.) Checking the proof it becomes clear that Θ is meant as an open subset of an Euclidean space, and that "uniformly $\sqrt{n}$-consistent" means "$\sqrt{n}$-consistent, uniformly on every compact subset of Θ". Since this is the only easily accessible place where this result can be found, it seems in order to mention that the proof of Theorem 1 is somewhat sketchy on p. 43. With applications in mind, the authors forgo proving the existence of consistent (rather than $\sqrt{n}$-consistent) estimator sequences for more general models.

The theorems mentioned above are more than "existence theorems": The estimator sequences are constructed explicitly. Yet, many steps in these constructions include arbitrary elements. While the maximum likelihood estimator is uniquely determined for every sample size (at least in the usual cases), the result of the foregoing construction for a particular sample is vague, which calls the operational significance of the asymptotic assertion in question.

5.5 Asymptotically Linear Estimator Sequences

Many estimator sequences with limit distribution are asymptotically linear. Asymptotically optimal estimator sequences in the sense of the Convolution Theorem are necessarily of this type.

Let $\mathfrak{P}$ be an arbitrary family of probability measures $P|\mathscr{A}$. For $P \in \mathfrak{P}$ set

$$L_*(P) := \left\{g \in L_2(P) : \int g dP = 0\right\}.$$

Let $\kappa : \mathfrak{P} \to \mathbb{R}^p$ be a functional. An estimator sequence $\kappa^{(n)} : X^n \to \mathbb{R}^p$ is *asymptotically linear* (for κ at P) if it admits a stochastic expansion of order 1, more precisely if there is a function $K(\cdot, P) \in L_*(P)^p$ such that

$$n^{1/2}(\kappa^{(n)} - \kappa(P)) - \tilde{K}(\cdot, P) \to 0 \quad (P^n) \tag{5.5.1}$$

with

$$\tilde{K}(\mathbf{x}_n, P) = n^{-1/2} \sum_{\nu=1}^{n} K(x_\nu, P).$$

In the common terminology, originating from the literature on robustness, K is called the *influence function*. The representation of $\kappa^{(n)}$, $n \in \mathbb{N}$, by (5.4.7) leads immediately to

$$P^n \circ n^{1/2}(\kappa^{(n)} - \kappa(P)) \Rightarrow N(0, \Sigma(P)) \tag{5.5.2}$$

with $\Sigma(P) = \int K(\cdot, P) K(\cdot, P)^\top dP$.

One could question whether an approximation of $n^{1/2}(\kappa^{(n)} - \kappa(P))$ by $\tilde{K}(\cdot, P)$ is operationally significant unless it holds in some sense locally uniformly. Bickel et al. (1993) [p. 201, Definition 2.2.6] require relation (5.5.1) to hold uniformly on compact (!) subsets of $\mathfrak{P}$, and they show that under this condition the estimator sequence converges regularly to the normal limit distribution with the covariance matrix $\Sigma(P)$. (Uniformity in (5.5.1) is not required by Witting and Müller-Funk 1995, p. 197, relation 6.147.)

Assume that κ has a *gradient* $\kappa^\bullet \in L_*(P)$ in the sense of Sect. 5.6,

$$n^{1/2}(\kappa(P_{n^{-1/2}g}) - \kappa(P)) \to \int \kappa^\bullet(x, P)g(x)P(dx)$$

for paths P_{tg} whose density p_{tg} fulfills a suitable representation $p_{tg}/p = 1 + tg + r_t$, $t \in (-\varepsilon, \varepsilon)$, with $g \in L_*(P)$. Then asymptotic linearity (5.5.1) holds with P replaced by $P_{n^{-1/2}g}$ iff

$$\tilde{K}(\cdot, P_{n^{-1/2}g}) = \tilde{K}(\cdot, P) - \int \kappa^\bullet(x, P)g(x)P(dx) + o(n^0, P^n). \tag{5.5.3}$$

Relation (5.5.3) holds in *parametric* families $\{P_\vartheta : \vartheta \in \Theta\}$, $\Theta \subset \mathbb{R}^k$, of probability measures under mild regularity conditions. It reads

$$\tilde{K}(\cdot, \vartheta + n^{-1/2}a) - \tilde{K}(\cdot, \vartheta) = \int K(x, \vartheta)\ell^\bullet(x, \vartheta)dP_\vartheta(dx) \cdot a + o(n^0, P_\vartheta^n). \tag{5.5.4}$$

To see this, let $K(\cdot, \vartheta) \in L_*(P_\vartheta)$ be suitably differentiable in ϑ with Jacobian $K^\bullet(\cdot, P_\vartheta)$. Then

$$\tilde{K}(\cdot, \vartheta + n^{-1/2}a) - \tilde{K}(\cdot, \vartheta) \to \int K^\bullet(\cdot, \vartheta)dP_\vartheta \cdot a \quad (P_\vartheta^n). \tag{5.5.5}$$

On the other hand,

$$\begin{aligned}
0 &= \int K(x, \vartheta + n^{-1/2}a)P_{\vartheta + n^{-1/2}a}(dx) \\
&= \int (K(x, \vartheta) + n^{-1/2}K^\bullet(x, \vartheta)a)(1 + n^{-1/2}\ell^\bullet(x, \vartheta)a)P_\vartheta(dx) + o(n^{-1/2}) \\
&= n^{-1/2}\int K^\bullet(x, \vartheta)P_\vartheta(dx) \cdot a + \int K(x, \vartheta)\ell^\bullet(x, \vartheta)P_\vartheta(dx) \cdot a + o(n^{-1/2}).
\end{aligned}$$

Hence

$$\int K^\bullet(x, \vartheta)P_\vartheta(dx) + \int K(x, \vartheta)\ell^\bullet(x, \vartheta)P_\vartheta(dx) = 0. \tag{5.5.6}$$

Together with (5.5.5) this implies (5.5.4).

Observe that relation (5.5.5) applied with $K(x, \vartheta) = \ell^\bullet(x, \vartheta)$ becomes

$$\int \left(\ell^{\bullet\bullet}(x, \vartheta) + \ell^\bullet(x, \vartheta)\ell^\bullet(x, \vartheta)^\top\right) P_\vartheta(dx) = 0.$$

With $L_{ij}(\vartheta) = \int \ell^{(ij)}(x, \vartheta) P_\vartheta(dx)$ and $L_{i,j}(\vartheta) = \int \ell^{(i)}(x, \vartheta)\ell^{(j)}(x, \vartheta) P_\vartheta(dx)$ this is the well known relation

$$L_{ij}(\vartheta) + L_{i,j}(\vartheta) = 0 \quad \text{for } i, j = 1, \ldots, k.$$

Relation (5.5.4) extends immediately to $K : X \times \Theta \to \mathbb{R}^k$. Observe that (5.5.4) implies that

$$\lim_{n\to\infty} P^n_{\vartheta+n^{-1/2}a} \circ \tilde{K}(\cdot, \vartheta + n^{-1/2}a)$$

does not depend on $a \in \mathbb{R}^k$. In fact, it holds with a replaced by a bounded sequence $a_n, n \in \mathbb{N}$.

Returning to the *nonparametric* setting, one could question whether asymptotic linearity with P replaced by $P_{n^{-1/2}g}$ is of much interest. Yet, at least in the case of parametric families, the relation

$$n^{1/2}(\kappa^{(n)} - \kappa(P_{\vartheta+n^{-1/2}a})) = \tilde{K}(\cdot, \vartheta + n^{-1/2}a) + o(n^0, P^n_\vartheta)$$

occurs by Bahadur's Lemma automatically for λ^p–a.a. ϑ of all rational a along some subsequence. Hence one cannot escape the question for conditions on $K(\cdot, P)$ which imply relation (5.5.3).

The results presented so far are based on asymptotic linearity of $\kappa^{(n)}$ and the differentiability of κ. If we include properties of $P^n_{n^{-1/2}g} \circ n^{1/2}(\kappa^{(n)} - \kappa(P_{n^{-1/2}a}))$, in our considerations, we obtain the following Proposition.

Proposition 5.5.1 *Assume that an estimator sequence of κ is at P asymptotically linear with influence function K. This estimator sequence is regular at P iff κ is differentiable at P, and K is a gradient of κ at P.*

Weaker versions of this result occur in Witting and Müller-Funk 1995, p. 422, Satz 6.201). That the influence function is a gradient is expressed in a disguised form as "Kopplungsbedingung (6.5.15)". For related results see also Bickel et al. 1993, p. 39, Proposition 2.4.3 A and p. 183, Theorem 5.2.3.

Proof Recall that $P^n \circ \tilde{K}(\cdot, P) \Rightarrow N(0, \Sigma(P))$. The relation

$$n^{1/2}(\kappa^{(n)} - \kappa(P)) = \tilde{K}(\cdot, P)) + o(n^0, P^n)$$

implies by Le Cam's Lemma that

$$P^n_{n^{-1/2}g} \circ n^{1/2}(\kappa^{(n)} - \kappa(P)) \Rightarrow N(\mu, \Sigma(P))$$

with $\mu = \int K(x, P)g(x)P(dx)$.

On the other hand,

$$n^{1/2}(\kappa^{(n)} - \kappa(P_{n^{-1/2}g})) \\ = n^{1/2}(\kappa^{(n)} - \kappa(P)) - n^{1/2}\big(\kappa(P_{n^{-1/2}g}) - \kappa(P)\big) + o(n^0, P^n). \qquad (5.5.7)$$

Therefore, $\kappa^{(n)}$ is regular iff the righthand side of (5.5.7) is asymptotically independent of g, i.e. iff

$$n^{1/2}(\kappa(P_{n^{-1/2}g}) - \kappa(P)) \to \int K(x)g(x)P(dx),$$

a relation equivalent to the differentiability of κ with derivative K. □

In the following we shall discuss how the representation of estimator sequences by (5.5.1) can be used to obtain asymptotically efficient estimator sequences. The technical problems are different for parametric, semiparametric and general families.

Parametric Families

As above we write $\ell^\bullet(x, \vartheta)$ for the vector $\ell^{(i)}(x, \vartheta)$, $i = 1, \ldots, k$, and $\ell^{\bullet\bullet}(x, \vartheta)$ for the matrix $\ell^{(ij)}(x, \vartheta)$, $i, j = 1, \ldots, k$, and we set $L(\vartheta)$ for the matrix with entries $L_{i,j}(\vartheta)$ and write $\Lambda(\vartheta) = L(\vartheta)^{-1}$.

Since ML estimators are not always easy to compute, Fisher (1925) [Sect. 5, pp. 707–709] suggests to approximate ML estimators by expanding the function $\sum_{\nu=1}^n \ell^\bullet(x_\nu, \vartheta)$ about the value $\vartheta = \vartheta^{(n)}(\mathbf{x}_n)$ of a $\sqrt{n}$-consistent preliminary estimator:

$$\sum_{\nu=1}^n \ell^\bullet(x_\nu, \vartheta) = \sum_{\nu=1}^n \ell^\bullet(x_\nu, \vartheta^{(n)}(\mathbf{x}_n)) + \sum_{\nu=1}^n \ell^{\bullet\bullet}(x_\nu, \vartheta)(\vartheta^{(n)}(\mathbf{x}_n) - \vartheta) + o(n^{-1/2}, P_\vartheta^n).$$

For the ML estimator $\hat{\vartheta}^{(n)}$, the likelihood equation $\sum_{\nu=1}^n \ell^\bullet(x_\nu, \hat{\vartheta}^{(n)}(\mathbf{x}_n)) = 0$ then leads to the approximation

$$\hat{\vartheta}^{(n)}(\mathbf{x}_n) = \vartheta^{(n)}(\mathbf{x}_n) + \Big(\sum_{\nu=1}^n \ell^{\bullet\bullet}(\mathbf{x}_\nu, \vartheta^{(n)}(\mathbf{x}_n))\Big)^{-1} \sum_{\nu=1}^n \ell^\bullet(x_\nu, \vartheta^{(n)}(\mathbf{x}_n)) \\ = \vartheta^{(n)}(\mathbf{x}_n) + \Lambda(\vartheta^{(n)}(\mathbf{x}_n))n^{-1} \sum_{\nu=1}^n \ell^\bullet(x_\nu, \vartheta^{(n)}(\mathbf{x}_n)) + o(n^{-1/2}, P_\vartheta^n).$$

Following his principle of "proof by example", Fisher carries this through for the Cauchy distribution, starting from the median as a preliminary estimator. In a more general context, the same idea occurs (independently?) in Le Cam (1956) [p. 139]: If $\{P_\vartheta : \vartheta \in \Theta\}$, $\Theta \subset \mathbb{R}^k$, is sufficiently regular (something close to LAN), then

$$\hat{\vartheta}^{(n)}(\mathbf{x}_n) := \vartheta^{(n)}(\mathbf{x}_n) + \Lambda(\vartheta^{(n)}(\mathbf{x}_n))n^{-1}\sum_{\nu=1}^{n} \ell^{\bullet}(x_\nu, \vartheta^{(n)}(\mathbf{x}_n))$$

is asymptotically linear with influence function $\Lambda(\vartheta)n^{-1/2}\sum_{\nu=1}^{n}\ell^{\bullet}(x_\nu, \vartheta)$, i.e.

$$n^{1/2}(\vartheta^{(n)}(\mathbf{x}_n) - \vartheta) = \Lambda(\vartheta)n^{-1/2}\sum_{\nu=1}^{n}\ell^{\bullet}(x_\nu, \vartheta) + o(n^0, P_\vartheta^n). \tag{5.5.8}$$

(His p. 138, Lemma 6, (47).)

The immediate consequence of relation (5.5.8): The limit distribution of the improved estimator sequence is $N(0, \Lambda(\vartheta))$, identical with the limit distribution of (consistent) ML-sequences.

The improvement procedure requires just $\sqrt{n}$-consistency of $\vartheta^{(n)}$, not more. It solves, by the way, also the problem of consistency connected with the ML method, a problem Fisher was not aware of.

Applications of the improvement procedure depend on ad hoc procedures for finding $\sqrt{n}$-consistent preliminary estimator sequences: The general existence theorems are hardly applicable. There are numerous papers investigating various aspects of the improvement procedure.

(i) First of all, the improvement procedure can be modified to

$$\hat{\vartheta}^{(n)}(\mathbf{x}_n) = \vartheta^{(n)}(\mathbf{x}_n) + \Lambda^{(n)}(\mathbf{x}_n)n^{-1}\sum_{\nu=1}^{n} \ell^{\bullet}(x_\nu, \vartheta^{(n)}(\mathbf{x}_n)), \tag{5.5.9}$$

with some sequence $\Lambda^{(n)} \to \Lambda(\vartheta)$ (P_ϑ^n), and one can ask whether

$$\Lambda^{(n)}(\mathbf{x}_n) = \Lambda(\vartheta^{(n)}(\mathbf{x}_n))$$

or

$$\Lambda^{(n)}(\mathbf{x}_n) = -n^{-1}\sum_{\nu=1}^{n} \ell^{\bullet\bullet}(x_\nu, \vartheta^{(n)}(\mathbf{x}_n))$$

works better in which examples.

(ii) Does iteration of the improvement procedure lead to asymptotically better estimator sequences? Perhaps, but this does not show up in an asymptotic analysis of first order: $\hat{\vartheta}^{(n)}$ defined by (5.5.9) is asymptotically optimal already.

(iii) How important is the influence of the preliminary estimators?

We forgo discussing such questions in more detail. In the following we take a closer look at the formal conditions required for the assertion that (5.5.9) implies (5.5.8). From (5.5.9) one obtains

$$n^{1/2}(\hat{\vartheta}^{(n)}(\mathbf{x}_n) - \vartheta) = n^{1/2}(\vartheta^{(n)}(\mathbf{x}_n) - \vartheta) + \Lambda^{(n)}(\mathbf{x}_n)n^{-1/2}\sum_{\nu=1}^{n} \ell^\bullet(x_\nu, \vartheta^{(n)}(\mathbf{x}_n))$$

Since $\Lambda^{(n)} \to \Lambda(\vartheta)$ (P_ϑ^n) and since $n^{-1/2}\sum_{\nu=1}^n \ell^\bullet(x_\nu, \vartheta)$ is stochastically bounded, relation (5.5.8) follows if

$$\begin{aligned}&n^{1/2}(\vartheta^{(n)}(\mathbf{x}_n) - \vartheta)\\&+\Lambda^{(n)}(\mathbf{x}_n)n^{-1/2}\sum_{\nu=1}^{n}\left(\ell^\bullet(x_\nu, \vartheta^{(n)}(\mathbf{x}_n)) - \ell^\bullet(x_\nu, \vartheta)\right) \to 0 \quad (P_\vartheta^n).\end{aligned}$$

Under conditions on the second derivative $\ell^{\bullet\bullet}$, as used, for instance, by Ibragimov and Has'minskii (1981) [p. 83, Theorem 8.1], one obtains (see (5.2.5)) that

$$n^{-1/2}\sum_{\nu=1}^{n}(\ell^\bullet(x_\nu, \vartheta^{(n)}(\mathbf{x}_n)) - \ell^\bullet(x_\nu, \vartheta)) = o(n^0, P_\vartheta^n),$$

provided $n^{1/2}(\vartheta^{(n)} - \vartheta)$ is stochastically bounded. Relation (5.5.9) follows under the usual condition that $\Lambda(\vartheta) = -L_{\bullet\bullet}(\vartheta)$.

Nonparametric Families

After asymptotic bounds for the concentration of estimator sequences in general families became available around 1975 (see Sect. 5.6), it was of topical interest if these bounds are attainable under general conditions.

In some cases, the bounds just confirmed that estimator sequences available in the literature are, in fact, asymptotically optimal (see e.g. Levit 1975, p. 732, Theorem 3.1).

The first general attempt at constructing asymptotically efficient estimator sequences in *nonparametric families* is due to Has'minskii and Ibragimov (1979). Their results refer to an arbitrary Gâteaux differentiable functional κ on a family $\mathfrak{P}$ of probability measures on $\mathbb{B}$ with smooth densities. The problem is to construct an estimator for κ with influence function K. They start from a preliminary estimator $P_{\mathbf{x}_{m_n}}$ of P, based on a subsample $\mathbf{x}_{m_n} = (x_1, \ldots, x_{m_n})$ of $\mathbf{x}_n = (x_1, \ldots, x_n)$. In relation (7), p. 44, they define the estimator

$$\kappa^{(n)}(\mathbf{x}_n) := \kappa(P_{\mathbf{x}_{m_n}}) + (n - m_n)^{-1} \sum_{\nu=m_n+1}^{n} K(x_\nu, P_{\mathbf{x}_{m_n}}). \tag{5.5.10}$$

Definition (5.5.10) is a version of an improvement procedure that uses *sample splitting*.

Semiparametric Families

Following the pioneering paper by Bickel (1982), there is now a large literature on the construction of estimator sequences for *semiparametric models*. The family of probability measures is now written as $\mathfrak{P} = \{P_{\vartheta,\eta} : \vartheta \in \Theta,\ \eta \in H\}$, with $\Theta \subset \mathbb{R}^k$ and H a topological space. The functional is $\kappa(P_{\vartheta,\eta}) = \vartheta$. To keep the following discussion transparent, we assume that $k = 1$. The problem is to construct an estimator sequence for the finite-dimensional parameter ϑ with influence function $K(\cdot, \vartheta, \eta)$.

The basic idea is the same as in the more general case of an nonparametric family $\mathfrak{P}$ and a differentiable functional κ on $\mathfrak{P}$, namely: To improve a given estimator sequence. For technical reasons, the details are slightly different. The starting point now is an estimator sequence $\vartheta^{(n)}$ of ϑ, which is $\sqrt{n}$-consistent in a locally uniform sense, i.e., $n^{1/2}(\vartheta^{(n)} - \vartheta_n)$ is stochastically bounded under $P^n_{\vartheta_n,\eta_n}$ whenever $P^n_{\vartheta_n,\eta_n}$ is stochastically bounded under $P^n_{\vartheta_n,\eta_n}$, whenever $P^n_{\vartheta_n,\eta_n}$ and $P^n_{\vartheta,\eta}$ are contiguous.

In spite of its more restricted character, semiparametric families comprise a plethora of special models, and the prospects for a general procedure working efficiently in each of these cases are not very favourable. Hence procedures working at the general level can hardly be more than existence theorems.

To illustrate the diversity of semiparametric models think of the problem of (i) estimating the center of symmetry for a family of probability measures on $\mathbb{B}$ with unknown symmetric densities, or (ii) of estimating the expectation of $A \to \int P_{\vartheta,\tau}(A)\Gamma(d\tau)$ where Γ is unknown. Dozens of further examples of special semiparametric models are presented in Bickel et al. (1993).

Let now $(\vartheta_n, \eta_n) \in \Theta \times H$ be such that $P^n_{\vartheta_n,\eta_n}$ and $P^n_{\vartheta,\eta}$ are mutually contiguous. If relation (5.5.1), applied for $\kappa(P_{\vartheta,\eta}) = \vartheta$ and $\kappa^{(n)} = \vartheta^{(n)}$, holds with $P_{\vartheta,\eta}$ and $P_{\vartheta_n,\eta}$, this implies

$$n^{1/2}(\vartheta_n - \vartheta) + n^{-1/2}\sum_{\nu=1}^{n}(K(x_\nu, \vartheta_n, \eta) - K(x_\nu, \vartheta, \eta)) = o(n^0, P^n_{\vartheta,\eta}). \quad (5.5.11)$$

If (5.5.1) holds with $P_{\vartheta_n,\eta}$ and P_{ϑ_n,η_n}, then

$$n^{-1/2}\sum_{\nu=1}^{n}(K(x_\nu, \vartheta_n, \eta_n) - K(x_\nu, \vartheta_n, \eta)) = o(n^0, P^n_{\vartheta,\eta}). \quad (5.5.12)$$

Both, (5.5.11) and (5.5.12), are necessary for the existence of estimator sequences admitting the influence function $K(\cdot, \vartheta, \eta)$ locally uniformly.

Klaassen (1987) was the first to consider the construction of asymptotically linear estimator sequences with an arbitrary influence function. He justifies condition (5.5.11) (corresponding to his condition (2.2), p. 1550) by the remark that "it often holds". He does not mention that it is necessarily true under the condition of local uniformity. Condition (5.5.11) occurs in Bickel et al. 1993, p. 395, (iv), as "smoothness condition" (!).

The paper of Schick (1986) is confined to $K(\cdot, \vartheta, \eta)$ being a canonical gradient. In this case, condition (5.5.11) follows from some kind of LAN-condition on the family (see p. 1142, relation (2.5)), an idea going back to Bickel (1982) [p. 670, relation (6.43)].

Condition (5.5.12) does not show up in the literature. Since this condition is necessary too, this requires an explanation. The results in Schick (1986), Klaassen (1987), Bickel et al. (1993) are based on the following condition (5.5.13) and (5.5.14): There exists an estimator sequence $(x_1, \ldots, x_m) \to \hat{K}_m(\cdot, \vartheta, x_1, \ldots, x_m)$ of the function $K(\cdot, \vartheta, \eta)$ with the following properties.

$$(x_1, \ldots, x_m) \to m^{1/2} \int \hat{K}_m(y, \vartheta_m, x_1, \ldots, x_m) P_{\vartheta_m,\eta}(dy) = o_p(m^0, P^m_{\vartheta,\eta}) \tag{5.5.13}$$

and

$$(x_1, \ldots, x_m) \to \int (\hat{K}_m(y, \vartheta_m, x_1, \ldots, x_m) - K(y, \vartheta_m, \eta))^2 P_{\vartheta_m,\eta}(dy) = o_p(m^0, P^m_{\vartheta,\eta}). \tag{5.5.14}$$

These conditions presume that $(P^m_{\vartheta_m,\eta})_{m\in\mathbb{N}}$ and $(P^m_{\vartheta,\eta})_{n\in\mathbb{N}}$ are mutually contiguous if $\vartheta_m - \vartheta = O(m^{-1/2})$. Conditions (5.5.13) and (5.5.14) imply that

$$(n - m_n)^{-1/2} \sum_{\nu=m_n+1}^{n} (\hat{K}_{m_n}(x_\nu, \vartheta_{m_n}, x_1, \ldots, x_{m_n}) - K(x_\nu, \vartheta_{m_n}, \eta)) = o_p(n^0, P^n_{\vartheta,\eta}) \tag{5.5.15}$$

for every sequence m_n such that $n - m_n \to \infty$ and $n^{-1}m_n$ is bounded away from 0.

If sequences $\hat{K}_m$ fulfilling (5.5.15) exist, then the necessary condition (5.5.12) is automatically fulfilled (use (5.5.15) with η and η_n, and replace $n - m_n$ by n).

That (5.5.15) follows from (5.5.13) and (5.5.14) can be seen as follows. Write

$$\begin{aligned}
&(n - m_n)^{-1/2} \sum_{\nu=m_n+1}^{n} (\hat{K}_{m_n}(x_\nu, \vartheta_{m_n}, x_1, \ldots, x_{m_n}) - K(x_\nu, \vartheta_{m_n}, \eta)) \\
&= (n - m_n)^{-1/2} \sum_{\nu=m_n+1}^{n} \Big(\hat{K}_{m_n}(x_\nu, \vartheta_{m_n}, x_1, \ldots, x_{m_n}) \\
&\quad - \int \hat{K}_{m_n}(y, \vartheta_{m_n}, x_1, \ldots, x_{m_n}) P_{\vartheta_{m_n},\eta}(dy) - K(x_\nu, \vartheta_{m_n}, \eta)\Big) \\
&\quad + (n - m_n)^{1/2} \int \hat{K}_{m_n}(y, \vartheta_{m_n}, x_1, \ldots, x_{m_n}) P_{\vartheta_{m_n},\eta}(dy).
\end{aligned} \tag{5.5.16}$$

The first term on the right-hand side of (5.5.16) converges stochastically to 0 because of (5.5.14), provided $n - m_n)$ tends to infinity. The second term converges to 0 because of (5.5.13) if $n^{-1}m_n$ is bounded away from 0. Both conditions are fulfilled if $m_n \sim n$.

Now we shall show how, under conditions (5.5.11) and (5.5.15), an estimator sequence with influence function $K(\cdot, \vartheta, \eta)$ can be constructed.

We first remark that relations (5.5.11) and (5.5.15) hold, in a modified form, with ϑ_m replaced by a $\sqrt{m}$-consistent estimator sequence $\vartheta^{(m)}$. This is true without any modifications if $(\vartheta^{(m)})_{n\in\mathbb{N}}$ is discretized. This was the choice in the papers by Bickel (1982) and Schick (1986). Alternatively, one might, following Klaassen (1987), use the splitting trick.

Let $1 < k_n < m_n < n$ be such that $n^{-1}k_n$ is bounded away from 0 and $n^{-1}m_n$ bounded away from 1. Relation (5.5.11) with $(x_1, \ldots, x_n)$ replaced by $(x_{m_n+1}, \ldots, x_n)$ yields

$$\begin{aligned}&(n-m_n)^{1/2}(\vartheta^{(m_n-k_n)}(x_{k_n+1},\ldots,x_{m_n})-\vartheta)\\&+(n-m_n)^{-1/2}\sum_{\nu=m_n+1}^{n}\big(K(x_\nu,\vartheta^{(m_n-k_n)}(x_{k_n+1},\ldots,x_{m_n}),\eta)-K(x_\nu,\vartheta,\eta)\big)\\&=o(n^0,P^n_{\vartheta,\eta}).\end{aligned}\tag{5.5.17}$$

Similarly, (5.5.16) implies

$$\begin{aligned}&(n-m_n)^{-1/2}\sum_{\nu=m_n+1}^{n}\big(\hat{K}_{k_n}(x_\nu,\vartheta^{(m_n-k_n)}(x_{k_n+1},\ldots,x_{m_n}),\mathbf{x}_{k_n})\\&\qquad\qquad-K(x_\nu,\vartheta^{(m_n-k_n)}(x_{k_n+1},\ldots,x_{m_n}),\eta)\big)\\&=o_p(n^0,P^n_{\vartheta,\eta}).\end{aligned}$$

Let now

$$\begin{aligned}\vartheta_1^{(n)}(\mathbf{x}_n)&:=\vartheta^{(m_n-k_n)}(x_{k_n+1},\ldots,x_{m_n})\\&+(n-m_n)^{-1}\sum_{\nu=m_n+1}^{n}\hat{K}_{k_n}(x_\nu,\vartheta^{(m_n-k_n)}(x_{k_n+1},\ldots,x_{m_n}),\mathbf{x}_{k_n}).\end{aligned}$$

We shall show that

$$\begin{aligned}&(n-m_n)^{1/2}(\vartheta_1^{(n)}(\mathbf{x}_n)-\vartheta)\\&=(n-m_n)^{-1/2}\sum_{\nu=m_n+1}^{n}K(x_\nu,\vartheta,\eta)+o(n^0,P^n_{\vartheta,\eta}).\end{aligned}\tag{5.5.18}$$

This can be seen as follows.

$$\begin{aligned}&(n-m_n)^{1/2}(\vartheta_1^{(n)}(\mathbf{x}_n)-\vartheta)\\&=(n-m_n)^{1/2}(\vartheta^{(m_n-k_n)}(x_{k_n+1},\ldots,x_{m_n})-\vartheta)\end{aligned}$$

$$+(n-m_n)^{-1/2}\sum_{\nu=m_n+1}^{n}\hat{K}_{k_n}(x_\nu,\vartheta^{(m_n-k_n)}(x_{k_n+1},\ldots,x_{m_n}),\mathbf{x}_{k_n})$$

$$=(n-m_n)^{-1/2}\sum_{\nu=m_n+1}^{n}K(x_\nu,\vartheta,\eta)$$

$$+(n-m_n)^{-1/2}\sum_{\nu=m_n+1}^{n}\big(\hat{K}_{k_n}(x_\nu,\vartheta^{(m_n-k_n)}(x_{k_n+1},\ldots,x_{m_n}),\mathbf{x}_{k_n})$$

$$-K(x_\nu,\vartheta^{(m_n-k_n)}(x_{k_n+1},\ldots,x_{m_n}),\eta)\big)+o(n^0,P^n_{\vartheta,\eta})$$

$$=(n-m_n)^{-1/2}\sum_{\nu=m_n+1}^{n}K(x_\nu,\vartheta,\eta)+o(n^0,P^n_{\vartheta,\eta}).$$

For an estimator sequence $\vartheta_2^{(n)}$ constructed in the same way as $\vartheta_1^{(n)}$, but with the roles of $x_1,\ldots,x_{m_n}$ and $x_{m_n+1},\ldots,x_n$ interchanged, we obtain in analogy to (5.5.18):

$$m_n^{1/2}(\vartheta_2^{(n)}(\mathbf{x}_n)-\vartheta)=m_n^{-1/2}\sum_{\nu=1}^{m_n}K(x_\nu,\vartheta,\eta)+o(n^0,P^n_{\vartheta,\eta}).$$

Hence

$$\hat{\vartheta}^{(n)}:=(1-n^{-1}m_n)\vartheta_1^{(n)}+n^{-1}m_n\vartheta_2^{(n)} \tag{5.5.19}$$

is asymptotically linear with influence function $K(\cdot,\vartheta,\eta)$.

This is a simplified version of the construction proposed by Klaassen (1987) [p. 1552, Theorem 2.1] and adopted by Bickel et al. (1993) [p. 395, Theorem 7.8.1].

Conditions (5.5.13) and (5.5.14) underlying the construction appear strong and artificial. It is an interesting aspect of Klaassen's paper that—under a uniform integrability condition on $K(\cdot,\vartheta,\eta)^2$—just these conditions are necessary (Klaassen 1987, p. 1553, Theorem 3.1, and Bickel et al. 1993, pp. 396/7, Theorem 7.8.2).

Remark The constructions presented above use separate estimators of ϑ and η, as in

$$\hat{K}_{k_n}(\cdot,\vartheta^{(m_n-k_n)}(x_{k_n+1},\ldots,x_{m_n}),x_1,\ldots,x_{k_n}).$$

In practical case, ϑ and η will be estimated simultaneously. Example: If ϑ is the center of a symmetric density, a preliminary estimator of ϑ will enter the estimator of the density based on $x_1,\ldots,x_{k_n}$. The theory requires this preliminary estimate to be based on $x_1,\ldots,x_{k_n}$ only, hence different from $\vartheta^{(m_n-k_n)}(x_{k_n+1},\ldots,x_{m_n})$. (It appears that this has been neglected in Bickel et al. 1993, pp. 400/1, Example 7.8.1.)

Remark As the result of a rather artificial construction, the estimate $\hat{\vartheta}^{(n)}(x_1,\ldots,x_n)$ defined by (5.5.19) will not be invariant under permutations of $(x_1,\ldots,x_n)$. However, the median of all permutations $\vartheta^{(n)}(x_{i_1},\ldots,x_{i_n})$ is permutation invariant and asymptotically linear with the same influence function (see Pfanzagl 1990, p. 7, Lemma 3.2).

The idea to construct an asymptotically optimal estimator sequence for semiparametric models by the improvement procedure was developed by Bickel (1982), independently of the paper by Has'minskii and Ibragimov (1979). (See p. 670, acknowledgement.) His construction is based on condition (5.5.14) (see p. 653, condition 3.5) and on the condition

$$\int \hat{K}_m(\xi, \vartheta_m, x_1, \ldots, x_m) P_{\vartheta_m,\eta}(d\xi) = 0,$$

a rather stringent version of (5.5.13). (To the reader who has trouble in finding this condition in Bickel's paper: It is part of "condition H" on p. 653.)

5.6 Functionals on General Families

Let $(X, \mathscr{A})$ be a measurable space, $\mathfrak{P}$ a family of probability measures $P|\mathscr{A}$, and $\kappa : \mathfrak{P} \to \mathbb{R}^k$ the functional to be estimated. The problem is to find an operationally meaningful concept of asymptotic optimality for estimator sequences. Of course, we encounter the same problems which were present already in the case of a parametric family $\mathfrak{P}$: that one has to find a suitable concept of local uniformity, and of an intrinsic bound for the asymptotic concentration of locally uniformly convergent estimator sequences.

If $\mathfrak{P}$ is an arbitrary family and $\kappa : \mathfrak{P} \to \mathbb{R}$ an arbitrary functional, one has no immediate standard for judging the quality of an estimator. Is the sample quantile asymptotically efficient if $\mathfrak{P}$ consists of all probability measures on $\mathbb{B}$ with continuous density? What is the asymptotic efficiency of the sample quantile if every density in $\mathfrak{P}$ is symmetric?

As another example, consider the problem of estimating

$$\kappa(P) := \int \Psi(x_1, x_2) P(dx_1) P(dx_2)$$

with $\Psi(x_1, x_2) = \Psi(x_2, x_1)$ and $\int \Psi(x_1, x_2)^2 P(dx_1) P(dx_2) < \infty$. A reasonable estimator is

$$(x_1, \ldots, x_n) \to n^{-2} \sum_{\nu=1}^{n} \sum_{\mu=1}^{n} \Psi(x_\nu, x_\mu).$$

But is this estimator asymptotically optimal? If it is known that P has a continuous λ-density, can one achieve an asymptotically better estimator sequence by estimating first the density of P by $p^{(n)}(\cdot, \mathbf{x}_n)$ and computing subsequently the estimator

$$\mathbf{x}_n \to \int \Psi(\xi_1, \xi_2) p^{(n)}(\xi_1, \mathbf{x}_n) p^{(n)}(\xi_2, \mathbf{x}_n) d\xi_1 d\xi_2.$$

Stein's Approach

Stein was the first scholar to strive for an answer to such questions. His paper (1956) is generally considered as pioneering for the problems of nonparametric bounds. Regrettably, Stein wrote his paper several years too early. He cites Le Cam (1953), but he ignores the problem of "superefficiency". Without a solid concept for the optimality in parametric families, he takes ML-sequences as asymptotically efficient straightaway, without entering the discussion about the asymptotic optimality of statistical procedures. Stein never came back to this problem when the adequate techniques were available. Recall that the idea to base the concept of asymptotic optimality on local uniformity as in Rao (1963) and Bahadur (1964) had not yet evolved, and that the optimality of ML-sequences based on such a solid concept of asymptotic optimality occurred first in Wolfowitz (1965) and Kaufman (1966).

The main idea in Stein's paper is to obtain the asymptotic bound for estimator sequences of a functional $\kappa : \mathfrak{P} \to \mathbb{R}$ at $P_0 \in \mathfrak{P}$ from the bound for estimator sequences in the "least favourable" parametric subfamily passing through P_0, say $\mathfrak{P}_0$. Assuming that ML-sequences are asymptotically optimal, the question remains which parametric subfamily is the least favourable one. With a solid concept of asymptotic optimality at our disposal, this is clear: $\mathfrak{P}_0$ is least favourable if the best possible estimator sequence which is regular in $\mathfrak{P}_0$ is also regular in $\mathfrak{P}$, or conversely: If an estimator sequence which is regular and optimal in $\mathfrak{P}$ is optimal in $\mathfrak{P}_0$. Without a precise concept of asymptotic optimality, Stein's argument (p. 187) remains necessarily vague.

> Clearly [!] a nonparametric problem is at least as difficult as any of the parametric problems

The reasoning underlying Stein's idea contains certain weak points: (i) He assumes that for parametric families the ML-sequence is asymptotically optimal. (ii) Moreover, it makes no sense to speak of the quality of an estimator sequence of $\kappa(P)$ at $P = P_0$ without reference to the class of competing estimator sequences. Hence it makes no sense to say what one would like to say, namely: That an estimator sequence is optimal on some family $\mathfrak{P}$ if it is optimal on some subfamily $\mathfrak{P}_0 \subset \mathfrak{P}$.

To illustrate Stein's approach, we consider the problem of estimating the quantile of an unknown symmetric distribution.

Example Let $\mathfrak{P}$ be the family of all symmetric distributions on $\mathbb{B}$ with a differentiable Lebesgue density p. The problem is to estimate the β-quantile $\kappa_\beta(P)$, defined by $P(-\infty, \kappa_\beta(P)] = \beta$. W.l.g. we assume $\beta \geq 1/2$. The β-quantile $\kappa^{(n)}$ of the sample is certainly a reasonable estimator; its limit distribution is normal with mean 0 and variance $\beta(1-\beta)/p(\kappa_\beta(P))^2$. Can the symmetry of the densities be utilized to obtain a better estimator? According to Stein's program, one has to find a parametric subfamily in which the estimation of the β-quantile is particularly difficult. Let $P_0 \in \mathfrak{P}$ be fixed. As starting point we consider the parametric family $\mathfrak{Q} := \{Q_\vartheta : \vartheta \in (-\varepsilon, \varepsilon)\}$ with Lebesgue density

$$q(x, \vartheta) = p_0(x-\vartheta)(1 + \vartheta\psi(x-\vartheta)), \quad p_0 \text{ symmetric about } 0. \tag{5.6.1}$$

The family $\{Q_\vartheta : \vartheta \in \mathbb{R}\}$ passes through P_0 for $\vartheta = 0$. The signed measure Q_ϑ is a probability measure (for ϑ sufficiently small) if

$$\inf_{x\in\mathbb{R}} \psi(x) > -\infty \tag{5.6.2}$$

and

$$\int \psi(x)p_0(x)dx = 0. \tag{5.6.3}$$

Since $Q_{n^{-1/2}a}$, $n \in \mathbb{N}$, is a sequence converging to P_0, condition (5.6.1) implies that Q_0 is a limit of $Q_{n^{-1/2}a} \circ n^{-1/2}(\kappa^{(n)} - \kappa_\beta(Q_{n^{-1/2}a}))$, $n \in \mathbb{N}$. In other words: The convergence of $\kappa^{(n)}$ is regular within the parametric family $\mathfrak{Q}$ in the sense of Sect. 5.8.

Q_ϑ is symmetric about ϑ if ψ is symmetric about 0. The aim is to choose ψ such that the estimation of the β-quantile in the family $\{Q_\vartheta : \vartheta \in (-\varepsilon, \varepsilon)\}$ becomes as difficult as possible. For the following notice that $\int_{-\infty}^{+\infty} \psi dP_0 = 0$ implies $\int_0^\infty \psi dP_0 = 0$ (since ψ and p_0 are symmetric). To simplify our notations, we write $q_\beta = \kappa_\beta(P_0)$. It is easy to check that

$$K(0) := \partial_\vartheta \kappa_\beta(Q_\vartheta)|_{\vartheta=0} = 1 - \int \Psi 1_{(0,q_\beta)} dP_0 \Big/ p_0(q_\beta).$$

Moreover, under suitable regularity conditions on ψ,

$$\partial\vartheta \log q(x,\vartheta)|_{\vartheta=0} = -\ell_0'(x) + \psi(x), \tag{5.6.4}$$

so that

$$L(Q_0) := \int (-\ell_0' + \psi)^2 dP_0 = \int (\ell_0')^2 dP_0 + \int \psi^2 dP_0.$$

Hence the intrinsic bound for the asymptotic variance of estimator sequences of the β-quantile in the family (5.6.1) is

$$\Big(1 - \int \Psi 1_{(0,q_\beta)} dP_0 \Big/ p_0(q_\beta)\Big)^2 \Big/ \Big(\int (\ell_0')^2 dP_0) + \int \psi^2 dP_0\Big). \tag{5.6.5}$$

The problem is to choose ψ symmetric and subject to the conditions (5.6.1) such that (5.6.5) is as large as possible. The solution of this task becomes more transparent if we represent $\psi = \lambda\psi_0$, with $\int \psi_0^2 dP_0 = 1$. Given ψ_0,

$$\Big(1 - \lambda \int \psi_0 1_{(0,q_\beta)} dP_0 \Big/ p_0(q_\beta)\Big)^2 \Big/ \Big(\int (\ell_0')^2 dP_0) + \lambda^2\Big)$$

attains its maximal value

$$\Big(\int (\ell_0')^2 dP_0\Big)^{-1} + \Big(\int \psi_0 1_{(0,q_\beta)} dP_0\Big)^2 \Big/ p_0(q_\beta)^2$$

for $\lambda = -\left(\int (\ell_0')^2 dP_0\right)^{-1} \int \psi_0 1_{(0,q_\beta)} dP_0 / p_0(q_\beta)$.

The problem now is to choose ψ_0, symmetric about 0, with $\int \psi_0 dP_0 = 0$ and $\int \psi_0^2 dP_0 = 1$ such that $(\int \psi_0 1_{(0,q_\beta)} dP_0)^2$ becomes as large as possible. The solution to this problem is (recall that $P_0(-\infty, q_\beta) = \beta$)

$$\psi_0(x) = (2(2\beta - 1)1 - \beta))^{-1/2}(1_{(0,q_\beta)}(|x|) - (2\beta - 1). \tag{5.6.6}$$

The least favourable subfamily is of the type (5.6.1) with Ψ given by (5.6.6). This leads to the variance bound

$$\left(\int (\ell_0')^2 dP_0\right)^{-1} + (\beta - 1/2)(1 - \beta)/p_0(q_\beta)^2. \tag{5.6.7}$$

Endowed with the concepts of "tangent space" and "gradient" introduced below, we now complete this example of bounds for the β-quantile of a symmetric distribution on $\mathbb{B}$. According to (5.6.4), the tangent cone of this family contains all functions $x \to -a\ell_0'(x) + \psi(x)$, where p_0 is any density symmetric about 0, and ψ a (differentiable) function fulfilling (5.6.2) and (5.6.3), which is symmetric about 0.A gradient of the β-quantile at P_0 is

$$\kappa_\beta^\bullet(x, P_0) = (\beta - 1_{(-\infty,q_\beta)}(x)) \Big/ p_0(q_\beta),$$

the canonical gradient is

$$\kappa_\beta^*(x, P_0) = -L(P_0)^{-1}\ell_0'(x) + \big(2\beta - 1 - 1_{(q(1-\beta),q(\beta))}(x)\big) \Big/ 2p_0(q_\beta).$$

The asymptotic variance bound $\int \kappa_\beta^*(x, P_0)^2 dP_0$ coincides with (5.6.7).

Remark With ψ_0 defined in (5.6.6), the probability measure Q_ϑ defined in (5.6.1) is not in $\mathfrak{P}$. Hence one needs to modify ψ_0 slightly in such a way that the density $q(\cdot, \vartheta)$ becomes differentiable, without disturbing the relevant local properties of Q_ϑ at $\vartheta = 0$.

The term (5.6.7) is an intrinsic bound obtained from one-parameter families of the particular type (5.6.1), and one cannot be sure that this is the largest bound resulting from one-parameter families passing through P_0, nor can one be sure that the largest bound obtained from all one-parameter subfamilies is attainable at P_0, locally uniformly in the full family $\mathfrak{P}$. All doubts whether (5.6.7) is the bound we were looking for can be removed by presenting a regular estimator sequence attaining this bound. □

This example illustrates what Stein could have done to demonstrate the power of his method: Look for a one-parameter subfamily in which the estimation of the functional is "most difficult", and show that the intrinsic bound for this subfamily is attainable in the full family $\mathfrak{P}$. What tein did is much less, and one can hardly refute Millar's verdict (1983, p. 261) that Stein's paper consists of "some obscure remarks".

This impression is due to the fact that Stein mingles the problems of "nonparametric bounds" and "adaptivity".

Most authors refer to Stein (1956) as a pioneering paper for the concept of nonparametric bounds. It is, in fact, a paper on adaptivity. Readers who have difficulties to understand Stein's enigmatic "algebraic lemma" (p. 189) might consult Bickel (1982) [p. 651, condition S] or Fabian and Hannan (1982) [p. 474, condition 8]. These authors offer simplified and more transparent versions of what Stein intended to express in his "algebraic lemma": That a certain orthogonality relation is necessary for the existence of "adaptive" procedures.

The examples in Stein's paper are all of the "adaptive" type, and in such cases it is superfluous to search for "least favourable parametric subfamilies": The intrinsic nonparametric bound is the intrinsic bound for the submodel where the nonparametric component is known.

The most interesting among Stein's examples is the estimation of the median of an unknown symmetric distribution. The estimation of an arbitrary β-quantile would have offered the possibility to show how the method of least favourable one-parameter subfamilies can be used to obtain a nonparametric bound. The restriction to the adaptive case $\beta = 1/2$ in Stein's paper deprives his example of its substance.

Tangent Sets

Shortly after Hájek's paper (1972) on bounds for the concentration of estimator sequences in parametric families, the problem of intrinsic bounds for functionals on general families was taken up in a series of papers by Levit (1974, 1975) and Koshevnik and Levit (1976). In these papers the ideas of Stein (1956) were transformed into a serviceable technique for determining intrinsic bounds, based on the concepts "tangent space" and "gradient".

To keep the following considerations transparent, we assume that the probability measures $P \in \mathfrak{P}$ are mutually absolutely continuous. We consider paths $P_t \in \mathfrak{P}$ for $t \in A = (-\varepsilon, \varepsilon)$ approximating P in the following sense: The density p_t of P_t can be represented as

$$t^{-1}(p_t/p - 1) = g + r_t \tag{5.6.8}$$

with $g \in L_*(P)$ and r_t "DCC-differentiable" as defined below.

The *tangent set* $T(P, \mathfrak{P})$ is the set of all $g \in L_*(P)$ occurring in a representation (5.6.8). By definition, $T(P, \mathfrak{P})$ is a cone. Applications sometimes require more properties of $T(P, \mathfrak{P})$. By definition, $T(P, \mathfrak{P})$ is closed under delations, i.e. $g \in T(P, \mathfrak{P})$ implies $tg \in T(P, \mathfrak{P})$ for every $t \in A$, neighbourhood of o. Throughout the following we assume that $T(P, \mathfrak{P})$ is linear and $\| \ \|_2$-closed.

Warning: Expressions for the (co-)variance bounds in the papers by Levit (1974) [p. 333] and Koshevnik and Levit (1976) [p. 744, Theorem 1] appear with the factor $1/4$. This is a consequence of the deviating definition of a Hellinger derivative g by

$$\lim_{t \to 0} \int \left(t^{-1}((p_t/p)^{1/2} - 1) - g\right)^2 dP = 0$$

rather than the natural (and now common) definition

$$\lim_{t\to 0}\int \Big(t^{-1}((p_t/p)^{1/2}-1)-\frac{1}{2}g\Big)^2 dP = 0.$$

Given a functional $\kappa : \mathfrak{P} \to \mathbb{R}$, Levit (1974) [p. 332] assumes a particular parametrization, namely $\kappa(P_{\vartheta_0+t}) = \kappa(P_{\vartheta_0}) + t$, so that $\kappa^{(\cdot)}(\vartheta_0) = 1$. In this case, the usual variance bound $1\Big/\int \ell^\bullet(\cdot,\vartheta_0)^2 dP_{\vartheta_0}$ changes to $\sigma^2(\vartheta_0) = 1\Big/\int g(\cdot,\vartheta_0)^2 dP_{\vartheta_0}$, and the least favourable parametric subfamily, leading to the largest asymptotic variance, is the one which minimizes $\int g(\cdot,\vartheta_0)^2 dP_0$, subject to the condition $\kappa(P_{\vartheta_0+t}) = \kappa(P_{\vartheta_0}) + t$.

We call the path p_t, $t \in A$, *DCC-differentiable* (at $t = 0$) with *derivative* g if $g \in L_*(P)$ and

$$p_t/p = 1 + tg + r_t \tag{5.6.9}$$

with r_t, $t \in A$, fulfilling the *degenerate convergence conditions* (DCC):

$$P\{|r_t| > \varepsilon t^{-1}\} = o(t^2) \quad \text{for every } \varepsilon > 0, \tag{5.6.10}$$

$$\int r_t 1_{\{|r_t|\le t^{-1}\}} dP = o(t), \tag{5.6.11}$$

$$\int r_t^2 1_{\{|r_t|\le t^{-1}\}} dP = o(t^o) \tag{5.6.12}$$

Theorem 5.6.1 *If p_t, $t \in A$, is DCC-differentiable with derivative g, then*

$$\sum_{\nu=1}^n \log(p_{n^{-1/2}a}(x_\nu)/p(x_\nu)) = an^{-1/2}\sum_{\nu=1}^n g(x_\nu) - \frac{1}{2}a^2\int g^2 dP_0 \quad (P_0^n). \tag{5.6.13}$$

In other words, (5.6.10)–(5.6.12) imply an LAN-condition. (In fact, $\int r_t^2 dP \to 0$ suffices.) For a proof see (Pfanzagl 1985, Proposition 1.2.7, p. 22 ff).

For parametric families $P_\vartheta : \vartheta \in \Theta$ with $\Theta \subset \mathbb{R}^k$ we set $P_{\vartheta+ta}$ in place of P_t and obtain

$$dP_{\vartheta+ta}/dP_\vartheta = 1 + ta^\top \ell^\bullet(\cdot,\vartheta) + tr_t.$$

Then

$$\sum_{\nu=1}^n \log(p(x_\nu,\vartheta+n^{-1/2}at)/p(x_\nu,\vartheta))$$
$$= n^{-1/2}ta^\top\sum_{\nu=1}^n \ell^\bullet(x_\nu) - \frac{1}{2}t^2a^\top\int \ell^\bullet(\cdot,\vartheta)\ell^\bullet(\cdot,\vartheta)^\top dP_\vartheta a \quad (P_\vartheta^n).$$

A representation (5.6.9) of the densities is natural from the intuitive point of view. Technically useful for the determination of the concentration bound is the LAN-condition (5.6.13). That (5.6.9) is almost necessary for (5.6.13) was suggested by Le Cam (1984, 1985). See Pfanzagl (1985) [p. 22, Proposition 1.2.7] for details. Let $H : \mathbb{R} \to \mathbb{R}$ be a function with continuous 2nd derivative in a neighbourhood of 0 such that

$$H(0) = 0 \quad \text{and} \quad H'(0) = 1. \tag{5.6.14}$$

We shall use the following

Lemma 5.6.2 *If*

$$h_t = tg + t^2 K + t r_t \tag{5.6.15}$$

with K a constant and r_t fulfilling (5.6.10)–(5.6.12), then

$$H(h_t) = tg + t^2 (K + \frac{1}{2} H''(0)\sigma^2) + t s_t \tag{5.6.16}$$

with s_t fulfilling (5.6.10)–(5.6.12).

This is a special case of Lemma 1.3.4 in Pfanzagl (1985) [pp. 30/1], applied with $a = b = 0$. See also the remark on p. 21 concerning the condition on H''. If G is the inverse of H in a neighbourhood of 0, then the conditions (5.6.14) on H imply $G(0) = 0$, $G'(0) = 1$ and $G''(0) = -H''(0)$. Hence, (5.6.16) and (5.6.15) are equivalent.

If relation (5.6.9) holds true, relation (5.6.15) is fulfilled with

$$h_t = p_t/p - 1 \quad \text{and} \quad K = 0.$$

Applied with $H(u) = \log(1 + u)$, relation (5.6.16) asserts that

$$\log p_t/p = H(h_t) = tg + t^2 \frac{1}{2} H''(0)\sigma^2 + t r_t.$$

Since $H''(0) = -1$, this is

$$\log p_t/p = tg - \frac{t^2}{2}\sigma^2 + t r_t. \tag{5.6.17}$$

To establish LAN for i.i.d. products one may use relation (5.6.17). Since

$$n^{-1/2} \sum_{\nu=1}^{n} r_{n^{-1/2}a}(x_\nu) \to 0$$

by the Degenerate Convergence Theorem, relation (5.6.13) follows.

Usually, the LAN-representation (5.6.13) is obtained from the assumption that $t \to p_t/p$ is Hellinger differentiable, i.e., that there exists a representation

$$(p_t/p)^{1/2} = 1 + \frac{t}{2}g + ts_t, \quad \text{with} \int s_t^2 dP \to 0. \tag{5.6.18}$$

We shall show that (5.6.18) is equivalent to (5.6.9) with $(r_t)_{t\in A}$ fulfilling conditions (5.6.10)–(5.6.12). Notice that (5.6.18) implies $\int g dP = 0$ and $\int g^2 dP < \infty$. According to Lemma 1.2.17 in Pfanzagl (1985) [p. 27], relation (5.6.18) is equivalent to the condition that $r_t := s_t + \frac{1}{8}t\sigma^2$ fulfills conditions (5.6.10)–(5.6.12). Hence

$$(p_t/p)^{1/2} = 1 + \frac{t}{2}g - \frac{1}{8}t^2\sigma^2 + tr_t \tag{5.6.19}$$

is equivalent to Hellinger differentiability of $t \to p_t/p$ with derivative g. As a consequence of (5.6.19), relation (5.6.15) is fulfilled with

$$h_t = 2((p_t/p)^{1/2} - 1) \quad \text{and} \quad K = -\frac{1}{4}\sigma^2.$$

Applied with $H(u) = u(1 + u/4)$, relation (5.6.16) asserts that

$$p_t/p - 1 = H(h_t) = tg + t^2\left(-\frac{1}{4}\sigma^2 + \frac{1}{2}H''(0)\sigma^2\right) + tr_t.$$

Since $H''(0) = 1/2$, this is relation (5.6.9). Hence conditions (5.6.9), (5.6.17), (5.6.18) and (5.6.19) are equivalent.

If condition (5.6.17) is the technically most convenient one, it is not so easy to justify from the intuitive point of view for a given family $\{P_\vartheta : \vartheta \in \Theta\}$. Pfanzagl (1982) is based on the more transparent condition (5.6.9) (see p. 23). Now generally accepted is condition (5.6.18), i.e. Hellinger differentiability. What makes this condition attractive is its provenance from a classical concept of differentiability. But is this really a strong argument if the underlying distance function is rather artificial? One reason for the general acceptance of Hellinger differentiability seems to be the prestige of Le Cam who started its use in (1969) (see e.g. p. 94, Théorème).

Gradients

The functional $\kappa : \mathfrak{P} \to \mathbb{R}$ is *differentiable* at P if there exists $\kappa^\bullet \in L_*(P)$ such that

$$t^{-1}(\kappa(P_t) - \kappa(P)) \to \int \kappa^\bullet g dP \tag{5.6.20}$$

for every $g \in T(P, \mathfrak{P})$ and every DCC-differentiable path P_t in $\mathfrak{P}$ with derivative g.

Any function $\kappa^\bullet \in L_*(P)$ fulfilling (5.6.20) is a *gradient* of the functional κ. A gradient in $T(P, \mathfrak{P})$ is unique. It is distinguished by the symbol κ^* and the name *canonical gradient*. It can be obtained as the projection of any gradient onto $T(P, \mathfrak{P})$.

The tangent space describes the local properties of the family $\mathfrak{P}$; the gradient describes the local properties of the functional κ. The Convolution Theorem (see Sect. 5.13) shows that the canonical gradient of the functional κ determines the bound for the concentration of regular estimator sequences.

Differentiability of the functional κ seems to be a natural condition for the existence of a reasonable estimator sequence. In fact, Bickel et al. (1993) [p. 183, Theorem 5.2.3] claim the "equivalence of regularity and differentiability", a result going back to van der Vaart (1988). This claim is based on the joint convergence of $(n^{1/2}(\kappa^{(n)} - \kappa(P)), n^{-1/2}\sum_{\nu=1}^{n} g(x_\nu))$ for any g in the tangent set. Without this popular but not operational assumption this is not true any more: Pfanzagl (2002a) [pp. 266–268] presents a one-parameter family and a non-differentiable functional admitting a regular estimator sequence with continuous limit distribution.

For parametric families $\{P_\vartheta : \vartheta \in \Theta\}$, $\Theta \subset \mathbb{R}^k$, it is natural to consider κ as a functional on Θ instead of a functional defined on the family of probability measures $\{P_\vartheta : \vartheta \in \Theta\}$ with $K(\vartheta) := \kappa(P_\vartheta)$. The canonical gradient of the functional κ may now be written as

$$\kappa^* = K^\bullet \Lambda(\vartheta)\ell^\bullet(\cdot, \vartheta)$$

with $K^\bullet$ the gradient K at ϑ.

With the introduction of "tangent cone" and "gradient" in Koshevnik and Levit (1976) [p. 742], the least favourable subfamily for a real-valued functional is expressed through the canonical gradient of κ. The tangent cone contains all information about the local structure of the family $\mathfrak{P}$ at P which is needed to obtain the intrinsic bound. Koshevnik and Levit (as well as Has'minskii and Ibragimov (1979), confine themselves to expressing the intrinsic bounds by Minimax Theorems, following Hájek (1972). They could have used the Convolution Theorem as well.

Levit (1974) [Sects. 2 and 3] illustrates the use of this approach by application to the functional

$$\kappa(P) := \int \Psi(x_1, \dots, x_m)P(dx_1)\dots P(dx_m),$$

a functional investigated by von Mises in (1947) and in some earlier papers. After some less satisfactory results of von Mises on the distribution of estimators of κ (see Filippova, 1962, p. 24, for critical remarks), the following result was obtained by Hoeffding (1948) in a paper "which was essentially completed before the paper by von Mises (1947) was published" (see p. 306). Specialized for one-dimensional functionals, Hoeffding's Theorem 7.4, p. 309, proves that the estimator

$$\kappa^{(n)}(x_1, \dots, x_n) := \binom{n}{m}^{-1} \Sigma\Psi(x_{i_1}, \dots, x_{i_m}), \tag{5.6.21}$$

with the summation extending over all $1 \le i_1 < \dots < i_m \le n$, is asymptotically distributed as $N(0, m^2\sigma^2(P))$, where

$$\sigma^2(P) := \int (\Psi_1(x, P) - \kappa(P))^2 P(dx),$$

with

$$\Psi_1(x, P) := \int \Psi(x, x_2, \dots, x_m) P(dx_2) \dots P(dx_m)$$

if Ψ is, w.l.g., assumed to be permutation-invariant. Levit (1974) [proof of Theorem 2.2] gives conditions which imply that $m(\Psi_1(\cdot, P) - \kappa(P))$ is the canonical gradient of κ, hence $m^2\sigma^2(P)$ the intrinsic variance bound. This proves that the estimator sequence defined by (5.6.21) is asymptotically optimal.

Le Cam's Lemmas

Le Cam's Theorem 2.1 (1960, p. 40) contains under point (6) the following assertion.

*Let P_n, P_n' be probability measures on $(X_n, \mathscr{A}_n)$ which are mutually contiguous. For any sequence $h_n : X_n \to \mathbb{R}$, the convergence of $P_n \circ (\log(dP_n'/dP_n), h_n)$ to a limit distribution M implies the convergence of $P_n' \circ (\log(dP_n'/dP_n), h_n)$ to the limit distribution with M-*density $(u, v) \to \exp[u]$.

By this result, assertions about the asymptotic performance of $(h_n)_{n\in\mathbb{N}}$ under $(P_n')_{n\in\mathbb{N}}$ can be obtained from assertions about the asymptotic performance of $(h_n)_{n\in\mathbb{N}}$ under $(P_n)_{n\in\mathbb{N}}$. It implies in particular the equivalence of

$$P_\vartheta^{(n)} \circ \log(dP^{(n)}_{\vartheta+c_n^{-1}a}/dP_\vartheta^{(n)}) \Rightarrow N\Big(-\frac{1}{2}a^\top L(\vartheta)a, a^\top L(\vartheta)a\Big)$$

and

$$P^{(n)}_{\vartheta+c_n^{-1}a} \circ \log(dP^{(n)}_{\vartheta+c_n^{-1}a}/dP_\vartheta^{(n)}) \Rightarrow N\Big(\frac{1}{2}a^\top L(\vartheta)a, a^\top(\vartheta)a\Big),$$

which is essential for certain asymptotic results on the concentration of estimator sequences obtained from the Neyman–Pearson Lemma (see Sect. 5.11).

Le Cam's beautiful Lemma inspired Hájek and Šidák (1967) [Sects. VI.1.2–VI.1.4] to introduce what they called Le Cam's 1st, 2nd and 3rd Lemma, an adaptation useful in their particular framework, and it is this version in which Le Cam's Lemma now usually occurs in the literature. Le Cam's original Lemma was extended to h_n with values in a metric space by Bickel et al. (1993) [p. 480, Lemma A.8.6]. Witting and Müller-Funk (1995) present the "three Lemmas", and two versions of Le Cam's original Lemma on p. 326 as Satz 6.138. See also Bening (2000) [Sect. A, pp. 149–156].

Since Le Cam's proof is opaque and the proof in Witting and Müller-Funk (with 5 references to auxiliary results) rather technical, we present the following lemma which shows the basic idea.

Lemma 5.6.3 *Let $(Y, \mathscr{B})$ be a topological space, endowed with its Borel algebra, and $Q_n|\mathscr{B}$, $n \in \mathbb{N}$, a sequence of probability measures converging weakly to* *a probability measure $Q|\mathscr{B}$. Let $q : Y \to [0, \infty)$ be a continuous function fulfilling $\int q dQ = 1$.*

If every $Q'_n|\mathscr{B}$ has Q_n-density q, then Q'_n, $n \in \mathbb{N}$, converges weakly to the probability measure Q' with Q-density q.

Proof By assumption, $\int f(v)Q_n(dv) \to \int f(v)Q(dv))$ for every bounded and continuous function $f : Y \to \mathbb{R}$. The assertion is that

$$\int f(v)q(v)Q_n(dv) \to \int f(v)q(v)Q(dv).$$

If q is bounded, $y \to f(v)q(v)$ is bounded and continuous; hence the assertion is proved. Since $\int q(v)Q_n(dv) = 1$ for $n \in \mathbb{N}$, and $\int q(v)Q(dv) = 1$, the assertion holds for q unbounded. (Hint: approach q by $0 \le q_\varepsilon \le q$ such that $\int (q - q_\varepsilon)dQ \le \varepsilon$ and $\int (q - q_\varepsilon)dQ_n < \varepsilon$ for $n \ge n_\varepsilon$.) □

Corollary 5.6.4 *Let P_n and P'_n, $n \in \mathbb{N}$, be probability measures on $(X, \mathscr{A})$, with $p_n \in dP'_n/dP_n$. Let $g_n : X \to \mathbb{R}^m$, $n \in \mathbb{N}$, be such that $P_n \circ (g_n, p_n) \Rightarrow Q|\mathbb{B}^m \times \mathbb{B}_+$. If $\int vQ(d(u, v)) = 1$, then $P'_n \circ (g_n, p_n)$, $n \in \mathbb{N}$, converges weakly to the probability measure Q' with Q-density $(u, v) \to v$.*

This implies in particular that $\int f(g_n)dP'_n \to \int f(u)vQ(d(u, v))$ for any bounded and continuous function f.

Addendum. *If Q has $\lambda^m \times \lambda$-density $(u, v) \to q(u, v)$, then $P'_n \circ g_n$, $n \in \mathbb{N}$, converges weakly to the probability measure with λ^m-density $u \to \int vq(u, v)dv$.*

Proof The corollary follows from Lemma (5.6.3), applied with $Q_n = P_n \circ (g_n, p_n)$ and $Q'_n = P'_n \circ (g_n, p_n)$, since

$$\int f(u, v)Q'_n(d(u, v)) = \int f(g_n, p_n)dP'_n = \int f(g_n, p_n)p_n dP_n = \int f(u, v)v dQ_n$$

for any measurable function $f : \mathbb{R}^m \times \mathbb{R}_+ \to \mathbb{R}$. The condition $\int q(u, v)Q(d(u, v)) = 1$ now becomes $\int vQ(d(u, v)) = 1$. □

The essential point of Le Cam's idea is to determine the limit distribution of a function S_n under P'_n from the limit distribution under P_n without knowing where this limit distribution comes from.

For functions $h, g \in L_*(P_0)$, an elementary computation shows that

$$P^n_{n^{-1/2}g} \circ \tilde{h}_n \Rightarrow N\left(hgdP_0, \int h^2 dP_0\right). \tag{5.6.22}$$

This relation is basic for the g-regularity of asymptotically linear estimator sequences with influence function h.

To illustrate Le Cam's idea from 1960 (Le Cam 1960), we prove the following generalization of (5.6.22), in which $\tilde{h}_n$ is replaced by a function S_n.

Theorem 5.6.5 *Let $S_n : X^n \to \mathbb{B}$ and $g : X \to \mathbb{B}$ be such that the limit distribution of $(S_n, \tilde{g}_n)$ under P^n is jointly normal $N(0, \Sigma)$ with covariance matrix*

$$\begin{pmatrix} \Sigma_{11} & \Sigma_{12} \\ \Sigma_{21} & \Sigma_{22} \end{pmatrix}.$$

Then the limit distribution of $(S_n, \tilde{g}_n)$ under $P^n_{n^{-1/2}g}$ is $N(\mu, \Sigma)$, with

$$\mu_1 = \Sigma_{12} \quad \text{and} \quad \mu_2 = \Sigma_{22}.$$

The second marginal of $N(\mu, \Sigma)$ i.e. the limit distribution of $\tilde{g}_n$ under $P^n_{n^{-1/2}g}$, is the well known $N(\Sigma_{22}, \Sigma_{22})$. Of interest is the first marginal, $N(\Sigma_{12}, \Sigma_{11})$, the limit distribution of S_n under $P^n_{n^{-1/2}g}$.

Applied for the special case $S_n = \tilde{h}_n$, the covariance matrix becomes

$$\Sigma_{11} = \int h^2 dP_0, \quad \Sigma_{22} = \int g^2 dP_0, \quad \Sigma_{12} = \int hg dP_0.$$

Proof According to (5.6.22), the density of $P^n_{n^{-1/2}g}$ with respect to P^n_0 can be approximated by $\exp[\tilde{g}_n - \frac{1}{2}\Sigma_{22}]$. Therefore the density of $P^n_{n^{-1/2}g} \circ \tilde{g}_n$ with respect to $P^n_0 \circ \tilde{g}_n$ is approximable by $v \to \exp[v\frac{1}{2}\Sigma_{22}]$. If $P^n_0 \circ (S_n, \tilde{g}_n) \Rightarrow Q$, then $P^n_{n^{-1/2}g} \circ (S_n, \tilde{g}_n) \to Q'$ with Q-density $(u, v) \to (u, \exp[v - \frac{1}{2}\Sigma_{22}])$. Hence, if $Q = N(0, \Sigma)$, this leads to $Q' = N(\mu, \Sigma)$ with $\mu_1 = \Sigma_{12}$, $\mu_2 = \Sigma_{22}$.

The conclusion from $N(0, \Sigma)$ to $N(\mu, \Sigma)$ requires an elementary but somewhat tedious computation. Hint: Rewrite the 2-dimensional normal distribution with covariance matrix Σ as

$$c \exp\Big[-\frac{1}{2}A_{11}u^2 + A_{12}uv - \frac{1}{2}A_{22}v^2\Big]$$

with

$$A_{11} = \frac{1}{(1-\rho^2)}, \quad A_{22} = \frac{1}{(1-\rho^2)\Sigma_{22}}, \quad A_{12} = \frac{\rho}{1-\rho^2} \cdot \frac{1}{(\Sigma_{11}\Sigma_{22})^{1/2}}.$$ □

5.7 Adaptivity

The estimator sequence $s_n^2(x) = n^{-1}\sum_{\nu=1}^n (x_\nu - \overline{x}_n)^2$ is asymptotically optimal for estimating σ^2 in the family $\mathfrak{P} = \{N(\mu, \sigma^2) : \mu \in \mathbb{R}, \sigma^2 > 0\}$. It is still asymptotically optimal in any of the subfamilies $\mathfrak{P}_\mu := \{N\mu, \sigma^2) : \sigma^2 > 0\}$, $\mu \in \mathbb{R}$. That means: Knowing μ does not help to obtain an estimator sequence for σ^2 which is asymptotically better than s_n^2. Within the realm of parametric families there are many such examples. (Another example is the estimation of ρ in the family

$\{N(\mu_1, \mu_2, \sigma_1^2\sigma_2^2, \rho) : \mu_i \in \mathbb{R},\ \sigma_i^2 > 0,\ \rho \in (-1, 1)\}$. Knowing that $\sigma_1^2 = \sigma_2^2$ does not help to obtain an estimator sequence asymptotically superior to the correlation coefficient. (See Pfanzagl 1982, p. 219, Example 13.2.4.)

This phenomenon of "adaptivity" did not find particular attention as long as it occurred in parametric families. It became a worthwhile subject of statistical theory as soon as it occurred in a general family. Let $\mathfrak{P}$ be the family of all distributions on $\mathbb{B}$ with a sufficiently smooth density; The problem is to estimate the median. Here is a natural idea as to how a "good" estimator for the median may be obtained. The simplest idea: Take the median of the sample, or: estimate the unknown density and take the median of the estimated density as an estimator of the "true" median. If the family of probability measures is large, no regular estimator sequence can be better than the sample median. The situation changes dramatically as soon as it is known that all densities are symmetric. Stein (1956) gave in his Sect. 4, pp. 190/1, a rough sketch of how "adaptive" tests for the position of the median of a symmetric density and for the difference in location and scale between two distributions of the same unknown shape could be obtained.

It took almost fifteen years until Stein's sketch was turned into a mathematically solid result. The (equivalent) two-sample problem was solved independently by van Eeden (1970) [p. 175, Theorem 2.1], Weiss and Wolfowitz (1970, Sect. 4, pp. 144/5) and Beran (1974) [Theorem 3.1, p. 70]. The papers by van Eeden and Beran are based on Hájek's (1962) theorem on adaptive rank tests. They also assert the existence of adaptive estimator sequences for the median of a symmetric distribution (their p. 180, Theorem 4.1, and p. 73, Theorem 4.1, respectively). See also the preliminary version of van Eeden's result, dating from (1968).

These basic papers were followed by a number of papers, doing what is usual in mathematics: To prove a stronger assertion under weaker assumptions. (See Fabian, 1973, Sacks, 1975, and Stone, 1975.) There is also a questionable paper by Takeuchi (1971): "We do not understand his argument" is what Weiss and Wolfowitz say (1970, p. 149). The most impressive of these results is Theorem 1.1 in Stone (1975) [p. 268] which asserts the existence of an adaptive translation and scale invariant estimator of the median without any regularity conditions on the density except for absolute continuity (and symmetry, of course).

Perplexing as the phenomenon of adaptivity in nonparametric families is, it obviously was not easy to find an appropriate conceptual framework. To discuss the results mentioned above we use the conceptual framework of "semiparametric" models introduced later by Bickel (1982). Let

$$\mathfrak{P} = \{P_{\vartheta,\eta} : \vartheta \in \Theta,\ \eta \in H\},$$

where $\Theta \in \mathbb{R}^k$ and $\eta \in H$, an abstract set. Intuitively speaking, an estimator sequence of ϑ is adaptive in $\mathfrak{P}$ if it is asymptotically as good as any optimal estimator sequence based on the knowledge of η. Yet, what is "the optimal estimator sequence of ϑ", based on the knowledge of η? The solution to this problem starts with Stein. Since Stein (1956) cites Le Cam (1953), he must have been familiar with the problem of superefficiency and the necessity of basing the concept of an optimal limit distribution on some kind of locally uniform convergence to this limit distribution. Yet, he did not care: "... in a sense, [the ML estimate] is the asymptotically best possible estimate" (p. 188). Similarly, his definition of adaptivity (p. 192) "... if it is as difficult when the form of the distribution is known as it is when the form of the distribution depends in a regular way [?] on an unknown parameter."

The succeeding authors, too, assume that the limit distribution of the ML-sequence (or the Cramér–Rao bound) are bounds for the quality of estimator sequences, without mentioning that such bounds are valid for "regular" estimator sequences only.

Most of the authors mentioned above are satisfied with estimator sequences $(\vartheta^{(n)})_{n\in\mathbb{N}}$ attaining for every $\eta \in H$ the optimal limit distribution $N(0, \sigma_\eta^2(\vartheta))$ and ignore the assumption of regularity (which is essential for the validity of $\sigma_\eta^2(\vartheta)$ as a bound in $\mathfrak{P}_\eta$.) None of these authors pays attention to the preconditions under which adaptivity is possible. In spite of the heuristic character of his paper, Stein arrives with his "algebraic Lemma" (Sect. 3, pp. 188–190) somehow at the conclusion that "orthogonality" is necessary for "adaptivity". Bickel presents Stein's "proof" that *"adaptivity" requires "orthogonality"* in an intelligible form (see 1982, p. 651). Bickel's presentation is repeated in Fabian and Hannan, 1982, p. 474, Theorem 9 and in Bickel et al. 1993, pp. 28/9.

To arrive at a precise concept of "adaptivity", let $\mathfrak{P}_\eta := \{P_{\vartheta,\eta} : \vartheta \in \Theta\}$, and let $N(0, \sigma_\eta^2(\vartheta))$ be the optimal limit distribution for estimator sequences of ϑ which are "ϑ-regular" within $\mathfrak{P}_\eta$. Bickel's definition of adaptivity (see 1982, p. 649) reads as follows. The estimator sequence $\vartheta^{(n)}$, $n \in \mathbb{N}$, is "adaptive" if it converges ϑ-regularly to $N(0, \sigma_\eta^2(\vartheta))$ for every $\eta \in H$, i.e. if

$$P^n_{\vartheta_n,\eta} \circ n^{1/2}(\vartheta^{(n)} - \vartheta_n) \Rightarrow N(0, \sigma_\eta^2(\vartheta)) \quad \text{for } \vartheta_n = \vartheta + n^{-1/2}a,\ a \in \mathbb{R}^k,\ \eta \in H.$$

Within this framework it would still be possible to construct ϑ-regular estimator sequences which are adaptive on a countable subset of H. To establish "orthogonality" as necessary for "adaptivity" is, therefore, impossible unless Bickel's definition of "adaptivity" is modified, e.g. by requiring continuity of $\eta \to \sigma_\eta^2(\vartheta)$.

Begun et al. (1983) [p. 438, Definition 3.1] suggest a more restrictive definition of "adaptivity". They require that

$$P^n_{\vartheta_n,\eta_n} \circ n^{1/2}(\vartheta^{(n)} - \vartheta_n) \Rightarrow N(0, \sigma_\eta^2(\vartheta))$$

for unspecified sequences $\vartheta_n \to \vartheta$ and $\eta_n \to \vartheta$ and $\eta_n \to \eta$. To justify their definition they do not say more than (see p. 438)

> Beran's result bolsters our feeling that [this definition] captures the local uniformity that should reasonably be required of adaptive estimates if [η] is unknown.

The same definition is accepted by Bickel et al. (1993) [p. 29, Definition 2.4.1]. It is, roughly speaking, equivalent to "(ϑ, η)-regularity", i.e. regularity w.r.t. $\mathfrak{P}$.

To make the connection between orthogonality and adaptivity more transparent, we consider the case that $\Theta \subset \mathbb{R}$. After all, the results for arbitrary families are derived from results for parametric subfamilies. Assume that $\vartheta^{(n)}$, $n \in \mathbb{N}$, is an estimator sequence which is adaptive and (ϑ, η)-regular. Adaptivity implies

$$P^n_{\vartheta,\eta} \circ n^{1/2}(\vartheta^{(n)} - \vartheta) \Rightarrow N(0, \sigma^2_\eta(\vartheta))$$

with $\sigma^2_\eta(\vartheta) = 1/I_\eta(\vartheta)$ and $I_\eta(\vartheta) = \int \ell^\bullet_\vartheta(x, \vartheta, \eta)^2 P_{\vartheta,\eta}(dx)$. Since $N(0, \sigma^2_\eta(\vartheta))$ is the optimal limit distribution for ϑ-regular estimator sequences in $\mathfrak{P}_\eta$, this implies that

$$n^{1/2}(\vartheta^{(n)} - \vartheta) = n^{-1/2} \sum_{\nu=1}^{n} \sigma^2_\eta(\vartheta)\ell^\bullet(x_\nu, \vartheta, \eta) \quad (P^n_{\vartheta,\eta}). \tag{5.7.1}$$

If $\eta_n \in H$ is a sequence such that $(P^n_{\vartheta,\eta_n})_{n\in\mathbb{N}} \ll (P^n_{\vartheta,\eta})_{n\in\mathbb{N}}$, this implies

$$n^{1/2}(\vartheta^{(n)} - \vartheta) = n^{-1/2} \sum_{\nu=1}^{n} \sigma^2_\eta(\vartheta)\ell^\bullet(x_\nu, \vartheta, \eta) \quad (P^n_{\vartheta,\eta_n}).$$

If $\vartheta^{(n)}$, $n \in \mathbb{N}$, is η-regular, i.e.,

$$P^n_{\vartheta,\eta_n} \circ n^{1/2}(\vartheta^{(n)} - \vartheta) \Rightarrow N(0, \sigma^2_\eta(\vartheta)), \tag{5.7.2}$$

then

$$P^n_{\vartheta,\eta_n} \circ n^{-1/2} \sum_{\nu=1}^{n} \sigma^2_\eta(\vartheta)\ell^\bullet(x_\nu, \vartheta, \eta) \Rightarrow N(0, \sigma^2_\eta(\vartheta)). \tag{5.7.3}$$

Yet,

$$P^n_{\vartheta,\eta+n^{-1/2}u} \circ n^{-1/2} \sum_{\nu=1}^{n} \sigma^2_\eta(\vartheta)\ell^\bullet(x_\nu, \vartheta, \eta) \Rightarrow N(u\sigma^2_\eta(\vartheta)L_{12}(\vartheta, \eta), \sigma^2_\eta(\vartheta)). \tag{5.7.4}$$

Hence (5.7.3) holds iff $L_{1,2}(\vartheta, \eta) = 0$, which is the orthogonality called for.

Conversely, if $(\vartheta^{(n)})_{n\in\mathbb{N}}$ is adaptive and ϑ-regular, we obtain from (5.7.1) and (5.7.4) under the condition $L_{1,2}(\vartheta, \eta) = 0$ the relations (5.7.3) and (5.7.2). Hence under this orthogonality condition *any* (in Bickel's sense) *adaptive and ϑ-regular estimator sequence is automatically also η-regular.* In a disguised form this occurs in Fabian and Hannan 1982, p. 474, Theorem 7.10.

Applied to the problem of estimating the parameter ϑ of a family with density $x \to p(x - \vartheta)$ this implies that any ϑ-equivariant (hence ϑ-regular) estimator sequence which is efficient for every p is "robust" in the sense that its limit distribution remains unchanged if the observations come from a density $x \to p_n(x - \vartheta)$ where p_n, $n \in \mathbb{N}$, tends to p from an orthogonal direction. Beran (1974, Remark on p. 74, 1978, p. 306, Theorem 4 and various other papers) seems to have been the first to exhibit estimator sequences that converge to an optimal limit distribution for a larger class of distributions (not just for the symmetric ones), and that this is related to "orthogonality". "This property can, we believe, be suitably re-expressed to apply generally" says Bickel (1982, p. 664, Remark 5.5), but apparently he never came back to this question.

With "asymptotic robustness" in mind, Beran makes in connection with the estimation of the median of a symmetric distribution a point of the fact that η-regularity (i.e., independence of the limit distribution from small deviations η_n) is not confined to symmetric distributions; relation (5.7.2) holds for *all* sequences η_n converging to η from a direction orthogonal to $\ell^\bullet(\cdot, \vartheta, \eta)$. Observe that this phenomenon is related to the regularity of all asymptotically linear estimator sequences (see Sect. 5.5).

So far we have discussed the phenomenon of adaptivity in its historical context. How does it present itself in a more general framework? Let $\mathfrak{P}$ be a general family of probability measures with tangent space $T(P, \mathfrak{P})$, and $\kappa : \mathfrak{P} \to \mathbb{R}$ a differentiable functional with canonical gradient $\kappa^*(\cdot, P) \in T(P, \mathfrak{P})$. According to the Convolution Theorem, the limit distribution of a regular estimator sequence cannot be better than $N(0, \sigma^2(P))$, with $\sigma^2(P) := \int \kappa^*(\cdot, P)^2 dP$. In the following we shall show that "optimality on a subfamily $\mathfrak{P}_0 \subset \mathfrak{P}$" is necessary and sufficient for "optimality on $\mathfrak{P}$" if $\kappa^*(\cdot, P)$ belongs to the tangent space of $\mathfrak{P}_0$.

(i) $\kappa^{(n)}$ is optimal on $\mathfrak{P}$ if it is $T(P, \mathfrak{P}_0)$-regular and asymptotically linear with influence function $\kappa^*(\cdot, P)$. Being regular on $\mathfrak{P}$, $\kappa^{(n)}$ is a fortiori regular on $T(P, \mathfrak{P}_0)$. It is optimal on $\mathfrak{P}_0$ iff $\kappa^*(\cdot, P)$ is in $T(P, \mathfrak{P}_0)$. Hence there is always a subfamily on which $\kappa^{(n)}$ is asymptotically optimal, namely the path converging to P from the direction $\kappa^*(\cdot, P)$.

(ii) If $\kappa^{(n)}$ is optimal in $\mathfrak{P}_0$, i.e. and) asymptotically linear with influence function $\kappa_0^*(\cdot, P)$, and hence $T(P, \mathfrak{P}_0)$-regular, then it is optimal on $\mathfrak{P}$ if it is $T(P, \mathfrak{P})$-regular. This is the case iff κ_0^* is a gradient in $T(P, \mathfrak{P})$, i.e. iff

$$\int \kappa_0^* g = \int \kappa^* g \quad \text{for } g \in T(P, \mathfrak{P}).$$

This is the case if $\kappa^*(\cdot, P) = \kappa_0^*(\cdot, P)$, i.e. if $\kappa^*(\cdot, P) \in T(P, \mathfrak{P}_0)$.

The idea that some kind of "orthogonality" is necessary for adaptivity goes back to Stein (1956). It occurs in the papers by Bickel and by Fabian and Hannan, mainly in connection with a special type of models: The estimation of the parameter ϑ for families $P_{\vartheta,\eta}$, with η known or unknown.

Does "orthogonality" also play a similar role in general models? Let $\mathfrak{P}$ be a general family, and $\kappa : \mathfrak{P} \to \mathbb{R}$ the functional to be estimated. Can we characterize the subfamilies $\mathfrak{P}_0$ on which the estimation of κ is as difficult as on $\mathfrak{P}$? The answer is affirmative if the restriction from $\mathfrak{P}$ to $\mathfrak{P}_0$ is based on a side condition of the following type: There is a differentiable functional $\gamma : \mathfrak{P} \to H$ such that

$$\mathfrak{P}_0(P) := \{\bar{P} \in \mathfrak{P} : \quad \gamma(\bar{P}) = \gamma(P)\}. \tag{5.7.5}$$

Observe that this corresponds to the restriction of $\mathfrak{P} = \{P_{\vartheta,\eta} : \vartheta \in \Theta, \eta \in H\}$ to $\mathfrak{P}_0 = \{P_{\vartheta,\eta_0} : \vartheta \in \Theta\}$ by the condition (5.7.5) with $\gamma(P_{\vartheta,\eta}) = \eta$. If $\gamma^*(\cdot, P)$ denotes the canonical gradient of the functional γ, the tangent space of this family is

$$T(P, \mathfrak{P}_0) = \left\{\gamma \in T(P, \mathfrak{P}) : \int \kappa^*(\cdot, P)\gamma dP = 0\right\}.$$

An estimator sequence which is optimal for κ on $\mathfrak{P}$ (i.e. $T(P_0, \mathfrak{P})$-regular and asymptotically linear with influence function $\kappa^*(\cdot, P)$), is optimal on the family $\mathfrak{P}_0$ iff

$$\kappa^*(\cdot, P) \in T(P, \mathfrak{P}), \quad \text{i.e.} \quad \int \gamma^*(\cdot, P)\kappa^*(\cdot, P)dP = 0.$$

This is the orthogonality condition which is necessary and sufficient for adaptivity, or: Optimal estimator sequences on $\mathfrak{P}$ cannot be improved if it is known that P_0 belongs to the smaller subfamily $\mathfrak{P}_0$.

In general, the knowledge that the "true" probability measure belongs to a given subfamily $\mathfrak{P}_0$, can be utilized to obtain an asymptotically better estimator sequence for $\kappa(P)$. But again, if $P \in \mathfrak{P}_0$ and

$$\kappa^*(\cdot, P) \in T(P, \mathfrak{P}_0),$$

then $\kappa^*(\cdot, P)$, the canonical gradient of κ in $\mathfrak{P}$, is also the canonical gradient in $\mathfrak{P}_0$, and $N(0, \sigma^2(P))$ (with the same $\sigma^2(P)$) is also the optimal limit distribution for $\mathfrak{P}_0$-regular estimator sequences. In this case, reducing the condition of $\mathfrak{P}$-regularity to $\mathfrak{P}_0$-regularity is not effective. A $\mathfrak{P}$-regular estimator sequence (which is a fortiori $\mathfrak{P}_0$-regular) cannot be replaced by a better $\mathfrak{P}_0$-regular estimator sequence.

Let now $\kappa_0^{(n)}$, $n \in \mathbb{N}$, be a $\mathfrak{P}_0$-regular estimator sequence which is optimal in $\mathfrak{P}_0$. This implies

$$n^{1/2}(\kappa_0^{(n)} - \kappa(P)) = \tilde{\kappa}_{0n}^*(\cdot, P) + o(n^0, P^n),$$

where κ_0^* is the canonical gradient of κ in $\mathfrak{P}_0$. If $\kappa^*(\cdot, P) \in T(P, \mathfrak{P}_0)$, then $\kappa_0^*(\cdot, P) = \kappa^*(\cdot, P)$, i.e.

$$n^{1/2}(\kappa_0^{(n)} - \kappa(P)) = \tilde{\kappa}_n^*(\cdot, P) + o(n^0, P^n),$$

This implies that the estimator sequence $\kappa_0^{(n)}$, $n \in \mathbb{N}$, is regular with respect to every $h \in T(P, \mathfrak{P})$: Any estimator sequence which is $\mathfrak{P}_0$-regular and optimal in $\mathfrak{P}_0$ is also $\mathfrak{P}$-regular (and therefore optimal in $\mathfrak{P}$).

For the purpose of illustration, we specialize the general considerations to a semiparametric model in which $\mathfrak{P}$ consists of all probability measures $P_{\vartheta,\eta}|\mathbb{B}$ with Lebesgue density

$$x \to p_\eta(x - \vartheta)$$

with $\vartheta \in \mathbb{R}$, where p_η is a density symmetric about 0. For this model, the tangent space $T(P_{\vartheta,\eta}, \mathfrak{P})$ consists of the functions of the form $a\ell'_\eta(x-\vartheta) + \Psi(x-\vartheta)$, where $\ell'_\eta(x) = \partial_x \log p_\eta(x)$ and Ψ is symmetric about 0 with $\int \Psi(x) P_{\vartheta,\eta}(dx) = 0$. The functional $\kappa(P_{\vartheta,\eta}) = \vartheta$ has in $\mathfrak{P}$ the canonical gradient $\kappa^*(\cdot, \vartheta, \eta) = \sigma(\eta)^2 \ell'_\eta (x-\vartheta)$ with $\sigma(\eta)^2 = (\int (\ell'_\eta)^2 dP_{\vartheta_0,\eta})^{-1}$.

If the density p_{η_0} is *known*, then the tangent space $T(P_{\vartheta,\eta_0}, \mathfrak{P}_{\eta_0})$ reduces to $\{a\ell'_{\eta_0}(\cdot - \vartheta) : a \in \mathbb{R}\}$, which still contains $x \to \sigma^2(\eta_0)\ell'_{\eta_0}(x-\vartheta)$, the canonical gradient of the functional $\kappa(P_{\vartheta,\eta}) = \vartheta$ in $T(P_{\vartheta,\eta_0}, \mathfrak{P})$. If an estimator sequence $\vartheta^{(n)}$, $n \in \mathbb{N}$, is ϑ-equivariant and optimal in $\mathfrak{P}_{\eta_0}$, it is regular in $\mathfrak{P}$, and even locally robust for certain directions, not necessarily in $T(P, \mathfrak{P})$, which are orthogonal to $x \to \ell'_{\eta_0}(x-\vartheta)$.

The Convolution Theorem provides precise information about the optimal limit distribution, based on the concepts of "*tangent space*" and "*gradient*". The concepts "tangent space" and "gradient" are somehow descendants of Stein's ideas. Yet it appears that something of Stein's intuitive ideas is lost, namely: An estimator sequence which is regular and optimal in the least favourable subfamily is also regular, hence also optimal, in the whole family.

To verify Stein's idea, and to elaborate on an aspect neglected by Stein, we apply his ideas to a "large" parametric family, and we study the performance of estimator sequences in one-parameter subfamilies. Let $\mathfrak{P} = \{P_\vartheta : \vartheta \in \Theta\}$, $\Theta \subset \mathbb{R}^k$, be an LAN family with tangent space $T(P_\vartheta, \mathfrak{P})$ spanned by $\ell^{(1)}(\cdot, \vartheta), \ldots, \ell^{(k)}(\cdot, \vartheta)$, i.e.

$$\log(dP^n_{\vartheta+n^{-1/2}a}/dP^n_\vartheta) = a^\top(\cdot, \vartheta)\tilde{\ell}^\bullet_n - \frac{1}{2}a^\top L(\vartheta)a + o(n^0, P^n_\vartheta).$$

We consider a one-parameter subfamily $\bar{P}_t := P_{\vartheta(t)}$, $t \in \mathbb{R}$, with $\vartheta(0) = \vartheta$. If the functions $\vartheta_i : \mathbb{R} \to \mathbb{R}$, $i = 1, \ldots, k$, have continuous derivatives ϑ'_i at $t = 0$, then at $t = 0$ the family $\tilde{\mathfrak{P}} := \{\bar{P}_t : t \in \mathbb{R}\}$ has the one-dimensional tangent space

$$T(P, \tilde{\mathfrak{P}}) = \left\{a \sum_{i=1}^k \vartheta'_i \ell^{(i)}(\cdot, \vartheta) : a \in \mathbb{R}\right\}$$

and the LAN-expansion

$$\log(d\bar{P}^n_{n^{-1/2}b}/dP^n_\vartheta) = b\sum_{i=1}^k \vartheta'_i \tilde{\ell}^{(i)}(\cdot, \vartheta) - \frac{1}{2}b^2 \sum_{i,j=1}^k \vartheta'_i \vartheta'_j L_{i,j}(\vartheta) + o(n^0, P^n_\vartheta).$$

We now consider estimation of $\tilde{\kappa}(t) := \kappa(\vartheta_1(t), \ldots, \vartheta_k(t))$ within the family $\tilde{\mathfrak{P}}$. According to the Convolution Theorem, the optimal estimator sequence among regular estimator sequences in this family has at $t = 0$ the stochastic expansion

$$n^{1/2}(\kappa^{(n)} - \tilde{\kappa}(0)) = n^{-1/2}\sum_{\nu=1}^n \kappa_0^*(x_\nu) + o(n^0, P^n_\vartheta)$$

with

$$\kappa_0^* = \Big(\sum_{i,j=1}^{k} \vartheta_i' \vartheta_j' L_{i,j}(\vartheta)\Big)^{-1} \sum_{i=1}^{k} \kappa^{(i)} \vartheta_i' \sum_{r=1}^{k} \vartheta_r' \ell^{(r)}(\cdot, \vartheta). \tag{5.7.6}$$

The pertaining asymptotic variance is

$$\Big(\sum_{i,j=1}^{k} \vartheta_i' \vartheta_j' L_{i,j}(\vartheta)\Big)^{-1} \Big(\sum_{i=1}^{k} \kappa^{(i)} \vartheta_i'\Big)^2.$$

The variance attains its maximal value $\sum_{r,s=1}^{k} \kappa^{(r)} \kappa^{(s)} \Lambda_{rs}(\vartheta)$ if

$$\vartheta_i' = \Big(\sum_{r,s=1}^{k} \kappa^{(r)} \kappa^{(s)} \Lambda_{rs}(\vartheta)\Big)^{-1} \sum_{i=1}^{k} \Lambda_{ij}(\vartheta) \kappa^{(j)}. \tag{5.7.7}$$

This follows from the Schwarz inequality if we rewrite $\sum_{i=1}^{k} \kappa^{(i)} \vartheta_i'$ as

$$\int \sum_{i=1}^{k} \vartheta_i' \ell^{(i)}(\cdot, \vartheta) \sum_{r,s=1}^{k} \Lambda_{rs}(\vartheta) \kappa^{(r)} \ell^{(s)}(\cdot, \vartheta) d P_\vartheta.$$

Hence the subfamily $\vartheta(t)$ with ϑ_i' given by (5.7.7) is the least favorable one.

Since $T(P, \tilde{\mathfrak{P}}) \subset T(P, \mathfrak{P})$, the minimal asymptotic variance of $T(P, \mathfrak{P})$-regular estimator sequences is "larger" than the minimal asymptotic variance of any one-dimensional subfamily, hence in particular larger than $\sum_{i,j=1}^{k} \kappa^{(i)} \kappa^{(j)} \Lambda_{ij}(\vartheta)$, the minimal asymptotic variance of the least favourable subfamily.

Now comes the point neglected by Stein: The estimator sequence which is optimal among the estimator sequences regular in the least favorable subfamily is $T(P, \mathfrak{P}$-regular, i.e., regular in the whole subfamily.

The canonical gradient κ_0^* for the subfamily $\{\bar{P}_t : t \in \mathbb{R}\}$ given by (5.7.6) reduces for the least favorable subfamily defined by (5.7.7) to

$$\kappa^* = \sum_{i,j=1}^{k} \Lambda_{ij}(\vartheta) \kappa^{(i)} \ell^{(j)}(\cdot, \vartheta).$$

Since

$$\kappa^{(r)} = \int \kappa^* \ell^{(r)}(\cdot, \vartheta) d P_\vartheta \quad \text{for } r = 1, \dots, k,$$

κ^* is also the canonical gradient of κ in $T(P, \mathfrak{P})$. Therefore any estimator sequence with stochastic expansion

$$n^{1/2}(\kappa^{(n)} - \kappa(\vartheta)) = n^{-1/2} \sum_{\nu=1}^{n} \kappa^*(x_\nu, \vartheta)$$

is $T(P, \mathfrak{P})$-regular.

5.8 "Regular" Estimator Sequences

With ML-sequences in mind one might think that regularity conditions on the family of probability measures is all one needs for assertions on the asymptotic performance of the estimator sequences. If one turns to arbitrary estimator sequences it becomes evident that meaningful results can be obtained only under conditions on the estimator sequences (such as uniform convergence) which are automatically fulfilled for ML-sequences under conditions on the family of probability measures.

The purpose of such regularity conditions on the estimator sequences is to make sure that the limit distribution provides information on the true distribution of the estimator for large samples. Ideally, this means that the convergence of

$$Q_P^{(n)} := P^{(n)} \circ c_n(\kappa^{(n)} - \kappa(P)), \quad n \in \mathbb{N},$$

to a limit distribution Q_P is uniform in $\mathfrak{P}$. Obviously such a property cannot be dealt with in the framework of local asymptotic theory. A possible way out is the resort to an asymptotic version of uniformity (based on local properties of $\mathfrak{P}$ and κ only) which is in some sense necessary for uniformity on $\mathfrak{P}$. Such a weak condition could be used to obtain results on the possible limit distributions (e.g. a bound for their concentration). Such results can be extended to obtain uniformity on $\mathfrak{P}$ for special models.

In the first papers by Rao and Wolfowitz for one-parameter families, the authors require uniformity on $\mathbb{R}$ or on a compact subset $\Theta \subset \mathbb{R}$, but they only use convergence of $Q^{(n)}_{\vartheta+n^{-1/2}a}$ to Q_ϑ for every $a \in \mathbb{R}$ (see Rao 1963, p. 197, proof of Lemma 2 and Wolfowitz 1965, p. 254, proof of Lemma 1).

Convergence of $Q^{(n)}_{\vartheta+n^{-1/2}a}$, $n \in \mathbb{N}$, to Q_ϑ was explicitly introduced by Hájek (1970) [p. 324] under the name "regular convergence". A similar condition had earlier been used by Bahadur (1964) [p. 1546]. It occurs in different variants: Ibragimov and Has'minskii (1981) [p. 151, Definition 9.1] with uniformity for $|a| \leq c$, for any $c > 0$; and with a replaced by a bounded sequence $(a_n)_{n \in \mathbb{N}}$ as in Bickel et al. 1993, p. 21, Definition 2.2.7.

Going back to the origins of "regular convergence", there is a technical reason why authors starting from the concept of "uniform convergence" change to $D(Q^{(n)}_{\vartheta_n}, Q_{\vartheta_0}) \to 0$ rather than $D(Q^{(n)}_{\vartheta_n}, Q_{\vartheta_n}) \to 0$, with $\vartheta_n = \vartheta_0 + n^{-1/2}a$ in the definition of a localized version of uniformity. Bickel (See e.g. et al., p. 18, Definition 2.2.3 for *uniformly* regular convergence, and p. 21, Definition 2.2.7 for *locally* regular convergence.) From the methodological point of view, "regular convergence" is the residual (Le Cam would, perhaps, say a ghost) of the operationally meaningful condition of "uniform convergence". It is convenient from the technical point of view, and it suffices to establish bounds for the concentration of estimator sequences.

Remark Hájek's concept of "regular convergence" might also be interpreted as a localized version of the idea to consider the performance of an estimator sequence $\vartheta^{(n)}$ within a one-dimensional subfamily $P_{\vartheta(t)} : t \in \mathbb{R}$, approaching P_ϑ from the direction $a \in \mathbb{R}^k$ if $t^{-1}(\vartheta_i(t) - \vartheta_i) \to a_i$ for $i = 1, \ldots, k$.

The choice of such a family is straightforward. If P_ϑ is sufficiently regular, the function $\vartheta(t)$ should be sufficiently smooth, say differentiable at $t = 0$. How should one choose a one-parameter subfamily, say $\{P_t : t \in A\}$ with P_t in a general family $\mathfrak{P}$? The essential point: For assertions about the asymptotic distribution of estimator sequences $\vartheta^{(n)}$ requires (in the i.i.d. case) an approximation to $(P_t^n)_{n\in\mathbb{N}}$ for t close to 0.

How could the idea of "regular convergence", defined for parametric families, be carried over to general families? Should we consider estimator sequences which are regularly convergent on all (sufficiently regular) subfamilies, or should we consider estimator sequences which converge regularly with respect to all directions in the tangent space $T(P, \mathfrak{P})$? (See Sect. 5.6 for details.)

A natural way to introduce such a tangent space is to start from a family of measures P_t, the P_0-density of which can be approximated by

$$p_t(x) = p_0(x)(1 + tg_t(x)).$$

If g_t converges to a function g_0 in a suitable sense (see (5.6.10)–(5.6.12)), then $\{(P_t^n)_{n\in\mathbb{N}}, t \in A\}$ fulfills an LAN-condition

$$dP^n_{n^{-1/2}t}/dP^n_0 = \exp\left[t\tilde{g}_0 - \frac{1}{2}t^2\sigma^2\right] \quad \text{with } \sigma^2 = \int g_0^2 dP_0.$$

(See Sect. 5.12 for details.)

Apart from its inherent relationship to uniform convergence, continuity of the limit distribution is of independent interest. It is surprising that few authors consider continuity of the limit distribution as some sort of regularity condition to be imposed on estimator sequences in general.

Since the continuity of $\vartheta \to Q_\vartheta^{(n)}$ is fundamental for the continuity of $\vartheta \to Q_\vartheta$, we state the following Theorem.

Theorem 5.8.1 *If $\vartheta \to P_\vartheta$ is continuous at ϑ_0 with respect to the sup-distance d, then $\vartheta \to Q_\vartheta^{(n)}$ is continuous at ϑ_0 for every n:*

$$\vartheta \to \vartheta_0 \quad \textit{implies} \quad \int f dQ_\vartheta^{(n)} - \int f dQ_{\vartheta_0}^{(n)} \to 0$$

for every bounded and continuous function $f : \mathbb{R}^k \to \mathbb{R}$.

Corollary 5.8.2 *If $\int f dQ_\vartheta^{(n)} \to \int f Q_\vartheta$ locally uniformly at ϑ_0 for all bounded and continuous $f : \mathbb{R}^k \to \mathbb{R}$, then $\vartheta \to Q_\vartheta$ is continuous at ϑ_0.*

Proof Theorem 5.8.1 and Lemma 5.3.9, applied with $\vartheta \to \int f dQ_\vartheta^{(n)}$ in place of $x \to h_n(x)$ and $\vartheta \to \int f dQ_\vartheta$ in place of $x \to h_0(x)$. □

Versions of Theorem 5.8.1 and Corollary 5.8.2 for general families are given in Pfanzagl (2003) [p. 109].

Regularly Attainable Limit Distributions are Continuous

In the following section, we consider functions $P \to Q_P$ from $\mathfrak{P}$ to the family of probability measures on $\mathbb{B}^m$. By continuity we mean that the map $P \to \int f dQ_P$ is continuous for every bounded and continuous function f, if $\mathfrak{P}$ is endowed with the sup-distance. In other words, that $d(P_n, P_0) \to 0$ implies $\int f dQ_{P_n} \to \int f dQ_{P_0}$. Recall that continuity of $P \to Q_P$ implies that $P \to \int f dQ_P$ is lower [upper] semicontinuous if f is bounded and lower [upper] semicontinuous. (Ash 2000, p. 122, Theorem 2.8.1.)

If $\mathfrak{P}$ is a parametric family $\{P_\vartheta : \vartheta \in \Theta\}$, $\Theta \subset \mathbb{R}^k$, we write Q_ϑ for Q_{P_ϑ}, and the continuity of $\vartheta \to Q_\vartheta$ follows if $\vartheta \to P_\vartheta$ is continuous with respect to the Euclidean distance in Θ and the sup-distance in $\mathfrak{P}$, i.e., if

$$\|\vartheta_n - \vartheta_0\| \to 0 \quad \text{implies} \quad d(P_{\vartheta_n}, P_{\vartheta_0}) \to 0.$$

For parametric families with $\Theta \subset \mathbb{R}$, the continuity of

$$\vartheta \to Q_\vartheta^{(n)} = P_\vartheta^n \circ n^{1/2}(\vartheta^{(n)} - \vartheta)$$

was first established by Rao (1963) [p. 196, Lemma 2(i)] under the unnecessarily restrictive condition that P_ϑ has a density such that $\vartheta \to p(x, \vartheta)$ is continuous for every $x \in X$. From this he inferred that a (normal) limit distribution of $Q_\vartheta^{(n)}$, $n \in \mathbb{N}$, is continuous if the convergence is uniform. Similar results occur in Wolfowitz (1965) [p. 254, Lemma 2].

Since continuity of the limit distribution is important in various connections, we present the following result on the continuity of $P \to Q_P^{(n)}$ in greater generality.

Lemma 5.8.3 *Let $\mathfrak{P}|\mathscr{A}$ be endowed with the sup-distance d. If $\kappa : \mathfrak{P} \to \mathbb{R}^k$ is continuous, then $P \to Q_P^{(n)}(B)$ is, for every $n \in \mathbb{N}$, lower [upper] semicontinuous if B is open [closed].*

The proof given in Pfanzagl (2003) [p. 109, Corollary 5.1] for the case $k = 1$ carries over to arbitrary k.

Theorem 5.8.4 *Let $\mathfrak{P}|\mathscr{A}$ be endowed with the sup-distance. If $\kappa : \mathfrak{P} \to \mathbb{R}^k$ is continuous, and $P^n \circ c_n(\kappa^{(n)} - \kappa(P))$, $n \in \mathbb{N}$, converges weakly to Q_P, uniformly on $\mathfrak{P}_0$, then $P \to Q_P$ is weakly continuous on $\mathfrak{P}_0$.*

In particular: $d(P_n, P_0) \to 0$ implies $Q_{P_n} \Rightarrow Q_{P_0}$.

The proof follows from Lemma 5.8.3.

5.9 Bounds for the Asymptotic Concentration of Estimator Sequences

The Convolution Theorem (Sect. 5.13) gives a satisfactory bound for the asymptotic concentration of "regular" estimator sequences. Before we present it, we give a survey of earlier attempts.

A Metaphysical Approach

We omit the attempts of Fisher at this problem that culminate in statements like (see Fisher 1925, p. 714):

> The efficiency of a statistic is the ratio of the intrinsic accuracy of its random sampling distribution to the amount of information in the data from which it has been derived.

The reader interested in Fisher's contributions to the problem of efficiency is referred to Pratt (1976).

The idea of a bound for the "accuracy" of estimators occurs already in Rao (1945) in connection with mean unbiased estimators in families admitting a complete sufficient statistic. The "accuracy" of the estimators is expressed by convex loss. The mathematics behind this result is unassuming (see Sect. 3.2 for details). A first attempt by Rao (1947) [p. 282, Theorem 2] to modify such (nonasymptotic) results to obtain the asymptotic optimality of ML-sequences is not satisfactory.

The way to an asymptotic bound (such as the Convolution Theorem) turned out demanding, both mathematically and methodologically. It was necessary to find a framework for the basic family of probability measures (say LAN), to find operational and effective conditions on the estimator sequence (like uniform convergence), and to find a concept of quality (say concentration) for limit distributions. It may be of interest to illustrate how troublesome the way to an asymptotic optimality concept had been.

When C.R. Rao embarked upon the problem of asymptotic efficiency—between 1961 and 1963 he published four papers on this subject which are distinct more by their titles than by their contents —the scene had changed drastically compared with the time of Fisher's endeavors to prove the optimality of ML-sequences. In 1953, Le Cam had presented various important results (Le Cam 1953). Adjusted to the present framework, some of these results may be rewritten as follows.

The Example of Hodges

One can always improve the asymptotic performance of a given estimator sequence $(\vartheta^{(n)})_{n\in\mathbb{N}}$ with $P^n_\vartheta \circ n^{1/2}(\vartheta^{(n)} - \vartheta) \Rightarrow N(0, \sigma^2(\vartheta))$ for some $\vartheta = \vartheta_0$ without changing its asymptotic performance for any other $\vartheta \in \Theta$. One just has to change the definition of $\vartheta^{(n)}(\mathbf{x}_n)$ to $\hat{\vartheta}^{(n)}_a(\mathbf{x}_n) := (1-a)\vartheta_0 + a\vartheta^{(n)}(\mathbf{x}_n)$ if $\vartheta^{(n)}(\mathbf{x}_n)$ is sufficiently close to ϑ_0, say $|\vartheta^{(n)}(\mathbf{x}_n) - \vartheta_0| < n^{-1/4}$. Then $(P^n_\vartheta \circ n^{1/2}(\hat{\vartheta}^{(n)}_a - \vartheta))_{n\in\mathbb{N}}$ converges to $N(0, \sigma^2(\vartheta))$ if $\vartheta \neq \vartheta_0$, and to $N(0, a^2\sigma^2(\vartheta_0))$ for $\vartheta = \vartheta_0$.

The risk $\int \ell(n^{1/2}(\vartheta^{(n)}_a - \vartheta))dP^n_\vartheta$, $n \in \mathbb{N}$, of superefficient estimators like $\hat{\vartheta}^{(n)}_a$ behaves irregularly in shrinking neighbourhoods of ϑ_0. This is the content of Le Cam (1953) [p. 327, Theorem 14], which may be written as follows.

(i) If

$$\limsup_{n\to\infty} \int \ell(n^{1/2}(\vartheta^{(n)} - \vartheta_0))dP^n_{\vartheta_0} < \int \ell dN(0, 1/I(\vartheta_0))$$

for some $\vartheta_0 \in \Theta$, then

$$\limsup_{n\to\infty} \sup_{|\vartheta-\vartheta_0|<\delta} \int \ell(n^{1/2}(\vartheta^{(n)} - \vartheta))dP^n_{\vartheta} > \int \ell dN(0, 1/I(\vartheta_0)).$$

(ii) If for some estimator sequence $\vartheta^{(n)}$, $n \in \mathbb{N}$,

$$\limsup_{n\to\infty} \int \ell(n^{1/2}(\vartheta^{(n)} - \vartheta))dP^n_{\vartheta} \leq \int \ell dN(0, 1/I(\vartheta)) \quad \text{for every } \vartheta \in \Theta, \tag{5.9.1}$$

then equality holds in (5.9.1) for λ-a.a. $\vartheta \in \Theta$. (Le Cam 1953, p. 314, Corollary 8.1.)

Though Rao cites Le Cam (1953) in each of his papers, he ignores the main results for this paper, except for the Hodges-example of superefficiency. His declared goal was to find a concept of asymptotic efficiency which excludes superefficiency, but he had no clear idea how this could be done. Ignoring the main corpus of Le Cam's paper, Rao started where Fisher had stopped almost forty years ago.

Rao (1961) [p. 537] suggested five equivalent (?) definitions of asymptotic efficiency. Definition (ii) of asymptotic efficiency requires $I_{\vartheta^{(n)}}(\vartheta) \to I(\vartheta)$. This definition is based on the inequality

$$I_{\vartheta^{(n)}}(\vartheta) \leq I(\vartheta) \tag{5.9.2}$$

with $I_{\vartheta^{(n)}}(\vartheta) := \int \left(q^\bullet_n(y, \vartheta)\big/q_n(y, \vartheta)\right)^2 Q^{(n)}_\vartheta(dy)$, which is, in Fisher's interpretation, the "amount of information per observation" contained in $\vartheta^{(n)}$. Here $q_n(\cdot, \vartheta)$ is the density of $Q^{(n)}_\vartheta := P^n_\vartheta \circ n^{1/2}(\vartheta^{(n)} - \vartheta)$. The inequality (5.9.2) is based on the fact that $q^\bullet_n(\cdot, \vartheta)/q_n(\cdot, \vartheta)$ is a conditional expectation of

$$(x_1, \ldots, x_n) \to \sum_{\nu=1}^{n} p^\bullet(x_\nu, \vartheta)/p(x_\nu, \vartheta),$$

given $n^{1/2}(\vartheta^{(n)} - \vartheta)$, with respect to P^n_ϑ. Doob (1936) [p. 415, Theorem 2] gave a precise but somewhat disorganized proof of this inequality. Rao's proof (1961, p. 534, Lemma 1 (iii)) is less precise, but more transparent.

Not being a master of concepts like "conditional expectation", Fisher had to resort to a notion like "summing over all $(x_1, \ldots, x_n)$ for which $T_n(x_1, \ldots, x_n)$ attains the same value". That Doob makes no use of "conditional expectations" is the more surprising since just two years later he presented a paper including a detailed chapter on "conditional probability" (1938, Sect. 3). This, by the way, is the paper notorious for its wrong theorem. On p. 96, Theorem 3.1, Doob tries to generalize Kolmogorov's theorem the existence of a regular conditional probability for probability measures on $(\mathbb{R}, \mathbb{B})$ to more general measurable spaces. In this paper he had overlooked that the proof requires some kind of compact approximability (which is, for instance, given in Polish spaces).

In Theorem 2, p. 535, Rao shows that equality in (5.9.2) follows if

$$n^{1/2}(\vartheta^{(n)} - \vartheta) - \big(\alpha(\vartheta) + \beta(\vartheta)\tilde{\ell}^{\bullet}(\cdot, \vartheta)\big) \to 0 \quad (P^n_\vartheta) \tag{5.9.3}$$

for some functions α and β (with $\alpha = 0$ in most of the subsequent papers). From then on Rao takes (5.9.3) as the definition of asymptotic efficiency. This corresponds to the principle that (see Rao 1961, p. 532)

> the efficiency of a statistic has to be judged by the degree to which the estimate provides an approximation to $[\tilde{\ell}^{\bullet}(\cdot, \vartheta)]$.

In 1963, p. 200, Rao praises his concept for being "... not explicitly linked with any loss functions" (Rao 1963). This is certainly true: It fails to reflect any meaningful property of the estimator sequence $(\vartheta^{(n)})_{n\in\mathbb{N}}$.

Remark One can all but admire Rao's courage to define 2nd order efficiency before he had established a plausible concept of 1st order efficiency. From a formal relation like $\tilde{\ell}^{\bullet}(\cdot, \vartheta) - \big(\alpha(\vartheta) + \beta(\vartheta)n^{1/2}(\vartheta^{(n)} - \vartheta) + \lambda(\vartheta)n^{1/2}(\vartheta^{(n)} - \vartheta)^2\big) \to 0$ (see Rao 1961, p. 532, 1962a, p. 49, 1963, p. 199), one can hardly expect a refined information about the performance of estimator sequences, let alone a result like "first order efficiency implies second order efficiency ". Le Cam (1974) [p. 233] politely says:

> The reader may also have noticed that we did not mention the concept introduced by C.R. Rao under the name of "second order efficiency".

Condition (5.9.3), occasionally with $\alpha = 0$, occurs in most of Rao's papers (1961, p. 537, Definition (iv), 1962a, p. 49, Definition 2.5). In 1963, p. 194, Definition 3C, Rao says

> It would be natural to define uniform first order efficiency as
>
> $$n^{1/2}(\vartheta^{(n)} - \vartheta) - \beta(\vartheta)\tilde{\ell}^{\bullet}(\cdot, \vartheta) \to 0$$
>
> without specifying the value of $\beta(\vartheta)$. It appears that if [this condition] is satisfied for various values of $\beta(\vartheta)$, then it is desirable to choose an estimator for which $\beta(\vartheta)$ is a minimum which is shown to be $(\int \ell^{\bullet}(\cdot, \vartheta)^2 dP_\vartheta)^{-1/2}$ [recte $(\int \ell^{\bullet}(\cdot, \vartheta)^2 dP_\vartheta)^{-1}$]

It escaped Rao's attention that under the condition of uniform convergence, the factor $\beta(\vartheta)$ is uniquely determined (as $1/I(\vartheta)$) (provided I is continuous). (Hint: Use that $\tilde{\ell}^{\bullet}(\cdot, \vartheta + n^{-1/2}a) - \tilde{\ell}^{\bullet}(\cdot, \vartheta) \to a \int \ell^{\bullet}(\cdot, \vartheta)^2 dP^n_\vartheta$, and that $P^n_{\vartheta+n^{-1/2}a}$ and P^n_ϑ are mutually absolutely continuous.) Rao's misconception of the role of uniform convergence is the more surprising since, in the very same paper (Rao 1963), he uses uniform convergence to establish (by means of the Neyman–Pearson Lemma) $1/I(\vartheta)$ as the optimal asymptotic variance (see p. 196, Lemma 2 (ii)).

There is a number of authors who mention relation (5.9.3) (together with its extension to second order efficiency) (Zacks 1971, p. 207 and p. 243; Schmetterer 1966, p. 416/7 and 1974, p. 341; Hájek 1971, p. 161, and 1972, p. 178; Ghosh and Subramanyam 1974, p. 331; Ibragimov and Has'minskii 1981, p. 102), but they make no use of it, nor do they question its interpretation, in particular the role of β.

Rao's Definition (5.9.3) requires basing the concept of asymptotic efficiency on $\rho(\vartheta)$, the asymptotic correlation between $n^{1/2}(\vartheta^{(n)} - \vartheta)$ and $\tilde{\ell}^{\bullet}(\cdot, \vartheta)$. Rao suggests to take $\rho(\vartheta)^2$ as a measure of asymptotic efficiency, hence $\rho(\vartheta) = 1$ as a criterion of optimality. Whatever "approximation by $\tilde{\ell}^{\bullet}(\cdot, \vartheta)$ of the estimate" means—the correlation between $n^{1/2}(\vartheta^{(n)} - \vartheta)$ and $\alpha(\vartheta) + \beta(\vartheta)\tilde{\ell}^{\bullet}(\cdot, \vartheta)$ is certainly not an adequate expression for this approximation, nor is this correlation in a meaningful relation to the asymptotic concentration of $n^{1/2}(\vartheta^{(n)} - \vartheta)$.

To see that neither the relation $I_{\vartheta^{(n)}}(\vartheta) \to I(\vartheta)$ nor $\rho(\vartheta) = 1$ is a meaningful concept of asymptotic efficiency, one might consider the example of the Hodges estimator $\hat{\vartheta}_a^{(n)}$. A somewhat tedious computation shows that the Lebesgue density of $N(\vartheta, 1)^n \circ \hat{\vartheta}_a^{(n)}$ is

$$q_n(y, \vartheta) := \begin{cases} n^{1/2}\varphi(n^{1/2}(y - \vartheta)) \\ \frac{1}{a}n^{1/2}\varphi(n^{1/2}(\frac{y}{a} - \vartheta)) \end{cases} \quad \text{if } |y| \begin{matrix} > \\ < \end{matrix} n^{-1/4}.$$

From this one easily obtains that

(i) $I_{\hat{\vartheta}_a^{(n)}}(\vartheta) \to I(\vartheta)$ for every $\vartheta \in \mathbb{R}$,
(ii) $\rho(\vartheta) = 1$ for every $\vartheta \in \mathbb{R}$.

On the other hand, we have, for $\vartheta = 0$, that $N(\vartheta, 1)^n \circ n^{1/2}(\hat{\vartheta}_a^{(n)} - \vartheta) \Rightarrow N(0, a^2)$, as well as $\int (n^{1/2}(\hat{\vartheta}_a^{(n)} - \vartheta))^2 dN(\vartheta, 1)^n \to a^2$.

Rao's "metaphysical" concepts of asymptotic efficiency were not generally accepted. Lindley (p. 68 in the discussion of Rao 1962a)

> Professor Rao follows in the footsteps of Fisher in basing his thesis on intuitive considerations of estimation that people, like myself, who lack such penetrating intuition, cannot aspire to.

Neyman (p. 90 in the discussion of Rao 1962b),

> I would like to request Professor Rao to explain his philosophical standpoint a little more clearly

Remark There is, by the way, an even more mystical concept of "optimality". Godambe (1960) [and many more papers] calls an estimating equation $\sum_{\nu=1}^{n} g(x_\nu, \vartheta) = 0$ with $\int g(\cdot, \vartheta) dP_\vartheta = 0$ "optimal" if the variance $\sigma^2(\vartheta)$ of $g(\cdot, \vartheta) / \int \partial_\vartheta g(\cdot, \vartheta) dP_\vartheta$ is minimal. Accordingly, estimators derived from an optimal estimating equation are "optimal"—by definition. In this sense, the estimating equation based on $g(\cdot, \vartheta) = \partial_\vartheta \log p(\cdot, \vartheta)$ is optimal, and this establishes the optimality of the ML-sequence (for every finite sample size!). One can hardly disagree with Hájek (1971) [p. 161], who says that

> Professor Godambe's suggestion how to prove the "optimality" of the maximum likelihood estimate for any finite n is ... not convincing enough.

Rao (1961) [p. 196, Lemma 2] gives conditions under which $P^n_\vartheta \circ n^{1/2}(\vartheta^{(n)} - \vartheta) \Rightarrow N(0, \sigma^2(\vartheta))$ locally uniformly implies that σ^2 is continuous and $\sigma^2(\vartheta) \geq 1/I(\vartheta)$. The inequality is proved using test theory. Though Rao considered the condition of uniform convergence as an important idea (see Sect. 5.6 for the question of priority), he did not fully exploit its impact. With a more subtle use of uniform convergence, he could have obtained the inequality $\sigma^2(\vartheta) \geq 1/I(\vartheta)$ earlier and without recourse to test theory. In (Rao 1961, p. 534, assumption 4, 1962b, p. 80, and 1963, p. 198) he assumes that $P^n_\vartheta \circ (n^{1/2}(\vartheta^{(n)} - \vartheta), \tilde{\ell}^\bullet(\cdot, \vartheta))$, $n \in \mathbb{N}$, converges to a normal limit distribution $N(0, \Sigma(\vartheta))$ of the form

$$\Sigma = \begin{pmatrix} \sigma^2 & \sigma_{12} \\ \sigma_{12} & I \end{pmatrix}, \tag{5.9.4}$$

with $I(\vartheta) := \int (\ell^\bullet(\cdot, \vartheta))^2 dP_\vartheta$.

Remark The existence of a limit distribution for $P^n_\vartheta \circ (n^{1/2}(\vartheta^{(n)} - \vartheta), \tilde{\ell}^\bullet(\cdot, \vartheta))$, $n \in \mathbb{N}$, is guaranteed if $\vartheta^{(n)}$, $n \in \mathbb{N}$, is regular (see van der Vaart 1991, p. 181, Theorem 2.1 and p. 198, Lemma A.1). Restrictive is the assumption that this joint distribution is normal, an assumption stronger than the asymptotic normality of $n^{1/2}(\vartheta^{(n)} - \vartheta)$. □

Assumption (5.9.4) is used in the somewhat mysterious Lemma 4, p. 198, to show that the correlation between $n^{1/2}(\vartheta^{(n)} - \vartheta)$ and $\tilde{\ell}^\bullet(\cdot, \vartheta)$ is unity iff $n^{1/2}(\vartheta^{(n)} - \vartheta) - \tilde{\ell}^\bullet(\cdot, \vartheta)/I(\vartheta) \to 0 \quad (P^n_\vartheta)$. A more subtle use of assumption (5.9.4) entails a much stronger result, namely: $n^{1/2}(\vartheta^{(n)} - \vartheta) - \tilde{\ell}^\bullet(\cdot, \vartheta)/I(\vartheta)$ and $\tilde{\ell}^\bullet(\cdot, \vartheta)$ are asymptotically stochastically independent. Though confined to asymptotically normal estimator sequences, this indicates what the essence of the Convolution Theorem is in general: The stochastic independence between $n^{1/2}(\vartheta^{(n)} - \vartheta) - \tilde{\ell}^\bullet(\cdot, \vartheta)/I(\vartheta)$ and $\tilde{\ell}^\bullet(\cdot, \vartheta)$ (see (5.13.2)). It would have been easy for Rao to prove, at a moderate level of rigor, that (5.9.4) implies

$$P^n_{\vartheta+n^{-1/2}a} \circ (n^{1/2}(\vartheta^{(n)} - \vartheta), \tilde{\ell}^\bullet(\cdot, \vartheta)) \Rightarrow N((a\sigma_{12}, a^2 I)^\top, \Sigma). \tag{5.9.5}$$

Beyond that, Rao (1963) cites Le Cam (1960), who provides in Theorem 2.1 (6), p. 40, the basis for a precise proof. (This theorem is, by the way, the source from which Hájek and Šidák (1967) [p. 208] extracted "Le Cam's 3rd Lemma".)

Relation (5.9.5) implies

$$P^n_{\vartheta+n^{-1/2}a} \circ n^{1/2}(\vartheta^{(n)} - (\vartheta + n^{-1/2}a)) \Rightarrow N(a(\sigma_{12} - 1), \sigma^2). \tag{5.9.6}$$

If the estimator sequence is regular (Rao even assumes uniform convergence on compact subsets of Θ), relation (5.9.6) implies $\sigma_{12} = 1$, an essential point missed by Rao. Relation (5.9.5) with $\sigma_{12} = 1$ implies

$$P_\vartheta \circ (n^{1/2}(\vartheta^{(n)} - \vartheta) - \tilde{\ell}^\bullet(\cdot, \vartheta)/I(\vartheta), \tilde{\ell}^\bullet(\cdot, \vartheta)) \Rightarrow N(0, \bar{\Sigma})$$

with

$$\bar{\Sigma} = \begin{pmatrix} \sigma^2 - I^{-1} & 0 \\ 0 & I \end{pmatrix},$$

which establishes I^{-1} as a lower bound for the asymptotic variance of regular (asymptotically normal) estimator sequences. At the same time, it asserts the asymptotic stochastic independence between $n^{1/2}(\vartheta^{(n)} - \vartheta) - \tilde{\ell}^\bullet(\cdot, \vartheta)/I(\vartheta)$ and $\tilde{\ell}^\bullet(\cdot, \vartheta)$, $n \in \mathbb{N}$.

Moreover, $\sigma_{12} = 1$ in (5.9.4) implies that $\rho(\vartheta)$, the asymptotic correlation between $n^{1/2}(\vartheta^{(n)} - \vartheta)$ and $\tilde{\ell}^\bullet(\cdot, \vartheta)$, is $I(\vartheta)^{-1/2}/\sigma(\vartheta)$. Hence

$$\rho(\vartheta)^2 = I(\vartheta)^{-1}/\sigma^2(\vartheta)$$

is an adequate measure for asymptotic efficiency. This justifies Rao's claim (1962b, p. 77, Definitions) for the case of regular, asymptotically normal estimator sequences.

First solid results

Intrinsic bounds for the asymptotic concentration of estimator sequences for a functional κ at $P_0 \in \mathfrak{P}$ depend on the local properties of $\mathfrak{P}$ and κ at P_0. As the example of Hodges convincingly demonstrated, regularity conditions on the family $\mathfrak{P}$ and on the functional κ are not enough to find a reasonable concept of "asymptotic efficiency": It would need conditions on the estimator sequence to avoid the "evil of superefficiency" (Has'minskii and Ibragimov 1979, p. 100).

The now common concept of "regularity" appears as a deus ex machina. It is, in fact, the outcome of the conceptual development that took place in the decade between 1960 and 1970. From the beginning it was clear that the convergence of $Q_\vartheta^{(n)} = P_\vartheta^n \circ n^{1/2}(\hat{\vartheta}^{(n)} - \vartheta)$ to a limit distribution Q_ϑ (which was usually assumed to be normal) needed to be "uniform" in some asymptotic sense. It was, however, not clear how this uniformity could be expressed adequately.

To show that ML-sequences are asymptotically optimal was the purpose of the asymptotic considerations; but there was no clear concept for the optimality of multivariate limit distributions. The outcome of these endeavors was that "regularity" in the sense of definition is the adequate expression for locally uniform performance of the estimator sequences. The maximal concentration on convex sets symmetric about the origin came out as quite a surprise.

Following the course of history, we start with the estimation of ϑ in one-parameter families $\{P_\vartheta : \vartheta \in \Theta\}$, $\Theta \subset \mathbb{R}$. The intention initially was to find an asymptotic bound for the concentration of estimators; a side result was that the limit distribution depends continuously on ϑ.

The first result is due to Rao: If $P_\vartheta^n \circ n^{1/2}(\vartheta^{(n)} - \vartheta) \Rightarrow N(0, \sigma^2(\vartheta))$ uniformly on compact subsets of Θ, then (Rao 1963, p. 196, Lemma 2(i)) the map $\vartheta \to \sigma^2(\vartheta)$ is continuous if the Lebesgue density $p(\cdot, \vartheta)$ of P_ϑ is a continuous function of ϑ, and furthermore (p. 196, Lemma 2(ii)) we have $\sigma^2(\vartheta) \geq 1/I(\vartheta)$. (Observe a misprint in Lemma 2(ii). Read $\geq$ instead of $\leq$.) According to Rao's Lemma 3, p. 197, the equality $\sigma^2(\vartheta) = 1/I(\vartheta)$ is achieved if

$$n^{1/2}(\vartheta^{(n)} - \vartheta) - \tilde{\ell}^\bullet(\cdot, \vartheta)/I(\vartheta) \to 0.$$

Rao's result, based on the continuity of

$$\vartheta \to P_\vartheta^n\{n^{1/2}(\vartheta^{(n)} - \vartheta) \leq t\} \quad \text{for } t \in \mathbb{R},$$

extends to general families $\mathfrak{P}$ endowed with the sup-metric d and estimator sequences $\kappa^{(n)}$ for a d-continuous functional $\kappa : \mathfrak{P} \to \mathbb{R}^p$ fulfilling

$$P^n \circ n^{1/2}(\kappa^{(n)} - \kappa(P)) \Rightarrow Q_P. \tag{5.9.7}$$

If the convergence in (5.9.7) is locally uniform at P_0, then $P \to Q_P$ is continuous at P_0 (by Proposition 5.3.5).

In addition to the continuity of the density as a function of ϑ, Rao requires a number of regularity conditions (see pp. 194/5) that made the Cramér type regularity conditions look moderate for some sequence $\vartheta_n \to \vartheta_0$. Weak convergence of $Q_\vartheta^{(n)}$ to Q_ϑ, uniformly on compact subsets of Θ as a remedy against superefficiency was suggested independently by C.R. Rao (1963) [p. 196, Lemma 2] and Wolfowitz (1963; see in this connection Wolfowitz 1965, p. 250, footnote 3). It was used as a condition in Kaufman (1966) [p. 157] and Inagaki (1970) [p. 3, (vi)]. These authors never give explicit reasons for requiring compactness. What seems to be natural is uniform convergence on open sets. To require uniform convergence on compact sets (with non-empty interior) is, perhaps, motivated by the fact that uniform convergence on (small) open sets implies uniform convergence on (large) compact sets, and that uniformity may fail at the boundary of open subsets of Θ. For a theory concerning general families $\mathfrak{P}$ it seems appropriate to assume uniform convergence on a given open set, or locally uniform convergence on $\mathfrak{P}$.

Wolfowitz (1965) considers—more generally—estimator sequences converging to some not necessarily normal limit distribution, say Q_ϑ. He shows that

$$Q_\vartheta(-t' < n^{1/2}(\vartheta^{(n)} - \vartheta) < t'') \leq N(0, 1/I(\vartheta))(-t', t'') \quad \text{for } t', t'' \geq 0$$

(see p. 251, relation (2.1)) for any estimator sequence such that $Q_\vartheta^{(n)}(-\infty, t], n \in \mathbb{N}$, converges to $Q_\vartheta(-\infty, t]$, uniformly in ϑ and t. Unlike Rao and Schmetterer he shows that ML-sequences converge in this sense to $N(0, 1/I(\vartheta))$ (see p. 253). Hence ML-sequences are optimal in the class of all estimator sequences fulfilling Wolfowitz's conditions of uniform convergence.

When Wolfowitz submitted his paper in 1965 he was still unaware that lower and upper medians are identical, a result which had been obtained in the meantime by Kaufman (see Wolfowitz 1965, p. 259, footnote). In our presentation of the results of Schmetterer and Wolfowitz this simplification has been taken into account. Surprisingly, Roussas (1972) still struggles with lower and upper medians (see p. 130).

Schmetterer (1966) claims that Theorem 2.2, p. 308, based on the condition of regular convergence, generalizes Rao's result. In fact, his assumption of regular convergence for every $\vartheta \in \Theta$ is, on a locally compact parameter set Θ, equivalent to Rao's assumption of uniform convergence on compact subsets. This seems to have escaped Schmetterer's attention. Schmetterer uses the condition of continuous convergence for two different purposes: To show that estimator sequences converge to $N(0, \sigma^2(\vartheta))$ continuously (less would have done) and to show that σ^2 is continuous (which follows from the fact that all limit distributions of uniformly convergent estimator sequences are continuous). Schmetterer requires (see p. 307) that $\vartheta \to \ell^\bullet(x, \vartheta)$ is continuous, a fact already familiar to Rao (1963) [p. 196, Lemma 2(i)] and Wolfowitz (1965) [p. 254, Lemma 2].

Still, the achievements of Rao and Wolfowitz are confined to one-parameter families. The next step followed soon: This was the paper by Kaufman (1966), which made all preceding results look poor. He shows that $n^{1/2}(\hat{\vartheta}^{(n)} - \vartheta)_{n\in\mathbb{N}}$ and $n^{1/2}(\vartheta^{(n)} - \hat{\vartheta}^{(n)})_{n\in\mathbb{N}}$ are asymptotically independent if the estimator sequence $\vartheta^{(n)}$, $n \in \mathbb{N}$ is sufficiently regular and $(\hat{\vartheta}^{(n)})_{n\in\mathbb{N}}$ is a ML-sequence. This discovery is the basis for an operational concept of multivariate optimality.

Some Properties of Limit Distributions

In this section we summarize properties of limit distributions which had been obtained by various authors as side results. The conditions in these papers are not easy to compare since continuity of the limit distribution usually occurs as a side result, and the regularity conditions are chosen with a different main result in mind.

Regularly attainable limit distributions have a Lebesgue density. Convolution products inherit certain properties of the factors: $Q_1 * Q_2$ is nonatomic or absolutely continuous with respect to the Lebesgue measure if at least one of the factors has this property (see e.g. Lukacs 1960, p. 45, Theorem 3.3.2). Hence limit distributions which are regularly attainable are, as a consequence of the Convolution Theorem, absolutely continuous with respect to the Lebesgue measure. Of historical interest are properties of limit distributions which were known prior to the Convolution Theorem.

Wolfowitz (1965) [p. 253, Lemma 1] asserts that uniformly attainable limit distributions have continuous distribution functions (for $\Theta \subset \mathbb{R}$). According to Kaufman (1966) [p. 173, Lemma 5.3 and p. 174, Lemma 5.4], they have a positive Lebesgue density for $\Theta \subset \mathbb{R}^k$.

The following more general result can be found in Pfanzagl (1994) [p. 229], Proposition 7.1.11): Let Θ be an open subset of $\mathbb{R}^k$, and $\kappa : \Theta \to \mathbb{R}^p$ a differentiable functional. Then

$$Q^{(n)}_{\vartheta+c_n^{-1}a} = P^{(n)}_{\vartheta+c_n^{-1}a} \circ c_n(\kappa^{(n)} - \kappa(\vartheta + c_n^{-1}a)) \Rightarrow Q_\vartheta$$

for every $a \in \mathbb{R}^k$ implies that $Q_\vartheta \ll \lambda^k$ if $(P^{(n)}_{\vartheta+c_n^{-1}a})_{n\in\mathbb{N}}$ is contiguous to $(P^{(n)}_\vartheta)_{n\in\mathbb{N}}$, for every $a \in \mathbb{R}^k$.

Locally uniformly attainable limit distributions are continuous. According to Rao (1963) [p. 197, Lemma 3(i)], continuity of $\vartheta \to Q_\vartheta$ follows from the continuity of $\vartheta \to Q_\vartheta^{(n)}$. In Lemma 2, pp. 254/5, Wolfowitz gives a less transparent proof, also based on the continuity of $\vartheta \to Q_\vartheta^{(n)}$, concealed in condition 4.15, p. 255.

Kaufman (1966) [p. 175, Lemma 5.5] shows that continuity of $\vartheta \to P_\vartheta$ with respect to the sup-distance suffices. Schmetterer (1966) [p. 308, Theorem 2.2] derives continuity of the (normal) limit distribution (i.e., continuity of $\vartheta \to \sigma^2(\vartheta)$) from continuous convergence of $Q_\vartheta^{(n)}(-\infty, t]$, $n \in \mathbb{N}$, to $N(0, \sigma^2(\vartheta)(-\infty, t]$ for every $t \in \mathbb{R}$, a condition slightly weaker but less convincing than uniform convergence.

A weaker condition for the continuity of limit distributions. In the following, $D(Q', Q'')$ denotes $|\int h dQ' - \int h dQ''|$, with a fixed bounded and continuous function f.

Continuity of $\vartheta \to Q_\vartheta$ at ϑ_0 follows if $D(Q_{\vartheta_n}^{(n)}, Q_{\vartheta_0}) \to 0$ for arbitrary sequences $\vartheta_n \to \vartheta_0$; the same relation for all sequences $(\vartheta_n)_{n\in\mathbb{N}}$ with $\limsup_{n\to\infty} c_n \|\vartheta_n - \vartheta_0\| < \infty$ is not strong enough. To start from the convergence of $Q_{\vartheta_n}^{(n)}$, $n \in \mathbb{N}$, to a fixed Q_{ϑ_0} is sanctioned by tradition, yet far from reasonable. Recall that locally uniform convergence is equivalent to $D(Q_{\vartheta_n}^{(n)}, Q_{\vartheta_n}) \to 0$ if $\vartheta_n \to \vartheta_0$. Since $\vartheta \to Q_\vartheta^{(n)}$ is continuous, continuity of $\vartheta \to Q_\vartheta$ follows by Lemma 5.8.3. A more subtle argument shows that $D(Q_{\vartheta_n}^{(n)}, Q_{\vartheta_n}) \to 0$ for all sequences $(\vartheta_n)_{n\in\mathbb{N}}$ fulfilling $\limsup_{n\to\infty} c_n \|\vartheta_n - \vartheta_0\| < \infty$ suffices to prove continuity of $\vartheta \to Q_\vartheta$ under additional regularity conditions. Continuity of $\vartheta \to Q_\vartheta^{(n)}$ plays no role in this argument. Applications of this theorem to parametric subfamilies of general families are straightforward.

The following Theorem is an improved version of the Proposition in Pfanzagl (1999a) [p. 72]. Condition (5.9.8) and condition (5.9.9) are fulfilled for LAN-families.

Theorem 5.9.1 *Let* $\{(P_\vartheta^{(n)})_{n\in\mathbb{N}} : \vartheta \in \Theta\}$, $\Theta \subset \mathbb{R}^k$, *be a family such that*

$$\limsup_{n\to\infty} \|a\|^{-1} d(P^{(n)}_{\vartheta_0 + c_n^{-1}a}, P^{(n)}_{\vartheta_0}) < \infty, \quad a \in \mathbb{R}^k, \tag{5.9.8}$$

for some sequence $(c_n)_{n\in\mathbb{N}} \uparrow \infty$ *fulfilling*

$$\lim_{n\to\infty} c_{n+1}/c_n = 1.$$

Let $\kappa : \Theta \to \mathbb{R}^p$, $p \le k$, *be a differentiable functional. Let* $Q_\vartheta^{(n)} := P_\vartheta^{(n)} \circ c_n(\kappa^{(n)} - \kappa(\vartheta))$. *Assume there is a family* $Q_\vartheta | \mathbb{B}^p$, $\vartheta \in \Theta$, *such that*

$$\lim_{n\to\infty} D(Q^{(n)}_{\vartheta_0 + c_n^{-1}a_n}, Q_{\vartheta_0 + c_n^{-1}a_n}) = 0 \tag{5.9.9}$$

for every sequence $(a_n)_{n\in\mathbb{N}}$, *converging to some* $a \neq 0$. *Then the following holds true: If* Q_{ϑ_0} *is nonatomic, then* $\vartheta \to Q_\vartheta$ *is continuous at* ϑ_0.

Proof We shall show that $\vartheta_m \to \vartheta_0$ implies $D(Q_{\vartheta_m}, Q_{\vartheta_0}) \to 0$. W.l.g. we assume that $\|\vartheta_m - \vartheta_0\| \downarrow 0$. Given $\varepsilon > 0$, let n_m be defined by $c_{n_m} \le \varepsilon/\|\vartheta_m - \vartheta_0\| < c_{n_m+1}$. Observe that $(n_m)_{m\in\mathbb{N}} \uparrow \infty$. For $a_m := (\vartheta_m - \vartheta_0)c_{n_m}$ we have

$$\varepsilon c_{n_m}/c_{n_m+1} < \|a_m\| \le \varepsilon. \tag{5.9.10}$$

Since $\|a_m\| \le \varepsilon$, there exists a convergent subsequence, say $(a_m)_{m\in\mathbb{N}_0} \to a$. Because of (5.9.10), $\|a\| = \varepsilon$. We have

$$D(Q_{\vartheta_m}, Q_{\vartheta_0}) \le D(Q_{\vartheta_m}, Q_{\vartheta_m}^{(n_m)}) + D(Q_{\vartheta_m}^{(n_m)}, Q_{\vartheta_0}).$$

Since $\vartheta_m = \vartheta_0 + c_{n_m}^{-1} a_m$, relation (5.9.9) implies $\lim_{n\in\mathbb{N}_0} D(Q_{\vartheta_m}, Q_{\vartheta_m}^{(n_m)}) = 0$. To simplify our notations, let

$$g_n(\cdot, \vartheta) := c_n(\kappa^{(n)} - \kappa(\vartheta)).$$

With these notations,

$$\begin{aligned}
&D(Q_{\vartheta_m}^{(n_m)}, Q_{\vartheta_0}) = D(P_{\vartheta_m}^{(n_m)} \circ g_{n_m}(\cdot, \vartheta_m), Q_{\vartheta_0})\\
&\le D(P_{\vartheta_m}^{(n_m)} \circ g_{n_m}(\cdot, \vartheta_m), P_{\vartheta_0}^{(n_m)} \circ g_{n_m}(\cdot, \vartheta_m)) + D(P_{\vartheta_0}^{(n_m)} \circ g_{n_m}(\cdot, \vartheta_m), Q_{\vartheta_0})\\
&\le d(P_{\vartheta_m}^{(n_m)}, P_{\vartheta_0}^{(n_m)}) + D(P_{\vartheta_0}^{(n_m)} \circ g_{n_m}(\cdot, \vartheta_m), Q_{\vartheta_0}).
\end{aligned}$$

Let J denote the Jacobian of κ at ϑ_0. By assumption,

$$g_{n_m}(\cdot, \vartheta_m) - g_{n_m}(\cdot, \vartheta_0) \to Ja \quad (P_{\vartheta_0}^{(n_m)}).$$

Writing $Q + Ja$ for $Q \circ (u \to u + Ja))$, we have

$$\begin{aligned}
&D(P_{\vartheta_0}^{(n_m)} \circ g_{n_m}(\cdot, \vartheta_m), Q_{\vartheta_0}) = D(P_{\vartheta_0}^{(n_m)} \circ g_{n_m}(\cdot, \vartheta_0) + Ja, Q_{\vartheta_0}) + o(n^0, P_{\vartheta_0}^{(n_m)})\\
&\le D(Q_{\vartheta_0}^{(n_m)} + Ja, Q_{\vartheta_0} + Ja) + D(Q_{\vartheta_0}, Q_{\vartheta_0} + Ja) + o(n^0, P_{\vartheta_0}^{(n_m)}),
\end{aligned}$$

hence

$$\limsup_{m\in\mathbb{N}_0} D(Q_{\vartheta_m}, Q_{\vartheta_0}) \le \limsup_{m\in\mathbb{N}_0} d(P_{\vartheta_m}^{(n_m)}, P_{\vartheta_0}^{(n_m)}) + D(Q_{\vartheta_0}, Q_{\vartheta_0} + Ja).$$

Since this relation holds for every $\varepsilon > 0$, we have

$$\lim_{m\in\mathbb{N}_0} D(Q_{\vartheta_m}, Q_{\vartheta_0}) = 0. \tag{5.9.11}$$

Since any subsequence of $\mathbb{N}$ contains a subsequence $\mathbb{N}_0$ fulfilling (5.9.11), the assertion follows. □

5.10 Regular Convergence and Continuity in the Limit

When Kaufman wrote his fundamental paper (1966) the problem was clear: There was an intuitively convincing method for the construction of estimators, the ML method. ML estimators were generally considered as "optimal". Yet, outstanding scholars had failed in their attempts to prove this optimality. In fact, nobody had an idea what optimality could mean for multivariate estimator sequences.

Kaufman gave regularity conditions for the (uniform) convergence of ML-sequences of a k-dimensional parameter ϑ to the (already well known) normal limit distribution with covariance matrix $\Lambda(\vartheta) = L(\vartheta)^{-1}$, and he showed that $\Lambda(\vartheta)$ is the optimal covariance matrix for estimator sequences converging uniformly on compact sets of Θ. His paper established the concept of multidimensional optimality of limit distributions as maximal concentration on sets that are convex and symmetric about the origin.

In his paper from 1970, Hájek approached the problem of asymptotic optimality of estimator sequences from a different point of view: Instead of presenting an estimator sequence and proving its asymptotic optimality, he started from "regularity" as the essential property of estimator sequences, and he gave a bound for the concentration of regular estimator sequences. (His Theorem refers to general LAN-sequences and is, therefore, not restricted to the i.i.d. case.)

In Kaufman's paper, the optimality of the ML-sequence follows from the fact that any convergent estimator sequence is a convolution product involving the limit distribution of the ML-sequence. Hájek's Theorem provides a bound for the concentration of regular estimator sequences; yet it remains to be shown that regular estimator sequences attaining this bound do exist.

The idea to replace "uniform convergence on compact subsets of Θ" by "regularity", i.e. by the convergence of $Q_\vartheta^{(n)}$ along sequences $\vartheta + n^{-1/2}a$ is now generally accepted. (See Ibragimov and Has'minskii 1970; Bickel et al. 1993; Witting and Müller-Funk 1995.) It was, in fact, regular convergence that was implicitly used by many authors. Starting from uniform convergence of $P_\vartheta^n \circ n^{1/2}(\vartheta^{(n)} - \vartheta)$ to a limit distribution Q_ϑ, they used convergence on paths $\vartheta + n^{-1/2}a$ only. (C.R. Rao, 1963, p. 196, Lemma 2, and Bahadur 1964, p. 1546, Proposition 1, and many more.) In test theory, the use of $P_{\vartheta+n^{-1/2}a}$ for approximations to the power function turned up quite early (see e.g. Eisenhart 1938, p. 32).

According to the Convolution Theorem, "regular convergence" is strong enough for obtaining a bound for the concentration of estimator sequences. Yet regularly convergent sequences $Q_\vartheta^{(n)}$, $n \in \mathbb{N}$, may have unpleasant properties.

(i) *Regular convergence of $Q_\vartheta^{(n)}$ to Q_ϑ at ϑ_0 does not imply that Q_ϑ is continuous at ϑ_0.*

That regular convergence is not strong enough to imply continuity of the limit distribution was first observed by Tierney (1987) [p. 430]. Beware of a misprint in line 20 of p. 430: Replace ϑ by ϑ_n in $\mathscr{L}(\sqrt{n}(T_n - \vartheta)|\vartheta_n))$. For $\mathfrak{P} = \{N(\vartheta, 1) : \vartheta \in \mathbb{R}\}$, and the estimator $\vartheta^{(n)}(\mathbf{x}_n) := (1 - n^{-1/2})\overline{x}_n + n^{-1/2}x_1$ if $|\overline{x}_n| < 1/\log n$

and $\vartheta^{(n)}(\mathbf{x}_n) := \overline{x}_n$ otherwise, one easily finds that $\lim_{n\to\infty} Q^{(n)}_{\vartheta_0+n^{-1/2}a}$ is $N(0,2)$ for $\vartheta_0 = 0$ if $a \in \mathbb{R}$, and $N(0,1)$ otherwise.

(ii) *Even if Q_ϑ is continuous at ϑ_0, regular convergence of $Q^{(n)}_\vartheta$ to Q_ϑ at ϑ_0 does not imply that $Q^{(n)}_{\vartheta_n} \to Q_{\vartheta_0}$ for arbitrary sequences $\vartheta_n \to \vartheta_0$.*
In spite of the fact that $Q^{(n)}_{\vartheta_0+n^{-1/2}a}(A) \to Q_\vartheta(A)$ for every set $A \in \mathbb{B}$ with boundary zero under Q_{ϑ_0}, it is not excluded that

$$Q^{(n)}_{\vartheta_n}(A) \to 0 \quad \text{for every bounded set } A \in \mathbb{B}^k$$

if the convergence of ϑ_n to ϑ_0 is too slow.

The more subtle question whether there are continuous limit distributions which can be attained regularly, but not uniformly, was answered to the affirmative by Fabian and Hannan (1982) by a (somewhat artificial) example in their Remark 3.2, p. 462.

Example Let $\mathbf{x}_n = (x_1, \ldots, x_n)$ be distributed as $N^n_\vartheta|\mathbb{B}^n$, with N_ϑ the normal distribution with mean ϑ and variance 1. Let

$$\vartheta^{(n)}(\mathbf{x}_n) := \begin{cases} \overline{x}_n \\ \overline{x}_n + n^{-1/2}u_n \end{cases} \quad \text{if } \overline{x}_n \begin{matrix} \le \\ > \end{matrix} n^{-1/2}\omega_n$$

with $\omega_n \uparrow \infty$ and $n^{-1/2}\omega_n \to 0$. We shall show that

$$P^n_{\vartheta_n} \circ n^{1/2}(\vartheta^{(n)} - \vartheta_n) \Rightarrow N_0 \quad \text{if} n^{1/2}\vartheta_n,\ n \in \mathbb{N}, \text{ is bounded} \tag{5.10.1}$$

and

$$P^n_{\vartheta_n} \circ (n^{1/2}(\vartheta^{(n)} - \vartheta_n) - u_n) \Rightarrow N_0 \quad \text{if} n^{1/2}\vartheta_n > 2\omega_n. \tag{5.10.2}$$

From this the assertions follow easily. (To prove the discontinuity of the regularly attainable limit distribution use $u_n = u$ for $n \in \mathbb{N}$.)

For convenience we introduce the random variable $\xi := n^{1/2}(\vartheta_n - \vartheta)$ which is under N^n_ϑ distributed as N_0. Hence the distribution of $n^{1/2}(\vartheta^{(n)}(x) - \vartheta)$ under N^n_ϑ is the same as the distribution of

$$H^{(n)}_\vartheta(\xi) := \begin{cases} \xi \\ \xi + u_n \end{cases} \quad \text{if } |n^{1/2}\vartheta + \xi| \begin{matrix} \le \\ > \end{matrix} \omega_n$$

under N_0. Since

$$N_0\{\xi : H^{(n)}_{\vartheta_n}(\xi) = \xi\} \to 1 \quad \text{if} n^{1/2}\vartheta_n, n \in \mathbb{N}, \text{ is bounded}$$

and

$$N_0\{\xi : H^{(n)}_{\vartheta_n}(\xi) = \xi + u_n\} \to 1 \quad \text{if} n^{1/2}\vartheta_n > 2\omega_n,$$

this implies (5.10.1) and (5.10.2). □

The example above presents a particular regularly attainable limit distribution which gives a misleading impression of the true distribution even for large sample sizes. If the basic family $\mathfrak{P}$ is large, then this may be necessarily so for any estimator sequence.

A first asymptotic impossibility result is due to Klaassen (1979). He shows (p. 853, Remark; see also Klaassen 1981, p. 62, Theorem 3.2.2) that for any sequence of equivariant and antisymmetric (hence median unbiasedmedian unbiased estimator) estimators of the median $\kappa(P)$ in the family $\mathfrak{P}$ of all symmetric densities with finite $\sigma^2(P) := \int (p'/p)^2 dP$,

$$\lim_{n\to\infty} \inf_{P\in\mathfrak{P}_0} P^n\{n^{1/2}(\kappa^{(n)} - \kappa(P))/\sigma(P) < t\} = 1/2 \quad \text{for every } t > 0,$$

if $\mathfrak{P}_0$ is a sufficiently large subset of $\mathfrak{P}$.

Ritov and Bickel (1990) use the estimation of $\kappa(P) = \int p(x)^2 dx$ in the family $\mathfrak{P}$ of all probability measures on $\mathbb{B}$ with λ-density p to illustrate the problems with uniformity in general families. In a slightly simplified version, their main result (Theorem 1, p. 926) reads as follows: For every $\varepsilon > 0$ there exists a compact subset $\mathfrak{P}_\varepsilon \subset (\mathfrak{P}, d)$ of diameter less than ε with the following property: For every estimator sequence $\kappa^{(n)}$, $n \in \mathbb{N}$, and every $\alpha > 0$, $\liminf_{n\to\infty} P^n\{|\kappa^{(n)} - \kappa(P)| > n^{-\alpha}\} > 0$ for some $P \in \mathfrak{P}_\varepsilon$. This result may be improved as follows (see Pfanzagl 2002a, p. 93): For any sequence $\varepsilon_n \downarrow 0$, the relation

$$\lim_{n\to\infty} P^n\{|\kappa^{(n)} - \kappa(P)| > \varepsilon_n\} = 1$$

holds for P in a dense subset of $(\mathfrak{P}, d)$.

The essential point in such nonexistence theorems is that the minimal asymptotic variance of a continuous functional $\kappa : \mathfrak{P} \to \mathbb{R}$, given by $\int \kappa^*(\cdot, P)^2 dP$, is discontinuous at some P_0. (See Pfanzagl 1982, p. 165, Corollary).

5.11 The Neyman–Pearson Lemma and Applications

The following version of the Neyman–Pearson Lemma refers to the case of mutually absolutely continuous probability measures $P_i|(X, \mathscr{A})$, $i = 1, 2$, with densities p_i.

Definition 5.11.1 A critical function φ_0 is of Neyman–Pearson type for $P_0 : P_1$ if

$$\varphi_0(x) = \begin{cases} 1 \\ 0 \end{cases} \text{if} \quad \frac{p_1(x)}{p_0(x)} \quad \text{for some } c > 0.$$

Definition 5.11.1 assumes nothing about the value of $\varphi_0(x)$ if $p_1(x) = c p_0(x)$.

It is easily seen that for any $\alpha \in (0, 1)$ there exists a critical function φ_0 of Neyman–Pearson type such that

$$\int \varphi_0(x) P_0(x) = \alpha.$$

Neyman–Pearson Lemma. *A critical function φ_0 with $\int \varphi_0(x) P_0(dx) \in (0, 1)$ is of Neyman–Pearson type for $P_0 : P_1$ iff for every critical function φ,*

$$\int \varphi(x) P_0(dx) = \int \varphi_0(x) P_0(dx)$$

implies

$$\int \varphi(x) P_1(dx) \leq \int \varphi_0(x) P_1(dx). \tag{5.11.1}$$

Addendum. *Equality in* (5.11.1) *holds iff*

$$\varphi(x) = \varphi_0(x) \quad \text{if } \varphi_0(x) \in \{0, 1\}.$$

Observe that the value of $\varphi(x)$ is irrelevant for x with $p_1(x) = cp_0(x)$.

According to Schmetterer (1974, footnote on p. 166), a first version of the Neyman–Pearson Lemma occurs in Neyman and Pearson (1936) [p. 207, relation (4)]. The basic idea of this Lemma occurs earlier in Neyman and Pearson (1933) [p. 300, p. 151, relation (25)].

In Wolfowitz's opinion, (Wolfowitz 1970, p. 767, footnote 4) "... the "Neyman–Pearson fundamental Lemma", which, no matter how "fundamental" it may be, is pretty trivial to prove and not difficult to discover".

Since the textbook by Schmetterer (1974) [Theorem 3.1, p. 166] needs four pages for the proof of the Neyman–Pearson Lemma, it is worthwhile to look for alternatives. Such can be found in Witting (1985) [p. 193, Satz 25] or Pfanzagl (1994) [p. 133, Lemma 4.3.3]. The basic idea of these proofs is the relation

$$(\varphi_0(x) - \varphi(x))(p_1(x) - cp_0(x)) \geq 0 \quad \text{for } x \in X,$$

which holds for every critical function φ.

The first optimality results for the asymptotic concentration of estimator sequences, obtained by C.R. Rao (1963), Bahadur (1964), Wolfowitz (1965), Schmetterer (1966) and Roussas (1968), are based on the Neyman–Pearson Lemma. Here is an outline of the basic idea. For a more general version see Theorem 5.11.4.

Definition 5.11.2 Let $\Theta \subset \mathbb{R}$. The estimator sequence $\vartheta^{(n)}, n \in \mathbb{N}$, is *asymptotically median unbiased* at ϑ_0 if

$$\limsup_{n\to\infty} P_{\vartheta_n}^{(n)}\{\vartheta^{(n)} \le \vartheta_n\} \le 1/2$$

and

$$\limsup_{n\to\infty} P_{\vartheta_n}^{(n)}\{\vartheta^{(n)} \ge \vartheta_n\} \le 1/2$$

for all sequences $\vartheta_n = \vartheta_0 + c_n^{-1}t$, with $|t| \le \varepsilon$, say.

Remark Let, more generally, $Q_n|\mathbb{B}$, $n \in \mathbb{N}$, be such that

$$\limsup_{n\to\infty} Q_n(-\infty, 0] \le 1/2 \quad \text{and} \quad \limsup_{n\to\infty} Q_n[0,\infty) \le 1/2.$$

Then $\lim_{n\to\infty} Q_n\{0\} = 0$, hence

$$\lim_{n\to\infty} Q_n(-\infty, 0] = 1/2 \quad \text{and} \quad \lim_{n\to\infty} Q_n[0,\infty) = 1/2.$$

Hint: $Q_n(-\infty, 0] + Q_n[0,\infty) = 1 + Q_n\{0\}$ implies

$$1 + \limsup_{n\to\infty} Q_n\{0\} \le \limsup_{n\to\infty} Q_n(-\infty, 0] + \limsup_{n\to\infty} Q_n[0,\infty) \le 1.$$

Assume LAN, i.e., the stochastic expansion

$$\Lambda_{nt} = \log(dP_{\vartheta_0+c_n^{-1}t}^{(n)}/dP_{\vartheta_0}^{(n)}) = t\Delta_n(\cdot,\vartheta_0) - \frac{1}{2}t^2L(\vartheta_0) + o(n^0, P_{\vartheta_0}^{(n)}).$$

Then asymptotic bounds for the concentration of asymptotically median unbiased estimator sequences can be obtained (as in C.R. Rao 1963, in the proof of his Lemma 2) by considering $\{\vartheta^{(n)} \le \vartheta_0 + c_n^{-1}t\}$ as a critical region for testing the hypothesis $P_{\vartheta_0+c_n^{-1}t}^{(n)}$ against the alternative $P_{\vartheta_0}^{(n)}$ and comparing this critical region with an asymptotically most powerful critical region

$$C_n(t) := \left\{\Lambda_{nt} \le \frac{1}{2}t^2L(\vartheta_0)^2\right\}.$$

We have

$$P_{\vartheta_0}^{(n)} \circ \Lambda_{nt} \Rightarrow N(-\frac{1}{2}t^2L(\vartheta_0)^2, t^2L(\vartheta_0)^2) \tag{5.11.2}$$

and, consequently, by Le Cam's 3rd Lemma,

$$P_{\vartheta_0+c_n^{-1}t}^{(n)} \circ \Lambda_{nt} \Rightarrow N(\frac{1}{2}t^2L(\vartheta_0)^2, t^2L(\vartheta_0)^2). \tag{5.11.3}$$

This implies

$$P_{\vartheta_0+c_n^{-1}t}^{(n)}(C_n(t)) \to 1/2$$

and

$$P_{\vartheta_0}^{(n)}(C_n(t)) \to \Phi(tL(\vartheta_0)).$$

Since

$$\limsup_{n\to\infty} P_{\vartheta_0+c_n^{-1}t}^{(n)}\{\vartheta^{(n)} \le \vartheta_0 + c_n^{-1}t\} \le 1/2 = \lim_{n\to\infty} P_{\vartheta_0+c_n^{-1}t}^{(n)}(C_n(t)),$$

an asymptotic version of the Neyman–Pearson Lemma implies

$$\limsup_{n\to\infty} P_{\vartheta_0}^{(n)}\{\vartheta^{(n)} \le \vartheta_0 + c_n^{-1}t\} \le \lim_{n\to\infty} P_{\vartheta_0}^{(n)}(C_n(t)) = \Phi(tL(\vartheta_0)).$$

Together with the corresponding relation with t replaced by $-t$, i.e.,

$$\limsup P_{\vartheta_0}^{(n)}\{\vartheta^{(n)} \ge \vartheta_0 - c_n^{-1}t\} \le \lim_{n\to\infty} \Phi(C_n(-t)) = \Phi(-tL(\vartheta_0)),$$

this implies for $t', t'' \ge 0$,

$$\limsup_{n\to\infty} P_{\vartheta_0}^{(n)}\{\vartheta_0 - c_n^{-1}t' < \vartheta^{(n)} \le \vartheta_0 + c_n^{-1}t''\} \le N(0, \Lambda(\vartheta_0))(-t', t''). \quad (5.11.4)$$

□

Observe that relation (5.11.4) refers to estimator sequences which are asymptotically median unbiased . Convergence to a limit distribution is not required.

As mentioned in Sect. 5.9, various authors presented bounds for the concentration of limit distributions under the assumption of uniform convergence, but they only use regular convergence to the median of the limit distribution. Though Bahadur (1964) [p. 1546] and Schmetterer (1966, p. 313, Remark 3.2) realize that their proofs use uniform convergence to the median of the limit distribution only, they are not surprised that this suffices to establish the limit distribution of ML-sequences as optimal, and they don't question what the apparently abundant assumption of uniform convergence to a limit distribution should be good for. Pfanzagl (1970) [p. 1502, Theorem 1] shows that median unbiasedness as such suffices for (5.11.4). It remains mysterious why this theorem is stated for sequences of exactly median unbiased estimators, though the proof obviously holds for estimator sequences which are asymptotically median unbiased only. Roussas (1972) still gives two separate theorems for sequences of median unbiased estimators (Lemma 5.1, p. 149), and estimator sequences converging to a limit distribution with median 0 (Theorem 5.1, p. 153 and Theorem 5.2, p. 154).

Pfanzagl (1972a) [p. 170, Theorem 5.2] gives conditions under which

$$\limsup_{n\to\infty} n^{1/2}\left(P_\vartheta^n\{\vartheta - n^{-1/2}t' < \vartheta^{(n)} < \vartheta + n^{-1/2}t''\} - N(0, \Lambda(\vartheta))(-t', t'')\right) < \infty.$$

With the same techniques, but under stronger regularity conditions, Michel (1974) [p. 207, Theorem] obtains the following result of Berry-Esseen type. If $(\vartheta^{(n)})_{n\in\mathbb{N}}$ is approximately median unbiased in the sense that for every compact $K \subset \Theta$ there is a_K such that, for $\vartheta \in K$,

$$P_\vartheta^n\{\vartheta^{(n)} < \vartheta\} \leq 1/2 + a_K n^{-1/2} \quad \text{and} \quad P_\vartheta^n\{\vartheta^{(n)} > \vartheta\} \leq 1/2 + a_K n^{-1/2},$$

then there is a'_K such that for every $\vartheta \in K$,

$$P_\vartheta^n\{\vartheta - n^{-1/2}t' < \vartheta^{(n)} < \vartheta + n^{-1/2}t''\} \leq N(0, \Lambda(\vartheta))(-t', t'') + a'_K n^{-1/2}.$$

Remark For some authors, the distinction between symmetric and arbitrary intervals (or loss functions) seems to be nothing to speak of; see Witting and Müller-Funk (1995) [p. 423]. Strasser (1985), presents a result concerning the concentration of median unbiased estimators on *arbitrary* intervals containing 0 on p. 162, Lemma 34.2, and a corresponding result for *symmetric* loss functions on p. 362, Lemma 72.1. Rüschendorf (1988) [p. 207, Satz 6.10] is on symmetric loss functions. □

To derive asymptotic median unbiasedness from regular or locally uniform convergence to a limit distribution poses no problems if the limit distribution is of the form $N(0, \sigma^2(\vartheta_0))$ (as in the papers by C.R. Rao, Schmetterer and Bahadur). Any estimator sequence converging regularly to $N(0, \sigma^2(\vartheta_0))$ is asymptotically median unbiased, and relation (5.11.4) can now be rewritten as

$$N(0, \sigma^2(\vartheta_0))(-t', t'') \leq N(0, \sigma_*^2(\vartheta_0))(-t', t'') \quad \text{for } t', t'' \geq 0,$$

which means

$$\sigma^2(\vartheta_0) \geq \Lambda(\vartheta_0).$$

(See C.R. Rao 1963, p. 196, Lemma 2(ii) and Bahadur 1964, p. 1546, Proposition 1; notice the misprint in Rao's Lemma 2(ii) which says $\sigma^2(\vartheta_0) \leq \Lambda(\vartheta_0)$.)

That regularly attainable limit distributions have, under the usual regularity conditions, a positive Lebesgue density and therefore a unique median was established by Kaufman (1966, p. 174, Corollary). Since this result was not yet available to Wolfowitz (1965), and unknown to Schmetterer (1966) and Roussas (1968 and 1972), these authors had to formulate a result with the median replaced by the lower or upper bound of the median interval, respectively.

Problems connected with the nonuniqueness of the median vanish if the optimality assertions are confined to intervals symmetric about 0. For regular one-parameter families, Weiss and Wolfowitz (1966) [p. 61] show that for any regularly attainable limit distribution Q_ϑ,

$$Q_\vartheta(-t, t) \leq N(0, \Lambda(\vartheta))(-t, t), \quad t > 0.$$

The argument in this paper is Bayesian, but the Neyman–Pearson Lemma suffices (see below).

The idea to obtain a bound from the Neyman–Pearson Lemma also works in "almost regular" cases. For shift parameter families, based on a probability measure with support (a, b), Akahira (1975) [II] presents in Theorem 3.1, p. 106, a result which implies, in particular, the following. Let

$$A := \lim_{x \downarrow a} (x-a)^{-1} p(x), \quad B := \lim_{x \uparrow b} (b-x)^{-1} p(x),$$
$$A' := \lim_{x \downarrow a} |p'(x)|, \quad B' := \lim_{x \uparrow b} |p'(x)|.$$

If these quantities are finite and positive, then the M.L. sequence of the location parameter, standardized by $c_n = n^{1/2}(\log n)^{1/2}$, is asymptotically normal with variance $\frac{1}{2}(A'^2/A + B'^2/B)$. According to Theorem 3.2, p. 107, this is the best possible limit distribution for estimator sequences $(\vartheta^{(n)})_{n\in\mathbb{N}}$ converging to their limit distribution Q_ϑ locally uniformly. Akahira's proof is based on the Neyman–Pearson Lemma. Roughly speaking, his Lemma 3.1, p. 102, implies the validity of conditions (5.11.2) and (5.11.3) with $c_n = n^{1/2}(\log n)^{1/2}$ and $L(\vartheta) = 2\left(A'^2/A + B'^2/B)\right)^{-1}$.

Though Akahira is aware of Woodroofe's related paper (1972) (see Akahira 1975, I, p. 26) he desists from discussing the connection of his results with that of Woodroofe and the pertaining optimality results of Weiss and Wolfowitz (1973).

By the same technique, bounds for asymptotically median unbiased estimator sequences may also be obtained for certain non-regular parametric families where the rate of convergence is $c_n = n$, and the optimal limit distribution is non-Gaussian. As a typical example we mention a result of Grossmann (1981) [p. 106, Corollary 4.3 and p. 107, Proposition 4.4]. Let $p|\mathbb{R}$ be a density such that $p(x) = 0$ for $x \le 0$ and

$$p(x) > 0 \text{ for } x > 0 \quad \text{with} \quad \lim_{x \downarrow 0} p(x) = A.$$

Let $P_\vartheta|\mathbb{B}$ be the distribution with density $x \to p(x - \vartheta)$. Then for any estimator sequence $(\vartheta^{(n)})_{n\in\mathbb{N}}$ fulfilling the condition of regular convergence with $c_n = n$,

$$\limsup_{n\to\infty} P^n_{\vartheta_0}\{\vartheta_0 - n^{-1}t' < \vartheta^{(n)} < \vartheta_0 + n^{-1}t''\} \le \frac{1}{2}\left(\exp[At'] - \exp[-At'']\right).$$

This bound is attained (for t', t'' sufficiently small) by the median unbiased estimator sequence $x_{1:n} - n^{-1}A^{-1}\log 2$.

Various examples of this kind (with $t' = t''$) can be found in Akahira (1982) and the references there. Notice that in non-regular cases the bounds obtained by the Neyman–Pearson Lemma are not necessarily attainable for every $t > 0$.

A Multidimensional Bound

By nature, results on median unbiased estimator sequences are restricted to one-dimensional functionals. They might, however, be used to derive some results on multidimensional functionals.

Let $\Theta \subset \mathbb{R}^k$. Assume that $\vartheta^{(n)} : X^n \to \mathbb{R}^k$, $n \in \mathbb{N}$, is an estimator sequence for ϑ which converges regularly to some limit distribution $Q_\vartheta|\mathbb{B}^k$. If this limit distribution is normal, say $Q_\vartheta = N(0, \Sigma(\vartheta))$, the estimator sequence $\kappa^{(n)} := a^\top \vartheta^{(n)}$, $n \in \mathbb{N}$, for $\kappa(\vartheta) = a^\top \vartheta$ is asymptotically normal: $P_\vartheta^n \circ c_n(\kappa^{(n)} - \kappa(\vartheta))$, $n \in \mathbb{N}$, converges regularly to $N(0, a^\top \Sigma(\vartheta)a)$. Hence $\kappa^{(n)}$ is asymptotically median unbiased, and relation (5.11.4), applied with $K = a$, leads to

$$N(0, a^\top \Sigma(\vartheta_0)a)(-t', t'') \leq N(0, a^\top \Lambda(\vartheta_0)a)(-t', t'') \quad \text{for } t', t'' \geq 0,$$

hence

$$a^\top \Sigma(\vartheta_0)a \geq a^\top \Lambda(\vartheta_0)a.$$

Since this relation holds for arbitrary $a \in \mathbb{R}^k$, the matrix $\Sigma(\vartheta_0) - \Lambda(\vartheta_0)$ is positive semidefinite. This argument occurs in Bahadur 1964, p. 1550, relation (26) and is repeated in Roussas (1968) [p. 255, Theorem 3.2 (iii)]. Both authors are satisfied with this formal relation between $\Sigma(\vartheta_0)$ and $\Lambda(\vartheta_0)$. Since Anderson's Theorem was well known at this time, (see in particular Anderson 1955, p. 173, Corollary 3) they could have turned this relation into the inequality

$$N(0, \Sigma(\vartheta_0))(C) \leq N(0, \Lambda(\vartheta_0))(C)$$

for all sets C which are convex and symmetric about 0, *a relation which anticipates the essence of the Convolution Theorem for the particular case of normal limit distributions.*

Remark The idea to conclude from the distribution of estimators $a^\top \vartheta^{(n)}$ to the distribution of $\vartheta^{(n)}$ occurs already in Kallianpur and Rao (1955) [p. 342], foreshadowed in a paper by Rao (1947) [p. 281, Corollary 1.1], which also contains the idea of an *intrinsic* bound for the asymptotic variance of (unbiased) estimator sequences. Surprisingly, it is not applied by Rao (1963). It was taken up by Roussas (1968, p. 255, Theorem 3 and 1972, p. 161, Theorem 7.1)—post festum. Though Roussas (1968) was aware of Kaufman's result from 1966 he obviously misunderstood its relevance (see Roussas 1968, p. 259, and 1972). Using the Neyman–Pearson Lemma, Roussas still struggles with upper and lower medians, a problem settled already by Kaufman (1966) [p. 174, Corollary]. It is not only the relevance of Kaufman's pre-Convolution Theorem which escaped Roussas' attention. In the very same book, he presents as Theorem 3.1, p. 136, a full version of the Convolution Theorem from which the above-mentioned results follow immediately.

Remark Observe that Roussas' paper deals with Markov chains fulfilling an LAN-condition. It seems to be the first paper at this level of generality, prior to Hájek (1970). A comparable result by Schmetterer (1966) [p. 316, Theorem 4.2] on Markov chains with a 1-dimensional parameter is technically less advanced. Schmetterer's paper was criticized by Roussas (1968) [p. 252] for being technically obsolete, using on p. 309 results by Daniels (1961) rather than Le Cam (1960) [p. 40, Theorem 2.1 (6)]. Roussas could have added that Schmetterer's Lemma 2.2, attributed by Schmetterer to Daniels (1961), occurs there (p. 156, relation (3.8)) with an incorrect proof. (Since this paper of Daniels is cited by many authors (Rao 1963, Bickel 1974, Lehmann and Casella 1998 and Le Cam in various places), it might be worth mentioning that this is not the only slip in Daniels' paper. See also G. Huber 1967, p. 211 and Williamson 1984.) □

Since Roussas avoids the use of Anderson's Theorem (for unknown reasons), the relation $\Lambda(\vartheta_0) \le \Sigma(\vartheta_0)$ between the covariance matrices (in the Löwner order) is used to derive relations between hyperplanes. For the case of an arbitrary multivariate limit distribution, Roussas (1968) [p. 257, Theorem 4] presents a result concerning the concentration about median hyperplanes.

Call $\{u \in \mathbb{R}^k : a^\top u = c\}$ a *median hyperplane* for the probability measure $Q|\mathbb{B}^k$ if

$$Q\{u \in \mathbb{R}^k : a^\top u \le c\} = Q\{u \in \mathbb{R}^k : a^\top u \ge c\} = 1/2.$$

The optimal limit distribution $N(0, \Lambda(\vartheta_0))$ is for any $a \in \mathbb{R}^k$ "more" concentrated about the median hyperplane $\{u \in \mathbb{R}^k : a^\top u = 0\}$ than any regularly attainable distribution Q_{ϑ_0} about its median hyperplane. More precisely: If $\{u \in \mathbb{R}^k : a^\top u = m_a(Q_{\vartheta_0})\}$ is a median hyperplane of Q_{ϑ_0}, then

$$\begin{aligned} &Q_{\vartheta_0}\{u \in \mathbb{R}^k : m_a(Q_{\vartheta_0}) - t' \le a^\top u \le m_a(Q_{\vartheta_0}) + t''\} \\ &\le N(0, \Lambda(\vartheta_0))\{u \in \mathbb{R}^k : -t' \le a^\top u \le t''\} \quad \text{for } t', t'' \ge 0. \end{aligned}$$

This result of Roussas generalizes the result of Wolfowitz (1965) [pp. 258/9, Theorem] from $k = 1$ to an arbitrary k. It is, in fact, a trivial consequence of the Convolution Theorem. Since $a^\top u$ is under Q_{ϑ_0} more spread out than under $N(0, \Lambda(\vartheta_0))$, it is (see 2.3.5) more concentrated—on arbitrary intervals—about its median under Q_{ϑ_0} than about its median under $N(0, \Lambda(\vartheta_0))$.

According to the Convolution Theorem, $N(0, K\Lambda K^\top)$ is the optimal limit distribution of regular estimator sequences for for a real-valued functional κ with gradient J. Since $J\Lambda\ell^\bullet(\cdot, \vartheta)$ is a gradient of κ, the stochastic expansion (5.11.6) implies that the convergence of any optimal median unbiased estimator sequence to $N(0, J\Lambda J^\top)$ is regular. Hence these estimator sequences are a fortiori optimal in the class of all regular estimator sequences for κ. It appears that this automatic regularity of asymptotically optimal median unbiased estimator sequences was overlooked in the papers by Rao, Wolfowitz, Schmetterer, etc.

A More Sophisticated Use of the Neyman–Pearson Lemma

The results presented so far are based on the Neyman–Pearson Lemma. Some of these results have later been obtained as a consequence of the Convolution Theorem—perhaps under stronger regularity conditions (such as regular convergence to a limit distribution, as opposed to asymptotic median unbiasedness).

In the following we mention two subtle consequences of the Convolution Theorem which, too, might have been obtained years earlier by an elementary proof, based on the Neyman–Pearson Lemma. Using an asymptotic version of the Neyman–Pearson Lemma, it can be shown that any estimator sequence which is asymptotically optimal among the asymptotically median unbiased estimator sequences is asymptotically linear with the optimal influence function. In this proof one has to use what has been ignored by the pioneers in their endeavor to establish the optimality of the ML-sequence, namely: That asymptotically optimal sequences of critical functions are asymptotically unique. The nonasymptotic version of this result dates from 1933. A contemporary textbook-reference could have been found in Schmetterer (1966) [Satz 3.1, pp. 202–206, in particular p. 204]. An asymptotic version of the Neyman–Pearson Lemma, particularly suitable for the subsequent results, is given in Lemma 5.11.7.

There is another result which is usually derived from the Convolution Theorem and which could have been obtained earlier by a proof based on the Neyman–Pearson Lemma: Any regularly attainable limit distribution (for estimators of a one-dimensional functional) is "more" spread out than the limit distribution determined by the LAN-condition. (See Sect. 2.8.)

The following results refer to parametric LAN-families $\{(P_\vartheta^{(n)})_{n\in\mathbb{N}} : \vartheta \in \Theta\}, \Theta \subset \mathbb{R}^k$. Since the value of ϑ remains fixed it will be omitted if there is no danger of confusion. Using the concepts developed in Sect. 5.13, these results can be easily extended to general families.

Theorem 5.11.3 *Assume that* $\{(P_\vartheta^{(n)})_{n\in\mathbb{N}} : \vartheta \in \Theta\}$, $\Theta \subset \mathbb{R}^k$, *is an LAN-family and* $\kappa : \Theta \to \mathbb{R}$ *a differentiable functional with gradient* J. *Then the following is true.*

(i) For any estimator sequence $\kappa^{(n)}$, $n \in \mathbb{N}$, *which is asymptotically median unbiased for* κ,

$$\limsup_{n\to\infty} P_\vartheta^{(n)}\{-t' \le c_n(\kappa^{(n)} - \kappa(\vartheta)) \le t''\} \le N(0, J\Lambda J^\top)(-t', t'') \quad \textit{for } t', t'' \ge 0. \tag{5.11.5}$$

(ii) If

$$\lim_{n\to\infty} P_\vartheta^{(n)}\{c_n|\kappa^{(n)} - \kappa(\vartheta)| \le t\} = N(0, J\Lambda J^\top)(-t, t) \quad \textit{for every } t > 0,$$

then

$$c_n(\kappa^{(n)} - \kappa(\vartheta)) - J\Delta_n \to 0 \quad (P_\vartheta^{(n)}). \tag{5.11.6}$$

Remark The scholars who tried to prove certain optimality properties of ML estimators by means of the Neyman–Pearson Lemma could have used an asymptotic expansion like (5.11.6) to show that an asymptotically normal estimator sequence with optimal marginals has the same (joint) limit distribution as the ML-sequences. This does, however, not yet establish ML-sequences as asymptotically optimal in some multivariate sense. Estimator sequences with marginals inferior to the marginals of the ML-sequences could, in principle, be asymptotically superior to the ML-sequences in some multidimensional sense. (See Sect. 5.13 for a discussion of this question.)

Proof of Theorem 4.11.3. (i) Let $a \in \mathbb{R}$ be fixed. If $\kappa^{(n)}$, $n \in \mathbb{N}$, is asymptotically median unbiased and κ is differentiable, the relation

$$\limsup_{n\to\infty} \int \varphi_n dP^{(n)}_{\vartheta+c_n^{-1}a} \le 1/2$$

is fulfilled for $\varphi_n = 1_{\{c_n^{-1}(\kappa^{(n)}-\kappa(\vartheta))\le Ja\}}$. Hence Corollary 5.11.8 below implies

$$\limsup_{n\to\infty} P^{(n)}_{\vartheta}\{c_n(\kappa^{(n)} - \kappa(\vartheta)) \le Ja\} \le \Phi((a^\top La)^{1/2}). \tag{5.11.7}$$

To make the upper bound as sharp as possible, one needs to minimize $a^\top La$ subject to the condition $Ja \ge t$. This is achieved by choosing $a = \bar{a}$ with

$$\bar{a} := \Lambda J^\top / J\Lambda J^\top. \tag{5.11.8}$$

Relation (5.11.7), applied with $\hat{a}$ given by (5.11.8), yields

$$\limsup_{n\to\infty} P^{(n)}_{\vartheta}\{c_n(\kappa^{(n)} - \kappa(\vartheta)) \le t\} \le \Phi(t(J\Lambda J^\top)^{-1/2}) \quad \text{for } t \ge 0.$$

Together with the corresponding relation for $t < 0$, this implies (5.11.5).

(ii) According to Corollary 5.11.8 applied with $\beta = 1/2$ and a replaced by $t\hat{a}$, the relation

$$\varphi_n = 1_{\{J\Delta_n < t\}}$$

holds for any critical function φ_n fulfilling

$$\int \varphi_n dP^{(n)}_{\vartheta+c_n^{-1}t\hat{a}} \to 1/2 \tag{5.11.9}$$

and

$$\int \varphi_n dP^{(n)}_{\vartheta} \to \Phi((J\Lambda J^\top)^{-1/2}). \tag{5.11.10}$$

If $\kappa^{(n)}$ is asymptotically median unbiased and optimal, then

$$\varphi_n = 1_{\{c_n(\kappa^{(n)} - \kappa(\vartheta)) \leq t\}}$$

has properties (5.11.9) and (5.11.10). Hence

$$1_{\{n^{1/2}(\kappa^{(n)} - \kappa(\vartheta)) \leq t\}} - 1_{\{J\Lambda\tilde{\ell}^\circ \leq t\}} \to 0 \quad \text{for } t \in \mathbb{R}.$$

By Lemma 5.11.6 below this implies

$$c_n^{-1}(\kappa^{(n)} - \kappa(\vartheta)) - J\Delta_n \to 0 \quad (P_\vartheta^{(n)}).$$

□

Proposition 5.5.1 implies that asymptotically median unbiased estimator sequences that attain at ϑ the maximal value $N(0, J\Lambda J^\top)(-t, t)$ for every $t > 0$, converge regularly to a limit distribution, which is $N(0, J\Lambda J^\top)$.

Leaving median unbiasedness aside, we turn to estimator sequences for $\kappa : \Theta \to \mathbb{R}$ with a regularly attainable limit distribution, say Q. The following Theorem 5.11.4 asserts that Q is more spread out than $N(0, J\Lambda J^\top)$. This assertion is weaker than the Convolution Theorem, since $Q \succeq N(0, J\Lambda J^\top)$ follows from $Q = N(0, J\Lambda J^\top) * R$ (see Sect. 2.8). It might be worth mentioning that the relation $Q \succeq N(0, J\Lambda J^\top)$ could thus have been obtained years before the Convolution Theorem.

Theorem 5.11.4 *Assume that* $\{(P_\vartheta^{(n)})_{n\in\mathbb{N}} : \vartheta \in \Theta\}$, $\Theta \subset \mathbb{R}^k$, *is an LAN-family and* $\kappa : \Theta \to \mathbb{R}$ *a differentiable functional with gradient* J. *Then* $Q_\vartheta \succeq N(0, J\Lambda J^\top)$ *for any regularly attainable limit distribution* Q_ϑ *of estimator sequences for* κ.

Proof. With $F(t) := Q_\vartheta(-\infty, t]$, let

$$\varphi_n(\cdot, a) := 1_{\{n^{1/2}(\kappa^{(n)} - \kappa(\vartheta + c_n^{-1}a)) \leq F^{-1}(\beta)\}}.$$

The relation $P^{(n)}_{\vartheta + c_n^{-1}a} \circ n^{1/2}(\kappa^{(n)} - \kappa(\vartheta + c_n^{-1}a)\} \Rightarrow Q_\vartheta$ implies

$$\int \varphi_n(\cdot, a) dP^{(n)}_{\vartheta + c_n^{-1}a} \to \beta$$

and

$$\int \varphi_n(\cdot, a) dP_\vartheta^{(n)} \to F(F^{-1}(\beta) + Ja).$$

Hence Corollary 5.11.8 implies

$$F(F^{-1}(\beta) + Ja) \leq \Phi(\Phi^{-1}(\beta) + (a^\top La)^{1/2}).$$

Applied with $a = td$ this leads to

$$F(F^{-1}(\beta)+t) \le \Phi(\Phi^{-1}(\beta)+t(J\Lambda J^\top)^{-1/2}). \tag{5.11.11}$$

By Lemma 5.11.5,

$$\Phi(\Phi^{-1}(\beta)+t\sigma^{-1}) = \Phi_\sigma(\Phi_\sigma^{-1}(\beta)+t)$$

where Φ_σ is the distribution function of $N(0,\sigma^2)$. Hence (5.11.11) may be rewritten as

$$F(F^{-1}(\beta)+t) \le \Phi_\sigma(\Phi_\sigma^{-1}(\beta)+t), \quad \text{with } \sigma^2 = J\Lambda J^\top.$$

This relation holds for every $\beta \in (0,1)$ and every $t \ge 0$. According to a well known Lemma on the spread order (see (2.4.3)), this implies $Q_\vartheta \succeq N(0, J\Lambda J^\top)$. □

Some Auxiliary Results

Lemma 5.11.5 *Let F and F_1 be increasing distribution functions of probability measures Q and Q_1, respectively. If*

$$F(F^{-1}(\beta)+t) \le F_1(F_1^{-1}(\beta)+t/\sigma) \quad \textit{for } \beta \in (0,1) \quad \textit{and } t \in \mathbb{R},$$

then Q is "more" spread out than $Q_\sigma := Q_1 \circ (u \to \sigma u)$.

Proof With F_σ denoting the distribution function of Q_σ, we have

$$F_1(F_1^{-1}(\beta)+t/\sigma) = F_\sigma(F_\sigma^{-1}(\beta)+t),$$

hence

$$F(F^{-1}(\beta)+t) \le F_\sigma(F_\sigma^{-1}(\beta)+t) \quad \text{for } \beta \in (0,1), t \in \mathbb{R}.$$

□

Lemma 5.11.6 *For $n \in \mathbb{N}$, let f_n, g_n be real functions defined on $(X, \mathscr{A})$, and $P_n|\mathscr{A}$ a probability measure.*

If

$$1_{\{f_n \le t\}} - 1_{\{g_n \le t\}} \to 0 \quad (P_n) \quad \textit{for } t \in \mathbb{R},$$

then

$$f_n - g_n \to 0 \quad (P_n).$$

Proof Since

$$\left|1_{\{f_n\le t\}} - 1_{\{g_n \le t\}}\right| = 1_{\{f_n \le t, g_n > t\}} + 1_{\{f_n > t, g_n \le t\}},$$

it suffices to prove that

$$\lim_{n\to\infty} P_n\{f_n > t, g_n \le t\} = 0 \quad \text{for every } t \in \mathbb{R}$$

implies

$$\lim_{n\to\infty} P_n\{f_n - g_n > \varepsilon\} = 0 \quad \text{for every } \varepsilon > 0.$$

If $f_n(x) - g_n(x) > \varepsilon$, the interval $(g_n(x) + \varepsilon/2, f_n(x) - \varepsilon/2)$ is nonempty, and there is $t \in \mathbb{Q}$ such that $g_n(x) + \varepsilon/2 < t < f_n(x) - \varepsilon/2$. Hence

$$\{x \in X : f_n(x) - g_n(x) < \varepsilon\} \subset \bigcup_{t\in\mathbb{Q}} \{x \in X : g_n(x) < t - \varepsilon/2,\ f_n(x) > t + \varepsilon/2\}.$$

Since

$$\limsup_{n\to\infty} P_n\{x \in X : g_n(x) < t - \varepsilon/2,\ f_n(x) > t + \varepsilon/2\}$$
$$\leq \lim_{n\to\infty} P_n\{x \in X : g_n(x) \leq t + \varepsilon/2,\ f_n > t + \varepsilon/2\} = 0,$$

the assertion follows. □

Lemma 5.11.7 *For $n \in \mathbb{N}$, let P_n and $P_n' \ll P_n$ be probability measures on a measurable space $(X, \mathscr{A})$. Let $q_n \in dP_n'/dP_n$. Assume that $P_n \circ q_n \Rightarrow Q|\mathbb{B}$, a non-degenerate probability measure on $\mathbb{B} \cap [0, \infty)$ fulfilling $\int u Q(du) = 1$, so that $P_n' \circ q_n \Rightarrow Q'$, with $\mathrm{id} \in dQ'/dQ$. (See also Le Cam's Lemmas.)*

Under these conditions, the following is true for any sequence of critical functions $\varphi_n : X \to [0, 1]$, $n \in \mathbb{N}$.

(i) For every $r > 0$,

$$\limsup_{n\to\infty} \int \varphi_n dP_n' \leq Q'[0, r] \tag{5.11.12}$$

implies

$$\limsup_{n\to\infty} \int \varphi_n dP_n \leq Q[0, r]. \tag{5.11.13}$$

(ii) If equality holds in (5.11.13) for some $r > 0$, then $\lim_{n\to\infty} \int \varphi_n dP_n = Q[0, r]$ and therefore

$$\varphi_n - 1_{\{q_n < r\}} \to 0 \quad (P_n). \tag{5.11.14}$$

Proof (i) For every $r > 0$,

$$(\varphi_n - 1_{\{q_n < r\}})(r - q_n) \leq 0. \tag{5.11.15}$$

Hence (5.11.13) follows from (5.11.12) after integration of (5.11.15) with respect to P_n.

(ii) If equality holds in (5.11.12) for some $r > 0$, relation (5.11.15) implies

$$\lim_{n\to\infty} \int |\varphi_n - 1_{\{q_n < r\}}| \cdot |r - q_n| dP_n = 0.$$

To show that this implies

$$\int |\varphi_n - 1_{\{q_n < r\}}| dP_n \to 0,$$

we assume, more generally, that

$$\int |\chi_n| \cdot |r - q_n| dP_n \to 0,$$

for some sequence $|\chi_n| \leq 1$, $n \in \mathbb{N}$.

Since

$$\int |\chi_n| dP_n = \int |\chi_n| 1_{\{|r-q_n|<\varepsilon\}} dP_n + \int |\chi_n| 1_{\{|r_n - q_n| \geq \varepsilon\}} dP_n \leq$$
$$P_n\{|r - q_n| < \varepsilon\} + \frac{1}{\varepsilon} \int |\chi_n| \cdot |r_n - q_n| dP_n, \tag{5.11.16}$$

relation (5.11.16) implies

$$\limsup_{n\to\infty} \int |\chi_n| dP_n \leq \limsup_{n\to\infty} P_n\{|r - q_n| < \varepsilon\}. \tag{5.11.17}$$

Since $P_n \circ q_n$, $n \in \mathbb{N}$, converges to a nonatomic distribution, the right-hand side of (5.11.17) may be made arbitrarily small by choice of $\varepsilon > 0$. This proves (5.11.14). □

Applied to an LAN-family $\{P_\vartheta^{(n)} : \vartheta \in \Theta\}$, $\Theta \subset \mathbb{R}^k$, with

$$\log(dP^{(n)}_{\vartheta + c_n^{-1}a} / dP^{(n)}_\vartheta) = a^\top \Delta_n(\cdot, \vartheta) - \frac{1}{2} a^\top L(\vartheta) a + o(n^0, P^{(n)}_\vartheta).$$

and

$$P^{(n)}_\vartheta \circ a^\top \Delta_n(\cdot, \vartheta) \Rightarrow N(0, a^\top L(\vartheta) a),$$
$$P^{(n)}_{\vartheta + c_n^{-1}a} \circ a^\top \Delta(\cdot, \vartheta) \Rightarrow N(a^\top L(\vartheta) a, a^\top L(\vartheta) a),$$

Lemma 5.11.7 yields the following Corollary.

Corollary 5.11.8 *Under the conditions specified above, the following holds true for any critical function* $\varphi_n : X^n \to [0, 1]$.

If

$$\limsup_{n\to\infty} \int \varphi_n dP^{(n)}_{\vartheta + c_n^{-1}a} \leq \beta \quad \textit{for some } a \in \mathbb{R}^k,$$

then

$$\limsup_{n\to\infty} \int \varphi_n dP^{(n)}_\vartheta \leq \Phi\big(\Phi^{-1}(\beta) + (a^\top L(\vartheta) a)^{1/2}\big).$$

Addendum. *If for some* $a \in \mathbb{R}^k$,

$$\int \varphi_n dP^{(n)}_{\vartheta+n^{1/2}a} \to \frac{1}{2}$$

and

$$\int \varphi_n dP^{(n)}_{\vartheta} \to \Phi((a^\top La)^{1/2}),$$

then

$$\varphi_n - 1_{\{a^\top \Delta_n < a^\top La\}} \to 0 \quad (P^{(n)}_{\vartheta}).$$

Proof (i) The assertion follows from Lemma 5.11.7(i), applied with $P_n = P^{(n)}_{\vartheta}$ and $P'_n = P^{(n)}_{\vartheta+c_n^{-1}a}$ and with $r_a = \exp[(a^\top La)^{1/2}\Phi^{-1}(\beta) + \frac{1}{2}a^\top La]$. Since

$$P^{(n)}_{\vartheta}\{\Lambda_{na} \le r_a\} - P^{(n)}_{\vartheta}\{a^\top \Delta_n) \le (a^\top La)^{1/2}\Phi^{-1}(\beta) + a^\top La\} \to 0,$$

we have

$$P^{(n)}_{\vartheta+c_n^{-1}a}\{\Lambda_{na} \le r_a\} \to \beta$$

and therefore

$$\limsup_{n\to\infty} P^{(n)}_{\vartheta}\{\Lambda_{na} < r_a\} \le \Phi\big(\Phi^{-1}(\beta) + (a^\top La)^{1/2}\big).$$

(ii) The addendum follows from Lemma 5.11.7(ii). □

Symmetric Optimality

We consider a parametric family $\{(P^{(n)}_{\vartheta})_{n\in\mathbb{N}} : \vartheta \in \Theta\}$, $\Theta \subset \mathbb{R}^k$ fulfilling an LAN-condition

$$\log(dP^{(n)}_{\vartheta+c_n^{-1}a}/dP^{(n)}_{\vartheta}) = a^\top \Delta_n(\cdot, \vartheta) - \frac{1}{2}a^\top L(\vartheta)a + o(n^0, P^{(n)}_{\vartheta}).$$

Asymptotic bounds for the concentration of estimator sequences for differentiable functionals $\kappa : \Theta \to \mathbb{R}^p$ are necessarily restricted to the concentration on symmetric sets if $p > 1$.

For functionals $\kappa : \Theta \to \mathbb{R}$ this is not necessarily so. Among median unbiased estimator sequences there may be estimator sequences which are asymptotically maximally concentrated on arbitrary intervals containing 0. For irregular models, estimator sequences with this strong optimum property do not necessarily exist. In such cases, optimality on intervals symmetric about 0 may be a useful surrogate.

A first result on symmetric optimality was presented by Weiss and Wolfowitz (1966) [p. 61]. To establish the asymptotic optimality of ML-sequences under Cramér-type regularity conditions they show that

$$Q_\vartheta(-t, t) \le N(0, \Lambda(\vartheta))(-t, t), \quad t > 0$$

for any regularly attainable limit distribution Q_ϑ. Since corresponding results with not necessarily symmetric intervals were already available at this time, one can only guess why they were interested in symmetric intervals. Perhaps their intention was to avoid trouble with the median. They use Bayesian arguments though a proof based on the Neyman–Pearson Lemma would have been simpler (see below).

The idea of symmetric optimality was taken up by Akahira and/or Takeuchi (see e.g. 1982), combined with the requirement of asymptotic median unbiasedness.

Is an optimality concept based on the concentration on symmetric intervals really required? In the regular cases considered above, there exist median unbiased estimator sequences which are optimal on all intervals containing 0. In certain non-regular models one encounters estimator sequences with an optimum property limited to symmetric intervals. Such limited optimum properties are of interest also from another point of view: They are neither related to the spread order nor to a convolution property.

Let $P_\vartheta^{(n)} \circ c_n(\vartheta^{n)} - \vartheta) \Rightarrow Q_\vartheta|\mathbb{B}^k$ with Q_ϑ nonatomic. The following proposition gives bounds for $Q_\vartheta(-t', t'')$ expressed by the asymptotic performance of

$$p_n(\cdot, \vartheta + c_n^{-1}t)/p_n(\cdot, \vartheta) \in dP^{(n)}_{\vartheta+c_n^{-1}t}/dP^{(n)}_\vartheta$$

under the assumption that the convergence to Q_ϑ is *regular* at $-t'$ and t'', i.e.,

$$P^{(n)}_{\vartheta+c_n^{-1}t} \circ c_n(\vartheta^{(n)} - (\vartheta + c_n^{-1}t)) \Rightarrow Q_\vartheta \quad \text{for } t = -t' \text{ and } t = t''.$$

Proposition 5.11.9 *Under the conditions indicated above,*

$$\begin{aligned} Q_\vartheta(-t', t'') \le \limsup_{n\to\infty} \big(&P^{(n)}_{\vartheta-c_n^{-1}t''}\{p_n(\cdot, \vartheta - c_n^{-1}t'') > p_n(\cdot, \vartheta + c_n^{-1}t')\} \\ &-P^{(n)}_{\vartheta+c_n^{-1}t'}\{p_n(\cdot, \vartheta - c_n^{-1}t'') > p_n(\cdot, \vartheta + c_n^{-1}t')\}\big). \end{aligned} \tag{5.11.18}$$

Proof By assumption,

$$P^{(n)}_{\vartheta+c_n^{-1}t'}\{\vartheta^{(n)} \le \vartheta\} = P^{(n)}_{\vartheta+c_n^{-1}t'}\{c_n(\vartheta^{(n)} - (\vartheta + c_n^{-1}t')) \le -t'\} \to Q_\vartheta(-\infty, -t')$$

and

$$P^n_{\vartheta-c_n^{-1}t''}\{\vartheta^{(n)} \le \vartheta\} = P^{(n)}_{\vartheta-c_n^{-1}t''}\{c_n(\vartheta^{(n)} - (\vartheta - c_n^{-1}t'')) \le t''\} \to Q_\vartheta(-\infty, t''),$$

hence

$$Q_\vartheta(-t', t'') = \lim_{n\to\infty} \int 1_{\{\vartheta^{(n)}\le\vartheta\}}(p_n(\cdot, \vartheta - c_n^{-1}t'') - p(\cdot, \vartheta + c_n^{-1}t'))d\mu_n.$$

Since

$$\left(1_{\{\vartheta^{(n)}\leq\vartheta\}} - 1_{\{p_n(\cdot,\vartheta-c_n^{-1}t'')>p_n(\cdot,\vartheta+c_n^{-1}t')\}}\right)$$
$$\left(p_n(\cdot,\vartheta-c_n^{-1}t'') - p_n(\cdot,\vartheta+c_n^{-1}t')\right) \leq 0,$$

this implies the assertion. □

This proposition, based on the Neyman–Pearson Lemma, offers another bound for the concentration of limit distributions, complementary to the bounds based on median unbiasedness and on the Convolution Theorem.

Recall the bound given in (5.11.4) for the concentration of asymptotically median unbiased estimator sequences,

$$\limsup_{n\to\infty} P_{\vartheta_0}^{(n)}\{-t' \leq c_n(\vartheta^{(n)} - \vartheta_0) \leq t''\} \leq N(0, \Lambda(\vartheta_0))(-t', t'').$$

Under equivalent regularity conditions (using (5.11.2) and (5.11.3) again), Proposition 5.11.9 yields the bound

$$\lim_{n\to\infty} P_{\vartheta_0}^{(n)}\{-t' \leq c_n(\vartheta^{(n)} - \vartheta_0) \leq t''\} \leq N(0, \Lambda(\vartheta_0))(-\frac{1}{2}(t'+t''), \frac{1}{2}(t'+t'')) \tag{5.11.19}$$

(now under the assumption that the lim on the left-hand side exists).

Whereas the bound given in (5.11.4) is attainable, this is not the case with the bound given in (5.11.19). Observe that

$$N(0,\sigma^2)(-t', t'') \leq N(0,\sigma^2)(-\frac{1}{2}(t'+t''), \frac{1}{2}(t'+t'')).$$

For $t' = t''(= t)$ the two bounds are the same,

$$\lim_{n\to\infty} P_{\vartheta_0}^{(n)}\{c_n|\vartheta^{(n)} - \vartheta_0| \leq t\} \leq N(0, \Lambda(\vartheta_0))(-t, t), \tag{5.11.20}$$

with two different interpretations, corresponding to the two different side conditions: median unbiasedness, and convergence to a limit distribution, respectively.

There is a third approach leading to the same bound, the Convolution Theorem, which requires regular convergence of $P_\vartheta^{(n)} \circ c_n(\vartheta^{(n)} - \vartheta)$, $n \in \mathbb{N}$, to some limit distribution Q_ϑ, and which implies, corresponding to (5.11.20), that

$$Q_{\vartheta_0}(-t, t) \leq N(0, \Lambda(\vartheta_0))(-t, t).$$

For multivariate LAN-families, the bounds provided by the Convolution Theorem are confined to the concentration on convex and symmetric sets. The following Theorem suggests how the same result may be obtained as a corollary to Proposition 5.11.9.

Theorem 5.11.10 *Let $C \subset \mathbb{R}^k$ be convex and symmetric about 0. If*

$$P^{(n)}_{\vartheta+c_n^{-1}a}\{c_n(\vartheta^{(n)} - (\vartheta + c_n^{-1}a)) \in C\}, \quad n \in \mathbb{N},$$

converges for every $a \in \mathbb{R}^k$ *to the same limit, then*

$$\lim_{n\to\infty} P^{(n)}_\vartheta\{c_n(\vartheta^{(n)} - \vartheta) \in C\} \le N(0, \Lambda(\vartheta))(C).$$

Observe that the set C is fixed. The proof follows from Proposition 5.11.9, applied with $a^\top\vartheta^{(n)}$, as an estimator of $a^\top\vartheta$, $a \in \mathbb{R}^k$.

Do we really need an independent concept of symmetric optimality? The following example presents a limit distribution which is (i) optimal on all symmetric intervals, and (ii) its optimality is not a consequence of the Convolution Theorem.

The operational significance of a representation $Q = \overline{Q} * R$ depends on properties of the factor $\overline{Q}$. In the case $k = 1$ this is the logconcavity of $\overline{Q}$, and the result is $Q(-t', t'') \le \overline{Q}(-t', t'')$ for $t', t'' \ge 0$. The following example presents two regularly attainable limit distributions $\overline{Q}$ and Q, such that $\overline{Q}(-t, t) \ge Q(-t, t)$ for every $t > 0$, but $Q(-t', t'') > \overline{Q}(-t', t'')$ for some $t', t'' > 0$. Since $\overline{Q}$ is logconcave, Q cannot be a convolution product involving the factor $\overline{Q}$. Hence the optimality of $\overline{Q}$ on all symmetric intervals is not a consequence of the Convolution Theorem.

Example For $\vartheta \in \mathbb{R}$ let P_ϑ be the probability measure with density $p(x, \vartheta) := 1_{[-1/2,1/2]}(x - \vartheta)$. By (5.11.18),

$$Q_\vartheta(-t', t'') \le 1 - e^{-(t'+t'')}. \tag{5.11.21}$$

Here $1 - e^{-(t'+t'')}$ is just a bound for $Q_\vartheta(-t', t'')$; equality in (5.11.21) for arbitrary t', t'' is impossible. For $t' = t'' = t$, we obtain

$$Q_\vartheta(-t, t) \le 1 - e^{-2t},$$

and this symmetric bound, $1 - e^{-2t}$, is attainable. For the estimator sequence

$$\hat\vartheta^{(n)}(\mathbf{x}_n) := \frac{1}{2}(x_{1:n} + x_{n:n}),$$

$P^n_\vartheta \circ n(\hat\vartheta^{(n)} - \vartheta)$ converges to the Laplace distribution with scale parameter $1/2$, hence

$$\lim_{n\to\infty} P^n_\vartheta\{n|\hat\vartheta^{(n)} - \vartheta| \le t\} = 1 - e^{-2t}.$$

The estimator sequence

$$\tilde\vartheta^{(n)}(\mathbf{x}_n) := x_{1:n} + \frac{1}{2} - n^{-1}\log 2$$

converges regularly to the limit distribution $\tilde{Q}$ with

$$\tilde{Q}(-t', t'') = \min\{1, \frac{1}{2}e^{t'}\} - \frac{1}{2}e^{-t''} \quad \text{for } -\log 2 < -t' \leq 0 < t'' < \infty.$$

We have $\tilde{Q}(-t, t) < \overline{Q}(-t, t)$ for $t > 0$ by Proposition 5.11.9 and $\tilde{Q}(-t', t'') > \overline{Q}(-t', t'')$ for t' close to $\log 2$ and t'' small. □

We conclude this section by some side results. Let, more generally, $\mathfrak{Q}$ denote a family of distributions on $\mathbb{B}$ which contains an element Q_0 with median 0 which is optimal on all symmetric intervals, i.e., for every $Q \in \mathfrak{Q}$,

$$Q(-t, t) \leq Q_0(-t, t) \quad \text{for } t > 0.$$

Then the following is true.

(i) If $\mathfrak{Q}$ contains an element which is minimal in the spread order , then Q_0 has this property.
(ii) If $\mathfrak{Q}$ contains an element $\overline{Q}$ which is minimal in the convolution order (i.e., $Q = \overline{Q} * R$ for every $Q \in \mathfrak{Q}$), then this is Q_0, provided $\int u^2 Q_0(du) < \infty$.

Proof (i) Assume that $\overline{Q}$ is minimal in the spread order. W.l.g. we may assume that $\overline{Q}$ has median 0. Then $\overline{Q}(-t', t'') \geq Q_0(-t', t'')$ for all $t', t'' \geq 0$. Since Q_0 is optimal on all symmetric intervals, $\overline{Q}(-t, t) = Q_0(-t, t)$ for $t > 0$. By Lemma 5.11.12 this implies $\overline{Q} = Q_0$.
(ii) Assume now that $\overline{Q}$ is minimal in the convolution order. Without assuming anything about $\overline{Q}$, it is not permitted to conclude that $\overline{Q}$ is also minimal in the spread order (since $\overline{Q} * R$ is not necessarily "more" spread out than $\overline{Q}$ in general). Yet, the relation $\overline{Q} = Q_0$ follows from Lemma 5.11.12. □

Lemma 5.11.11 *If $Q_0(-t, t) \geq Q(-t, t)$ for $t > 0$, and $Q_0(-t', t'') \leq Q(-t', t'')$ for $t', t'' \geq 0$, then $Q = Q_0$.*

Proof We have

$$\begin{aligned} Q(-t'', t') + Q(-t', t'') &= Q(-t', t') + Q(-t'', t'') \\ &\leq Q_0(-t', t') + Q_0(-t'', t'') = Q_0(-t'', t') + Q_0(-t', t''). \end{aligned}$$

If $Q(-t'', t') \geq Q_0(-t'', t')$ and $Q(-t', t'') \geq Q_0(-t', t'')$, this implies $Q(-t', t'') = Q_0(-t', t'')$ for arbitrary $t', t'' \geq 0$, whence $Q = Q_0$. □

Lemma 5.11.12 *Assume that $\mathfrak{Q}$ contains an element Q_0 which is optimal in the sense that for every $Q \in \mathfrak{Q}$,*

$$Q(-t, t) \leq Q_0(-t, t) \quad \text{for } t > 0. \tag{5.11.22}$$

Assume that $\int w^2 Q_0(dw) < \infty$. *If* $\mathfrak{Q}$ *contains an element which is minimal in the convolution order, then* Q_0 *has this property.*

Proof Assume that $Q_0 = \overline{Q} * R$. It is easily seen that $\int w^2 Q_0(dw) < \infty$ implies $\int u^2 \overline{Q}(du) < \infty$ and $\int v^2 R(dv) < \infty$. Hence $\int u \overline{Q}(du)$ exists, and $\overline{Q}$ may be chosen such that $\int u \overline{Q}(du) = 0$. This implies

$$\int w^2 Q_0(dw) = \int (u+v)^2 \overline{Q}(du) R(dv) = \int u^2 \overline{Q}(du) + \int v^2 R(dv).$$

Relation (5.11.22) with $\overline{Q}$ in place of Q implies $\int u^2 \hat{Q}(du) \geq \int w^2 Q_0(dw)$. Hence $\int v^2 R(dv) = 0$, and $\hat{Q} = Q_0$. □

Bounds for Asymptotically Mean Unbiased Estimator Sequences

The definition of asymptotic median unbiasedness is straightforward. In the case of asymptotic mean unbiasedness one encounters the problem that $\int \vartheta^{(n)}(\mathbf{x}_n) P_\vartheta^n(d\mathbf{x}_n)$ does not necessarily exist. Even if this expectation exists, the natural definition, $\lim_{n\to\infty} \int \vartheta^{(n)}(\mathbf{x}_n) P_\vartheta^n(d\mathbf{x}_n) = \vartheta$, is not strong enough to be technically useful. Moreover, it does not entail some sort of local uniformity. Therefore, the theory of asymptotically mean unbiased estimator sequences is based on the following somewhat artificial definition, which goes back to Chernoff (1956) [p. 12], at least.

For $u > 0$ let

$$L_u[y] := y 1_{[-u,u]}(y), \quad y \in \mathbb{R}.$$

($L_u[y] := y 1_{[-u,u]}(y) + u(1_{(-\infty,-u)} + 1_{(u,\infty)})$ serves the same purpose.)

An estimator sequence $(\vartheta^{(n)})_{n\in\mathbb{N}}$ is at ϑ_0 *asymptotically mean unbiased* with rate $(c_n)_{n\in\mathbb{N}}$ if for every sequence if $\vartheta_n = \vartheta_0 + c_n^{-1} a$, $a \in \mathbb{R}$,

$$\lim_{u\to\infty} \limsup_{n\to\infty} \left| \int L_u\big(c_n(\vartheta^{(n)} - \vartheta_n)\big) d P_{\vartheta_n}^n \right| = 0.$$

If the family $\{P_\vartheta : \vartheta \in \Theta\}$, $\Theta \subset \mathbb{R}$, fulfills an LAN-condition and if $(\vartheta^{(n)})_{n\in\mathbb{N}}$ is mean unbiased with rate $c_n = n^{1/2}$, then

$$\lim_{u\to\infty} \liminf_{n\to\infty} \int L_u[n^{1/2}(\vartheta^{(n)} - \vartheta_0)]^2 d P_{\vartheta_0}^n \geq \Lambda(\vartheta_0). \tag{5.11.23}$$

(See Pfanzagl 2001, p. 507, Theorem 3.1, improving an earlier result by Liu and Brown (1993). Notice a misprint: c_n in (3.6) has to be replaced by $n^{1/2}$.)

Of course, relation (5.11.23) follows from the Convolution Theorem if $P_{\vartheta_n}^n \circ n^{1/2}(\vartheta^{(n)} - \vartheta_n)$, $n \in \mathbb{N}$ converges to a limit distribution. The essence of this theorem is that the local uniformity condition refers to mean unbiasedness only; convergence to a limit distribution is not required.

5.12 Asymptotic Normality: Global and Local

The idea to take the LAN-condition as a basis for asymptotic results developed slowly. The starting point was the paper by Wald (1943), suggesting the idea to approximate a k-parameter family of i.i.d.-products for large n by a family of k-dimensional normal distributions, and to obtain an approximate solution for the original problem (say an optimal test) from the solution for the approximating normal distribution. The most interesting result in this paper seems to be Lemma 2, p. 443, asserting the existence of a map $W_n : \mathbb{B}^n \to \mathbb{B}^k$ such that

$$\left|P_\vartheta^n(B) - N(0, \Lambda(\vartheta))(W_n(B))\right| \to 0 \qquad (5.12.1)$$

uniformly in ϑ and $B \in \mathbb{B}^n$.

In this relation, $W_n(B)$ is something like $\vartheta^{(n)}(B)$, with $\vartheta^{(n)} : X^n \to \mathbb{R}^k$ a ML estimator. With W_n a set-transformation, the representation (5.12.1) is not easy to deal with (though Wald presents various applications to asymptotic test theory).

If Wolfowitz (1952) [p. 10] says that Wald wrote papers "without too much thought of elegance": this paper by Wald (containing 57 pages) is a convincing example. Probably, nobody ever had the patience to study it in detail.

Global Asymptotic Normality

Wald's approach was superseded by the following result of Le Cam (1956) [p. 140, Theorem 1]:

Under suitable regularity conditions on the family $\{P_\vartheta|\mathscr{A} : \vartheta \in \Theta\}$, $\Theta \subset \mathbb{R}^k$, dominated by $\mu|\mathscr{A}$, there exists an estimator sequence $(\vartheta^{(n)})_{n\in\mathbb{N}}$ and a sequence of probability measures $Q_{n,\vartheta}|\mathscr{A}^n$ with μ^n-density

$$K_n(\vartheta)h_n 1_{B_n}(\vartheta^{(n)} - \vartheta)\exp\Big[-\frac{1}{2}n(\vartheta^{(n)} - \vartheta)^\top L(\vartheta)(\vartheta^{(n)} - \vartheta)\Big] \qquad (5.12.2)$$

such that

$$d(P_\vartheta^n, Q_{n,\vartheta}) \to 0, \qquad (5.12.3)$$

uniformly for ϑ in compact subsets of Θ.

The regularity conditions correspond to a Taylor expansion of order 2 for $\vartheta \to \log p(x, \vartheta)$. Le Cam's proof of this assertion involves the construction of an asymptotically efficient estimator sequence under rather weak conditions. (See Sect. 5.5.)

Since the density of $Q_{n,\vartheta}$ is of the type $h_n(\mathbf{x}_n)g_n(\vartheta^{(n)}(\mathbf{x}_n), \vartheta)$, the function $\vartheta^{(n)}$ is sufficient for $\{Q_{n,\vartheta} : \vartheta \in \Theta\}$; naturally, $(\vartheta^{(n)})_{n\in\mathbb{N}}$ is called "asymptotically sufficient" for $\{(P_\vartheta^n)_{n\in\mathbb{N}} : \vartheta \in \Theta\}$, $n \in \mathbb{N}$. Observe that relation (5.12.3) implies, in fact, an asymptotic property of $(\vartheta^{(n)})_{n\in\mathbb{N}}$ stronger than "sufficiency". Sufficiency of an estimator $\tilde\vartheta^{(n)}$ for $\{P_\vartheta^n : \vartheta \in \Theta\}$ in the usual sense means that—in the present framework—any $P_\vartheta^n(B_n)$ can be expressed through $\tilde\vartheta^{(n)}$ as $\int M_n(\tilde\vartheta^{(n)}, B_n)dP_\vartheta^n$. The asymptotic sufficiency as explained above means that $P_\vartheta^{(n)}(B_n)$ can be approximated by $Q_{n,\vartheta}(B_n)$, without randomization.

Relation (5.12.3) is used in the papers by Kaufman (1966) and Inagaki (1970). Kaufman (p. 170, Theorem 4.3) gives an independent proof of relation (5.12.3), with $Q_{n,\vartheta}$ similar to (5.12.2). Under his strong regularity conditions, the ML estimator may be taken for $\vartheta^{(n)}$.

In Michel and Pfanzagl (1970, p. 188, Theorem) it was shown that the factor $1_{B_n}(\vartheta^{(n)}(\mathbf{x}_n) - \vartheta)$ in definition (5.12.2) may be omitted. In Pfanzagl (1972b) [p. 177, Theorem 1] it was shown that for every compact subset $K \subset \Theta$ there exists a constant a_K such that

$$\sup_{\vartheta \in K} d(P_\vartheta^n, Q_{n,\vartheta}) \le a_K n^{-1/2}.$$

According to Pfanzagl (1972c), approximations of the type (5.12.2) with

$$\sup_{\vartheta \in K} n^{1/2} d(P_\vartheta^n, Q_{n,\vartheta}) \to 0$$

are impossible, even in a simple case like $p(x, \vartheta) = \vartheta^{-1} \exp[-\vartheta^{-1} x]$, $x > 0$, with $\Theta = (0, \infty)$.

After the general LAN -condition (see below) had been firmly established as a technically useful tool for asymptotic theory, it was natural to reconsider "global asymptotic normality", originally confined to i.i.d., at this more general level. This was done in papers by Milbrodt (1983), Droste (1985) and Pfanzagl (1995). For $n \in \mathbb{N}$ let $(X_n, \mathscr{A}_n)$ be a measurable space. The family of sequences of distributions $\{(Q_{n,\vartheta})_{n\in\mathbb{N}} : \vartheta \in \Theta\}$, $\Theta \subset \mathbb{R}^k$ is *quasi-normal* if $Q_{n,\vartheta}$ has, with respect to some σ-finite measure, a density

$$\mathbf{x}_n \to K_n(\vartheta) \exp\Big[-\frac{1}{2} c_n^2 (\vartheta^{(n)}(\mathbf{x}_n) - \vartheta)^\top L(\vartheta)(\vartheta^{(n)}(\mathbf{x}_n) - \vartheta)\Big]$$

with $K_n(\vartheta) \to 1$.

The family of sequences $\{(P_\vartheta^{(n)})_{n\in\mathbb{N}} : \vartheta \in \Theta\}$ is called *asymptotically normal* if there exists a quasi-normal family such that

$$d(P_\vartheta^{(n)}, Q_{n,\vartheta}) \to 0.$$

It can be shown that $\{(P_\vartheta^{(n)})_{n\in\mathbb{N}} : \vartheta \in \Theta\}$ is asymptotically normal iff it fulfills an LAN-condition with $\Delta_n(\cdot, \vartheta) = L(\vartheta) c_n(\vartheta^{(n)} - \vartheta)$, and with a remainder term fulfilling $r_n(\cdot, \vartheta, a_n) \to 0$ $(P_\vartheta^{(n)})$ for every bounded sequence $(a_n)_{n\in\mathbb{N}}$. This result makes clear that an LAN-condition with a remainder term fulfilling $r_n(\cdot, \vartheta, a) \to 0$ $(P_\vartheta^{(n)})$ for every $a \in \mathbb{R}^k$, is not strong enough to imply asymptotic normality. Milbrodt (1983) [p. 406, Theorem 2.7] proves that LAN implies asymptotic normality under a stronger condition on the remainder term in the LAN-condition, namely $\sup_{|a|\le\varepsilon} |r_n(\cdot, \vartheta, a)| \to 0$ $(P_\vartheta^{(n)})$. The result with a remainder term $r_n(\cdot, \vartheta, a_n) \to 0$

which establishes the equivalence is due to Droste (1985) [p. 47, Satz 5.5]. Pfanzagl (1995) [p. 117, Theorem] asserts the same result with an improved proof. The papers mentioned above consider, in fact, a locally uniform version of this equivalence. Since Θ is an open subset of $\mathbb{R}^k$, the results hold then also uniformly on compact subsets of Θ; hence the name "global asymptotic normality".

That intuitively equivalent conditions on the remainder term like $r_n(\cdot, \vartheta, a_n) \to 0$ and $r_n(\cdot, \vartheta, a) \to 0$ are of decisive influence on an asymptotic result is somewhat irritating.

Global asymptotic normality of i.i.d.-families was used by Kaufman and Inagaki in the proof of the Convolution Theorem. The extension of global asymptotic normality from i.i.d. models to LAN-families is an interesting, nontrivial achievement of asymptotic theory. From the technical point of view the proof of the Convolution Theorem is easier if it uses the LAN-condition immediately (rather than taking the intuitively inverting digression via global asymptotic normality).

Local Asymptotic Normality

Unlike Kaufman (1966) and Inagaki (1970), Hájek starts in his papers (Hájek 1970, 1972) not from relation (5.12.2), but from the following LAN-condition:

LAN-condition. Let $(X_n, \mathscr{A}_n)$, $n \in \mathbb{N}$, be a sequence of measurable spaces. For $\vartheta \in \Theta \subset \mathbb{R}^k$ let $P_\vartheta^{(n)}$ be a sequence of probability measures such that

$$\log(dP^{(n)}_{\vartheta+c_n^{-1}a}\Big/dP_\vartheta^{(n)}) = a^\top \Delta_n(\cdot, \vartheta) - \frac{1}{2}a^\top L(\vartheta)a + r_n(x, \vartheta, a) \qquad (5.12.4)$$

with

$$P^{(n)}_{\vartheta+c_n^{-1}a} \circ \Delta_n(\cdot, \vartheta) \Rightarrow N(a^\top L(\vartheta), L(\vartheta)) \qquad (5.12.5)$$

and

$$r_n(\cdot, \vartheta, a) \to 0 \quad (P_\vartheta^{(n)}) \quad \text{for every } a \in \mathbb{R}^k. \qquad (5.12.6)$$

(For variants of condition (5.12.6) see Pfanzagl 1994, p. 264.)

Under suitable regularity conditions on the family $\{P_\vartheta : \vartheta \in \Theta\}$, relation (5.12.4) holds for $(X_n, \mathscr{A}_n) = (X^n, \mathscr{A}^n)$ and $P_\vartheta^{(n)} = P_\vartheta^n$ with

$$\Delta_n(\mathbf{x}_n, \vartheta) = n^{-1/2}\sum_{\nu=1}^n \ell^\bullet(x_\nu, \vartheta) \quad \text{and} \quad L(\vartheta) = \int \ell^\bullet(\cdot, \vartheta)\ell^\bullet(\cdot, \vartheta)^\top dP_\vartheta.$$

More on sufficient conditions for LAN can be found in Sect. 5.6.

In regular cases, the LAN-condition holds with $\Delta_n(\mathbf{x}_n, \vartheta) = n^{-1/2}\sum_{\nu=1}^n \ell^\bullet(x_\nu, \vartheta)$; in "almost regular" cases usually with $\Delta_n(\mathbf{x}_n, \vartheta) = c_n^{-1}\sum_{\nu=1}^n \ell^\bullet(x_\nu, \vartheta)$ and $c_n = n^{1/2}L(n)$ with L slowly varying.

As an example we mention the location parameter family generated by the density $p(x) = 2x\exp[-x^2]1_{(0,\infty)}(x)$, where LAN holds with

$$\Delta_n(\mathbf{x}_n, \vartheta) = n^{-1/2}(\log n)^{-1/2} \sum_{\nu=1}^{n} \big(2(x_\nu - \vartheta) - (x_\nu - \vartheta)^{-1}\big).$$

(See Pfanzagl 2002b, p. 484, Example 1. See also Ibragimov and Has'minskii 1973, p. 249, Theorem 1, and 1981, p. 134, Theorem 5.1.)

There are, however, almost regular models with $\Delta_n(\mathbf{x}_n, \vartheta) = \sum_{\nu=1}^{n} h_n(x_\nu, \vartheta)$, where h_n cannot be replaced by $c_n^{-1}\ell^\bullet$. (See Pfanzagl 2002b, p. 481, Proposition 2.2 and Example 2.)

Proposition 5.12.1 *The central sequence Δ_n, $n \in \mathbb{N}$, is unique up to a term $o_p(n^o)$ only. It may always be chosen such that*

$$dP^{(n)}_{\vartheta+c_n^{-1}a}/dP^{(n)}_\vartheta - \exp\Big[a^\top \Delta_n(\cdot, \vartheta) - \frac{1}{2}a^\top L(\vartheta)a\Big] \to 0 \quad (P^{(n)}_\vartheta). \tag{5.12.7}$$

This is, in certain instances, a convenient alternative to (5.12.4).

Proof To simplify our notations, let

$$g_n := \exp\Big[a^\top \Delta_n(\cdot, \vartheta) - \frac{1}{2}a^\top L(\vartheta)a\Big]$$

and

$$\delta_n := \exp[r_n(\cdot, \vartheta, a)]. \tag{5.12.8}$$

Relation (5.12.6) implies

$$\delta_n \to 1 \quad (P^{(n)}_\vartheta).$$

As a consequence of (5.12.5),

$$\int \delta_n g_n dP^{(n)}_\vartheta = 1 \quad \text{for } n \in \mathbb{N}. \tag{5.12.9}$$

According to a Lemma of Hájek (1970) [p. 327, Lemma 1], Δ_n may be replaced by a truncated version such that, in addition to (5.12.4),

$$\int g_n dP^{(n)}_\vartheta \to 1. \tag{5.12.10}$$

(According to Hájek, the convergence in (5.12.10) holds even uniformly on compact subsets of a, but this is not needed here.)

Since

$$P^{(n)}_\vartheta \circ \Delta_n(\cdot, \vartheta) \Rightarrow N(0, L(\vartheta))$$

and

$$\int \exp[a^\top u - \frac{1}{2}a^\top L(\vartheta)a] N(0, L(\vartheta))(du) = 1,$$

there are, for every $\varepsilon > 0$, numbers M_ε and n_ε such that

$$\int g_n 1_{[0,M_\varepsilon]}(g_n) dP_\vartheta^{(n)} > 1 - \varepsilon \quad \text{for } n \geq n_\varepsilon.$$

Together with (5.12.10) this implies

$$\int g_n 1_{(M_\varepsilon,\infty)}(g_n) dP_\vartheta^{(n)} < \varepsilon \quad \text{for } n \geq n'_\varepsilon. \tag{5.12.11}$$

Relation (5.12.7) follows if we show that

$$(\delta_n - 1) g_n \to 0 \quad (P_\vartheta^{(n)}). \tag{5.12.12}$$

By (5.12.9) and (5.12.10),

$$\int (\delta_n - 1) g_n dP_\vartheta^{(n)} \to 0. \tag{5.12.13}$$

We shall show that

$$\int (\delta_n - 1) 1_{[0,1)}(\delta_n) g_n dP_\vartheta^{(n)} \to 0. \tag{5.12.14}$$

Together with (5.12.13) this implies

$$\int (\delta_n - 1) 1_{(1,\infty)}(\delta_n) g_n dP_\vartheta^{(n)} \to 0,$$

so that

$$\int |\delta_n - 1| g_n dP_\vartheta^{(n)} \to 0, \tag{5.12.15}$$

which implies (5.12.12).

We have (use (5.12.11)

$$0 \geq \int (\delta_n - 1) 1_{[0,1]}(\delta_n) g_n dP_\vartheta^{(n)} \geq M_\varepsilon \int (\delta_n - 1) 1_{[0,1]}(\delta_n) dP_\vartheta^{(n)} - \varepsilon$$

for $n \geq n'_\varepsilon$. Since $\delta_n \to 1$ implies

$$\int (\delta_n - 1) 1_{[0,1]}(\delta_n) dP_\vartheta^{(n)} \to 0,$$

relation (5.12.14) follows. □

The LAN-condition may be considered as a localized version of global asymptotic normality (in the sense of (5.12.2)). It describes the local properties of the family $\{(P_\vartheta^{(n)})_{n\in\mathbb{N}} : \vartheta \in \Theta\}$ in a technically convenient way, sufficient for an asymptotic theory of first order. How difficult is was, even for leading statisticians of this time, to see the importance of conditions like local or global asymptotic normality, and to understand the meaning of the resulting intrinsic bounds (as shown in the Convolution Theorem and the Minimax Theorem) can be seen from C.R. Rao's contribution to the discussion of Hájek's survey paper (1971, p. 160).

Remark LAN is usually attributed to Le Cam. Various conditions closely related to (5.12.4) occur in Le Cam 1960 (see p. 46 "system A"; p. 51 "asymptotically differentiable"; p. 57 "differentially asymptotically normal"), and in several subsequent papers. Notice that "asymptotiquement dérivable" in Le Cam (1969) [p. 61] is something different from "asymptotically differentiable" in (1960, p. 46). In these papers, Le Cam's interest is confined to technical details of his LAN-variants. More surprising than Le Cam's concentration on technicalities is the fact that he never tried to apply LAN-conditions to statistical problems: It was left to Hájek (1970, 1972) to extract from the LAN-profusion in Le Cam's papers the technically useful version (5.12.4), and to show that it provides a base for convolution- and minimax-theorems. Though Le Cam's LAN-variants refer to families of general sequences $(P_\vartheta^{(n)})_{n\in\mathbb{N}}$, his examples deal with product measures only. The first instance of a LAN-model outside this realm is due to Roussas (1965) [p. 979, Theorem 3.1] who proves LAN for ergodic Markov-processes. The first examples of LAN-families with a rate other than $n^{1/2}$ occur in Le Cam (1969) [pp. 108–112]. The most interesting among these examples is the location parameter family generated $p(x) = C(\alpha)\exp[-|x|^\alpha]$, $\alpha > 0$, a family first studied by Prakasa Rao (1968). This family is differentiable in quadratic mean for $\alpha > 1/2$, but not any more for $\alpha = 1/2$, the "almost regular" case with the rate $n^{1/2}(\log n)^{1/2}$ mentioned above. Le Cam's comments on this boundary case are rather scarce. "On vérifie alors sans difficulté que la famille est encore asymptotiquement gaussienne..." Nothing about the central sequence Δ_n, and nothing about the asymptotic variance; it remained that vague also in later publications (Le Cam 1986, p. 59) or Le Cam and Yang (1990, Example 2, pp. 111–114. Notice a misprint: In the relation $C(\alpha) = \alpha/2\Gamma(1/\alpha)$, the factor α is missing.)

5.13 The Convolution Theorem

Kaufman's Paper

In 1966, the optimality of ML estimators for i.i.d. observations was still an open problem. Since Le Cam's paper (1953) it was clear that there are problems with the concept of an optimal limit distribution, and it took about 10 years to find a solution in the restriction to estimator sequences which attain their limit distribution (locally) uniformly (Rao 1963, Wolfowitz 1965).

In the fundamental paper by Kaufman (1966) and the related paper by Inagaki (1970), the main result reads as follows: Under suitable regularity conditions, the ML-sequence $(\hat{\vartheta}^{(n)})_{n\in\mathbb{N}}$ converges on the family $\{P_\vartheta : \vartheta \in \Theta\}$, $\Theta \subset \mathbb{R}^k$, uniformly on compact subsets of Θ to $N(0, \Lambda(\vartheta))$, where $\Lambda(\vartheta)$ is the inverse of the matrix $L(\vartheta) := \int \ell^\bullet(\cdot, \vartheta)\ell^\bullet(\cdot, \vartheta)^\top dP_\vartheta$.

Leaving the relation to the ML-sequence aside (which was the main subject of interest in this time), the essence of the Convolution Theorem can be described as follows:

If $P_\vartheta^n \circ (n^{1/2}(\vartheta^{(n)} - \vartheta)$, $n \in \mathbb{N}$, converges to $Q_\vartheta|\mathbb{B}^k$, *uniformly on* Θ, then

$$Q_\vartheta(C) \leq N(0, \Lambda(\vartheta))(C) \tag{5.13.1}$$

for every C which is convex and symmetric about 0.

This follows easily from the main result in these papers, namely that $n^{1/2}(\vartheta^{(n)} - \hat{\vartheta}^{(n)})$ and $n^{1/2}(\hat{\vartheta}^{(n)} - \vartheta)$ are, under $(P_\vartheta^n)_{n\in\mathbb{N}}$, asymptotically stochastically independent (Kaufman 1966, p. 164, Lemma 3.4, and Inagaki 1970, p. 8, Lemma 2.3). By Anderson's Theorem, this implies (5.13.1), which is Kaufman's Theorem 2.1, p. 157. Inagaki presents this result in a more elegant make up, as a Convolution Theorem (p. 10, Theorem 3.1):

$$Q_\vartheta = N(0, \Lambda(\vartheta)) * R_\vartheta. \tag{5.13.2}$$

There can be no question that Kaufman could have arrived at the convolution version, but he preferred the direct way via Anderson's Theorem (see p. 166).

The use of Anderson's Theorem was not obvious from the beginning. In the abstract of Kaufman's paper from 1965, the optimality assertion (5.13.1) was confined to ellipsoids C which are concentric to the concentration ellipsoid of $N(0, \Lambda(\vartheta))$.

Compared with later versions of the Convolution Theorem, Kaufman's paper is overcrowded with regularity conditions on the parametric family $\{P_\vartheta : \vartheta \in \Theta\}$, $\Theta \subset \mathbb{R}^k$. These regularity conditions are to ensure the usual properties of the ML-sequence. Moreover, they are needed to prove that the ML-sequence is "asymptotically sufficient" in the following sense: There exist sequences of functions $g_n(\cdot, \vartheta)$, $\vartheta \in \Theta$, and h_n on X^n such that uniformly on compact subsets of Θ,

$$\int \left| \prod_{\nu=1}^{n} p(x_\nu, \vartheta) - h_n(\mathbf{x}_n) g_n(\hat{\vartheta}^{(n)}(\mathbf{x}_n), \vartheta) \right| \mu_n(d\mathbf{x}_n) \to 0.$$

Kaufman's proof has a clear underlying idea: For asymptotic considerations, the sequence of i.i.d.-products P_ϑ^n, $n \in \mathbb{N}$, can be replaced by a family, say $Q_\vartheta^{(n)}$, with μ_n-density $\mathbf{x}_n \to h_n(\mathbf{x}_n) g_n(\hat{\vartheta}^{(n)}(\mathbf{x}_n), \vartheta)$. Since $\hat{\vartheta}^{(n)}$ is sufficient for the family $\{Q_\vartheta^{(n)} : \vartheta \in \Theta\}$, any estimator $\vartheta^{(n)}$ can be obtained from $\hat{\vartheta}^{(n)}$ by randomization, and this suggests that $\vartheta^{(n)}$ is less accurate than $\hat{\vartheta}^{(n)}$. Plausible as this idea is, it is difficult to carry through in a mathematically precise way. (For this purpose, $(\hat{\vartheta}^{(n)})_{n\in\mathbb{N}}$ and

$(\vartheta^{(n)})_{n\in\mathbb{N}}$ have to be replaced by asymptotically equivalent sequences etc.) Moreover, as later proofs show, the idea that $\vartheta^{(n)}$ is a randomization of $\hat{\vartheta}^{(n)}$ is dispensable.

Kaufman's paper is concerned with the asymptotic optimality of ML-sequences. It cannot serve as a model for general convolution theorems. Yet, it contains an idea which is basic for convolution theorems in general: The representation of an estimator as the sum of two (asymptotically) stochastically independent terms, in Kaufman's paper (see p. 158).

$$n^{1/2}(\vartheta^{(n)} - \vartheta) = n^{1/2}(\hat{\vartheta}^{(n)} - \vartheta) + n^{1/2}(\vartheta^{(n)} - \hat{\vartheta}^{(n)}), \tag{5.13.3}$$

where $\hat{\vartheta}^{(n)}$ is a modified version of the ML estimator.

Relation (5.13.3) may remind the reader of a similarly looking relation due to Fisher (1925) [p. 706], namely

$$n^{1/2}(\vartheta^{(n)} - \vartheta) = n^{1/2}(\hat{\vartheta}^{(n)} - \vartheta) + n^{1/2}(\vartheta^{(n)} - \hat{\vartheta}^{(n)}), \tag{5.13.4}$$

where $\vartheta^{(n)}$ is an arbitrary unbiased estimator. If $\hat{\vartheta}^{n}$ is an unbiased estimator of minimal variance, the two terms on the right-hand side of (5.13.4) are uncorrelated (but not necessarily stochastically independent). See also Sect. 7 in Fisher (1922).

Even if Kaufman's proof is not easy to follow, it is the first mathematically acceptable proof of a version of relation (5.13.1). (It does not diminish the value of Kaufman's paper that he uses in one place, p. 168, a Theorem of Bergström (1945), the proof of which is not correct; see Bergström 1974. Inagaki's proof from (1970) is just a streamlined version of Kaufman's proof. (At a surface inspection, Inagaki's paper looks much nicer with respect to the regularity conditions, but this is just because of his generous "we use the same regularity conditions ... as in Kaufman".) Weiss and Wolfowitz (1966) [Sect. 5] suggest for Kaufman's result an alternative proof which they consider as "simple and perspicuous".

Considering the fact that Kaufman was a student of Wolfowitz, and that his paper appeared almost simultaneously with the paper by Wolfowitz (1965) confined to univariate families, which, for its part, uses results from Kaufman's paper, one could think that Kaufman did not do more than generalizing Wolfowitz's result from one-parameter to k-parameter families. This would be totally wrong. Wolfowitz's result is—like the results of C.R. Rao, Roussas and Schmetterer (see Sect. 5.11)—based on the Neyman–Pearson Lemma. Kaufman's result is based on Le Cam's idea of "asymptotic sufficiency". See Le Cam (1956, 1964).

How was Kaufman's result greeted by the scientific community? Schmetterer (1956 and 1974), familiar with a fundamental preliminary version of Wolfowitz (1965) ignores Kaufman altogether. Le Cam, whose paper (1956) was a cornerstone for Kaufman, did not react until Hájek's paper appeared in 1970 (Hájek 1970). Roussas, who was coauthor of the paper on the convolution theorem by Roussas and Soms, published (1973), confines himself to the statement (see Roussas 1972 (!), p. 131):

> Kaufman [1966] has generalized Wolfowitz's result for the case that Θ is k-dimensional.

Hajek's Convolution Theorem

Shortly after Kaufman's Theorem from (1966), Hájek (1970) published a general Convolution Theorem . In Kaufman's paper the limit distribution of uniformly convergent estimator sequences had been shown to be a convolution product involving the limit distribution of the ML-sequence. A general Convolution Theorem establishes that the limit distribution of any uniformly convergent sequence is a convolution product—of what? This question is answered implicitly in Hájek's paper. He presumes that the basic family of probability measures fulfills an LAN-condition (see his p. 324, relation 2).

The Convolution Theorem asserts that the limit distribution of any convergent estimator sequence is a convolution product with the limit distribution of $P_\vartheta^n \circ \Delta_n(\cdot, \vartheta)$, i.e. that $P_\vartheta^n \circ \Delta_n(\cdot, \vartheta)$ is an intrinsic bound for the concentration of everywhere convergent estimator sequences. There is, however, no general result that guarantees that this bound is attained by an estimator sequence. What is the meaning of "attained" in this connection? After all the concept of "regular convergence" is technically useful but not intuitively convincing. A mathematically consistent result requires that there should be regular estimator sequences attaining this bound. From an intuitive point of view one would like to have an estimator sequence attaining this bound that converges on compact subsets of Θ.

The asymptotic optimality of $P_\vartheta^n \circ \Delta_n(\cdot, \vartheta)$ would be a natural consequence of the representation

$$n^{1/2}(\vartheta^{(n)} - \vartheta) = \Delta_n(\cdot, \vartheta) + n^{1/2}(\vartheta^{(n)} - \vartheta) - \Delta_n(\cdot, \vartheta) \tag{5.13.5}$$

with asymptotically stochastically independent terms $\Delta_n(\cdot, \vartheta)$ and $n^{1/2}(\vartheta^{(n)} - \vartheta) - \Delta_n(\cdot, \vartheta)$. Surprisingly, this argument does not occur in connection with general functions of the parameters explicitly until Bickel et al. (1993) (where it occurs at various levels of generality: Theorem 2.3.1, p. 24, Theorem 3.3.2, p. 63, Theorem 5.2.2, p. 182, etc.).

Moreover, Hájek's proof of his Convolution Theorem on pp. 324/5, is not optimal in its presentation. It is, therefore, not surprising that it was superseded soon by the now common proof suggested by Bickel and worked out in detail by Roussas and Soms (1973; see the reference on p. 28, and Theorem 5.1, pp. 34/5).

Bickel's idea to base the proof of the Convolution Theorem on characteristic functions was presented in detail in Roussas and Soms (1973, p. 28, and Theorem 5.1, pp. 34/5). It appears in various textbooks (Bickel et al. 1993, Theorem 3.3.2, pp. 63/4, Pfanzagl 1994, Theorem 8.4.1, pp. 278/9; Witting and Müller-Funk (1995, Satz 6.211, pp. 433/4).

What distinguishes Hájek's approach from that of Kaufman and Inagaki is (i) the use of an LAN-condition, extending the applicability from i.i.d. families to more general ones, (ii) the idea of an intrinsic bound for the asymptotic accuracy of estimator sequences, a bound which is determined by the local structure of the family, as described by the LAN-condition, and which makes no reference to a particular (asymptotically efficient) estimator sequence, and (iii) the technical details of the proof are concerned with limit distributions, not with the estimators themselves. (What has not survived from Hájek's approach are Bayesian techniques.) The comments by C.R. Rao (p. 160) and Godambe (p. 159) on Hájek's paper from (1971) show how progressive Hájek's approach has been.

Obviously, Hájek considers his Convolution Theorem as a generalization of Kaufman's Theorem. Yet, he is rather terse about the consequences of his theorem for the optimality of ML-sequences.

Remark Some authors name the "Convolution Theorem" after Hájek and Le Cam. This eponymy takes into account that it was Le Cam who had blazed the trail for Hájek's version of the Convolution Theorem by his papers on asymptotic normality, local and global, mainly in Le Cam (1956, 1960 and 1966). If it is true that Hájek's publication of the Convolution Theorem (1970) came to Le Cam as "a bolt out of the blue" (a personal communication of Beran, cited in van der Vaart 2002, p. 643) this confirms that Le Cam had, at this time, was not focused on what his fellow statisticians considered as relevant problems. In his last papers prior to Hájek's path-breaking paper from 1970, Le Cam (1969, entitled *Théorie asymptotique de la décision statistique*, still plays around with various subtle points related to "asymptotic normality". The emphasis of this paper can be illustrated by the references which include Alexiewicz (differentiation of vector valued functions), Dudley (convergence of Baire measures), Pettis (differentiation in Banach spaces), Saks (theory of the integral), but not the paper by Kaufman (1966), which was based on Le Cam (1956). If Le Cam says that he could prove the Convolution Theorem immediately after he had seen it (Yang 1999, p. 236), this confirms his ability to perceive the abstract structure of a problem (the local translation invariance, in this case). Two years after Hájek's paper, Le Cam, using the techniques of Le Cam (1964), offers an abstract version of the Convolution Theorem (Le Cam 1972, p. 256, Proposition 8) which asserts the existence of a transition between the limit distribution of a distinguished statistic and an arbitrary limit distribution. In Proposition 10, pp. 257/6, he gives conditions under which this transition may be represented as a convolution. In Sect. 6 he indicates certain applications of this result, including convolutions with an exponential factor, cases which are ignored by Hájek. Ibragimov and Has'minskii obtained for the same models the same results by elementary techniques. (See Ibragimov and Has'minskii (1981), p. 278, Theorem 5.2. Hint: In relation 5.5, the letter n must be replaced by u, a misprint dating from the Russian original, 1979, p. 370.)

We shall not follow Le Cam's path to more and more abstract versions, a path which ends up with the idea that the essence of the Convolution Theorem was there long before statisticians had any notion of it: No, it is not the unpublished thesis of Boll (1955); it is a paper by Wendel (1952) on the representation of bounded linear transformations on integrable functions on locally compact groups with right invariant Haar measure. Such transformations can be represented by a convolution if they commute with all operations of left multiplication. (See the reference in Le Cam 1994, p. 405 or 1998, p. 27.) □

The first Convolution Theorems refer to estimator sequences of $\vartheta \in \Theta \subset \mathbb{R}^k$. For a coherent theory of asymptotic efficiency, a convolution theorem for estimator sequences $\kappa^{(n)}$ of $\kappa(\vartheta)$ and on functionals $\kappa(P)$) is indispensable. Without such a general Convolution Theorem it is impossible to say that $\kappa(\vartheta^{(n)})$ is asymptotically optimal for $\kappa(\vartheta)$ if $\vartheta^{(n)}$ is asymptotically optimal for ϑ, that $\vartheta^{(n)} = (\vartheta_1^{(n)}, \ldots, \vartheta_p^{(n)})$ is asymptotically optimal for $\vartheta = (\vartheta_1, \ldots, \vartheta_p)$ if each component $\vartheta_i^{(n)}$ is asymptotically optimal for ϑ_i, etc. Such a coherent theory, based on local properties of the family $\mathfrak{P}$ and the functional κ leads to an inherent concept of asymptotic optimality. Convolution Theorems for functionals (rather than parameters) appear in Pfanzagl 1982, p. 158, Theorem 9.3.1, van der Vaart (1988, p. 20, Theorem 2.4), a Le Cam-type Convolution Theorem in Strasser (1985) [p. 199, Theorem 38.26], Bickel et al. (1993) [p. 63, Theorem 3.3.2 and p. 182, Theorem 5.2.2.]

Lehmann and Casella (1998) present an asymptotic estimation theory without the Convolution Theorem. In Theorem 6.5.1b, p. 463, they give conditions under which the ML-sequence $\hat{\vartheta}^{(n)}$ is asymptotically normal with variance $\Lambda(\vartheta)$. The problem is to show that $N(0, \Lambda(\vartheta))$ is optimal in some sense. The authors try to answer this question in Theorem 6.5.1. Since their relation 5.10 is an immediate consequence of $P_\vartheta^n \circ n^{1/2}(\hat{\vartheta}^{(n)} - \vartheta) \Rightarrow N(0, \Lambda(\vartheta))$, it remains unclear where the asymptotic efficiency asserted in Theorem 6.5.1c should come from. The reference to Theorem 6.3.10, p. 449 and to Definition 6.2.4, p. 439 is of no help. Theorem 6.3.10 refers to univariate parametric families, and Definition 6.2.4 is motivated by the relation $\sigma^2(\vartheta) \geq \Lambda(\vartheta)$ which holds "under some additional restriction" for estimator sequences with limit distribution $N(0, \sigma^2(\vartheta))$.

With this presentation, the authors withhold two important messages from the reader:

(i) Asymptotic optimality is restricted to estimator sequences converging to a limit distribution locally uniformly in some sense.
(ii) There is an operationally meaningful concept of multivariate optimality (which applies, in particular, to ML-sequences).

Hence the reader Lehmann and Casella's book gets no complete picture of the asymptotic optimality of ML-sequences—a problem in the focus of statistical theory since the twenties, and with a satisfying answer available since Kaufman's paper from 1966.

The textbook by Witting and Müller-Funk 1995 (with 803 pages on asymptotic theory), might have been the appropriate forum to present a coherent theory of multivariate asymptotic optimality. They forgave this chance. Though they present a detailed proof of the Convolution Theorem as Satz 6.211 (with a reference to Hájek 1970, followed by a proof based on Bickel's idea), it appears that the authors misunderstood the relevance of this theorem for proving the asymptotic optimality of ML-sequences. In Satz 6.35, p. 202, they show that the ML-sequence converges to $N(0, \Lambda(\vartheta))$, in Satz 6.203, p. 423 they claim the asymptotic optimality of the ML-sequence. This claim is based on Satz 6.202, p. 423 which asserts that $\Lambda(\vartheta)$ is minimal in the Löwner-order—among all normal limit distributions of regular and asymptotically linear [!] estimator sequences. Asymptotic linearity is a widespread and technically useful property of estimator sequences. Yet one might prefer an optimality concept for estimator sequences which is based solely on an operationally significant property. The Convolution Theorem, presented in Satz 6.211, p. 433, would have provided the appropriate framework to establish the optimality of the ML-sequence among all regular estimator sequences of the parameters. Estimators for functionals $\kappa : \mathbb{R}^k \to \mathbb{R}^p$ flare up, here and there, like an ignis fatuus. On p. 175 the authors suggest for this purpose a curious estimator (which is, in their opinion, ML "im erweiterten Sinne"). On p. 424 the problem of estimating $\kappa(\vartheta)$ is raised again. At this point they should have noticed that an appropriate answer requires a Convolution Theorem for estimators of $\kappa(\vartheta)$. What they offer in this connection, Satz 6.204, p. 424 is on functionals κ which are bijective and differentiable. In part b) of this Theorem they just compute the distribution of $\kappa(\vartheta^{(n)})$ if $\vartheta^{(n)}$ is asymptotically linear and efficient, but they say nothing about the optimality of these estimator sequences.

A reader who has studied section their 6.5 on "local asymptotic efficiency" of almost 50 pages will discover that he has not learned how to answer a simple question like: "Which is the asymptotically best estimator sequence for the coefficient of variation in the family $\{N(\mu, \sigma^2)^n : \mu > 0, \sigma^2 > 0\}$?"

So far we have followed the first steps towards a convolution theorem for estimators of the parameters. Now we turn to the general problem of a convolution theorem for a general functional defined on an arbitrary family of probability measures $\mathfrak{P}$. Let now $\kappa : \mathfrak{P} \to \mathbb{R}^p$ be a functional that is differentiable at P with canonical gradient $\kappa^*(\cdot, P)$. The presentation (5.13.5) suggests to consider for an estimator $\kappa^{(n)} : X^n \to \mathbb{R}^p$ the analogous presentation

$$n^{1/2}(\kappa^{(n)} - \kappa(P)) = \tilde{\kappa}^*(\cdot, P) + \big(n^{1/2}(\kappa^{(n)} - \kappa(P)) - \tilde{\kappa}^*(\cdot, P)\big). \qquad (5.13.6)$$

To prove that $P^n \circ n^{1/2}(\kappa^{(n)} - \kappa(P))$ is asymptotically a convolution product with $P^n \circ \tilde{\kappa}_n^*(\cdot, P)$, it remains to be shown that $\tilde{\kappa}^*(\cdot, P)$ and $n^{1/2}(\kappa^{(n)} - \kappa(P)) - \tilde{\kappa}^*(\cdot, P)$ are asymptotically stochastically independent.

In fact, a more general result holds true. The sequence $n^{1/2}(\kappa^{(n)} - \kappa(P)) - \tilde{\kappa}^*(\cdot, P)$ is asymptotically stochastically independent of any $\tilde{g}$ with $g \in T(\mathfrak{P})^p$. This follows immediately from Proposition 5.13.1.

The proofs of various versions of the Convolution Theorem, too, in Bickel et al. (1993), are based on the presentation (5.13.6). (See p. 24, Theorem 2.3.1, relation 3, pp. 63/4, Theorem 3.3.2, relations 26 and 27, p. 182, Theorem 5.2.2, relation 21.) The proofs of all these convolution theorems follow the same pattern, due to Bickel, as presented in Roussas and Soms (1973). Observe that for parametric families a much simpler proof could be obtained from Proposition 5.13.1.

In the following we present a result for estimators $(\kappa_1^{(n)}, \ldots, \kappa_p^{(n)})$ of a functional $(\kappa_1(P), \ldots, \kappa_p(P))$ on a general family of probability measures with tangent space $T(P)$.

Proposition 5.13.1 *If* $(\kappa_1^{(n)}, \ldots, \kappa_p^{(n)})$ *is regular with respect to the linear space spanned by* $g_1, \ldots, g_k \in T(P)$, *then* $\tilde{g}_1, \ldots, \tilde{g}_k$ *and*

$$n^{1/2}(\kappa_1^{(n)} - \kappa_1(P)) - \tilde{\kappa}_1^*(\cdot, P), \ldots, n^{1/2}(\kappa_p^{(n)} - \kappa(P)) - \tilde{\kappa}_p^*(\cdot, P)$$

are asymptotically stochastically independent.

Proof The multidimensional version follows from the one-dimensional one, applied with

$$\kappa = \sum_{i=1}^{p} a_i \kappa_i, \quad \kappa^{(n)} = \sum_{i=1}^{p} a_i \kappa_i^{(n)}, \quad g = \sum_{i=1}^{p} a_i g_i.$$

In the following proof we write κ^* for $\kappa^*(\cdot, P)$. Since the sequence of probability measures

$$P^n \circ \big(\tilde{g}, \tilde{\kappa}^*, n^{1/2}(\kappa^{(n)} - \kappa(P))\big), \quad n \in \mathbb{N},$$

is tight, Prohorov's Theorem implies the existence of a sequence $\mathbb{N}_0$ and of a probability measure $\Pi|\mathbb{B} \times \mathbb{B} \times \mathbb{B}$ such that

$$P^n \circ \big(\tilde{g}, \tilde{\kappa}^*), n^{1/2}\kappa^{(n)} - \kappa(P))\big)_{n \in \mathbb{N}_0} \Rightarrow \Pi|\mathbb{B} \times \mathbb{B} \times \mathbb{B}.$$

Since the limits turn out to be independent of $\mathbb{N}_0$ the reference to $\mathbb{N}_0$ will be omitted throughout. By assumption, the marginal on the first two components of Π is the normal distribution $N(0, \Sigma(P))$ with

$$\Sigma_{11} = \int u^2 \Pi(d(u, v, w)), \quad \Sigma_{12} = \int uv\Pi(d(u, v, w)), \quad \Sigma_{22} = \int v^2 \Pi(d(u, v, w)).$$

Since $\kappa^{(n)}$ is regular with respect to $ag + b\kappa^*$, there is a probability measure $M|\mathbb{B}$ such that

$$P^n_{n^{-1/2}(ag+b\kappa^*)} \circ n^{1/2}(\kappa^{(n)} - \kappa(P_{n^{-1/2}(ag+b\kappa^*)})) \Rightarrow M \quad \text{for} \quad a, b \in \mathbb{R}.$$

Since

$$n^{1/2}\big(\kappa(P_{n^{-1/2}(ag+b\kappa^*)}) - \kappa(P)\big) \to \int \kappa^*(ag + b\kappa^*)dP = a\Sigma_{12} + b\Sigma_{22},$$

the regularity of $\kappa^{(n)}$ implies that

$$P^n_{n^{-1/2}(ag+b\kappa^*)} \circ \big(n^{1/2}(\kappa^{(n)} - \kappa(P)) - a\Sigma_{12} - b\Sigma_{22}\big) \Rightarrow M \quad \text{for} \quad a, b \in \mathbb{R}.$$

This implies that

$$\int H(w - a\Sigma_{12} - b\Sigma_{22})$$

$$\exp\Big[(au+bv)) - \frac{1}{2}(a^2\Sigma_{11} + 2ab\Sigma_{12} + b^2\Sigma_{22})\Big]\Pi(d(u,v))$$
$$= \int H(w)\Pi(dw) \quad \text{for} \quad a, b \in \mathbb{R} \tag{5.13.7}$$

for every bounded and continuous function H. With $H(w) = \exp[itw]$ we obtain from (5.13.7) that

$$\int \exp\Big[it(w - a\Sigma_{12} - b\Sigma_{22})\Big]$$
$$\exp\Big[(au+bv) - \frac{1}{2}(a^2\Sigma_{11} + 2ab\Sigma_{12} + b^2\Sigma_{22})\Big]\Pi(d(u,v))$$
$$= \int \exp[itw]\Pi(dw) \quad \text{for} \quad a, b \in \mathbb{R}. \tag{5.13.8}$$

Since the left-hand side of (5.13.8) is a holomorphic function of a, b, relation (5.13.8) holds for all $a, b \in \mathscr{C}$. From (5.13.8), applied with $a = is$ and $b = -it$ we obtain

$$\int \exp[isu + it(w - v)]\Pi(d(u,v,w))$$
$$= \exp\Big[-\frac{1}{2}s^2\Sigma_{11} + t^2\Sigma_{22}\Big]\int \exp[itw]\Pi(dw). \tag{5.13.9}$$

Applied with $t = 0$, relation (5.13.9) yields

$$\int \exp[isu]\Pi(du) = \exp\Big[-\frac{1}{2}s^2\Sigma_{11}\Big]$$

which is the characteristic function of $N(0, \Sigma_{11})$. Applied with $s = 0$, relation (5.13.9) yields

$$\int \exp[it(w-v)]\Pi(d(v,w)) = \exp[t^2\Sigma_{22}]\int \exp[itw]\Pi(dw).$$

Inserting this relation into (5.13.9), we obtain

$$\int \exp[isu + it(w-v)]\Pi(d(u,v,w)) = \int \exp[isu]\Pi(du)\int \exp[it(w-v)]\Pi(d(v,w)),$$

which implies the stochastic independence between u and $w - v$ under Π. □

A Convolution Theorem for General Families

Theorem 5.13.2 *Let $\kappa : \mathfrak{P} \to \mathbb{R}^p$ be a differentiable functional, i.e., there exists a canonical gradient $\kappa^*(\cdot, P)$ such that*

$$n^{1/2}(\kappa(P_{n^{-1/2}g}) - \kappa(P)) \to \int \kappa^*(\cdot, P)g\,dP \quad \textit{for} \quad g \in T(P, \mathfrak{P}).$$

Let $\kappa^{(n)}$, $n \in \mathbb{N}$, be a regular estimator sequence for $\kappa(P)$, i.e.

$$M := \lim_{n\to\infty} P^n_{n^{-1/2}g} \circ n^{1/2}(\kappa^{(n)} - \kappa(P_{n^{-1/2}g}))$$

is the same for every $g \in T(P, \mathfrak{P})$. Then M is a convolution product of the factor $N(0, \Sigma(P))$, with

$$\Sigma(P) = \int \kappa^*(\cdot, P)\kappa^*(\cdot, P)^\top dP, \tag{5.13.10}$$

and the limit distribution of

$$(R^{(n)}(\cdot, P) = n^{1/2}(\kappa_i^{(n)} - \kappa(P)) - \tilde{\kappa}_i^*(\cdot, P).$$

By Anderson's Theorem this implies

$$N(0, \Sigma(P))(C) \geq \Pi_0(C) \quad \text{for} \quad C \in \mathscr{C}_p. \tag{5.13.11}$$

Equality in (5.13.11) holds for every $C \in \mathscr{C}_p$ iff

$$R_i^{(n)}(\cdot, P) \to 0\ (P^n) \quad \text{for} \quad i = 1, \ldots, p,$$

i.e.

$$n^{1/2}(\kappa_i^{(n)} - \kappa_i(P)) = \tilde{\kappa}_i^*(\cdot, P) + o(n^0, P^n),$$

hence $\kappa_i^{(n)}$ is asymptotically linear with influence function $\kappa_i^*(\cdot, P)$. By Proposition 5.5.1, such estimators are automatically regular.

Proof of the Theorem. Since

$$n^{1/2}(\kappa_i^{(n)} - \kappa_i(P)) = \tilde{\kappa}_i^*(\cdot, P) + R_i^{(n)}(\cdot, P) + o(n^0, P^n),$$

we obtain from Proposition 5.13.1 that $\tilde{g}$ and $R^{(n)}(\cdot, P)$ are asymptotically stochastically independent for every $g \in T(P, \mathfrak{P})$. Applied with $g = \alpha_i\kappa_i^*(\cdot, P)$ this yields the asymptotic independence between $\tilde{\kappa}^*(\cdot, P)$ and $R^{(n)}(\cdot, P)$. Hence the limit distribution of $n^{1/2}(\kappa^{(n)} - \kappa(P))$ is a convolution product between the limit distribution of $\kappa^*(\cdot, P)$ and that of $R^{(n)}(\cdot, P)$. □

The Convolution Theorem Applied to Parametric Families

If $\mathfrak{P} = \{P_\vartheta : \vartheta \in \Theta\}$, $\Theta \subset \mathbb{R}^k$, then the tangent space is (under suitable regularity conditions on the densities) the linear subspace of $L_*(P)$ spanned by the components of $\ell^\bullet(\cdot, \vartheta)$. The canonical gradient of the functional $\kappa(\vartheta) := \vartheta$ is $\Lambda(\vartheta)\ell^\bullet(\cdot, \vartheta)$, and the optimal limit distribution of regular estimators of ϑ is $N(0, \Lambda(\vartheta))$. Regular estimator sequences $(\vartheta^{(n)})$, $n \in \mathbb{N}$, attaining this limit distribution have the stochastic expansion

$$n^{1/2}(\vartheta^{(n)} - \vartheta) = \Lambda(\vartheta)\tilde{\ell}^\bullet(\cdot, \vartheta) + o(n^0, P_\vartheta).$$

If $\kappa : \Theta \to \mathbb{R}^p$ is continuously differentiable with Jacobian

$$J(\vartheta) = (\kappa_i^{(j)}(\vartheta))_{i=1,\dots,p;j=1,\dots,k},$$

it suggests itself to use $\kappa^{(n)} := \kappa(\vartheta^{(n)})$ as an estimator of $\kappa(\vartheta)$. If $\vartheta^{(n)}$ is regular for ϑ, then $\kappa(\vartheta^{(n)})$ is also regular by the expansion

$$n^{1/2}(\kappa(\vartheta^{(n)} - \kappa(\vartheta)) = J(\vartheta^{(n)})n^{1/2}(\vartheta^{(n)} - \vartheta) + o(n^0, P_\vartheta^n). \tag{5.13.12}$$

If $P_\vartheta^n \circ n^{1/2}(\vartheta^{(n)} - \vartheta)$ is asymptotically normal with covariance matrix $\Sigma(\vartheta) = \Lambda(\vartheta)$ (the optimal limit distribution), then $n^{1/2}(\kappa(\vartheta^{(n)} - \kappa(\vartheta))$ has (as a consequence of the asymptotic expansion (5.13.12)) the normal limit distribution with covariance matrix $J(\vartheta)\Lambda(\vartheta)J(\vartheta)^\top$. One might suspect that this limit distribution is optimal among *all* regular estimator sequences of $\kappa(\vartheta)$ (not just among estimator sequences of the type $\kappa(\vartheta^{(n)})$. To answer such a question, we need a concept of asymptotic optimality for estimators of a functional $\kappa(\vartheta)$. Since the canonical gradient of $\kappa(\vartheta)$ in the tangent space spanned by the components of $\ell^\bullet(\cdot, \vartheta)$ is

$$\kappa^*(\cdot, \vartheta) = J(\vartheta)\Lambda(\vartheta)\ell^\bullet(\cdot, \vartheta),$$

this leads to $N(0, \Sigma(\vartheta))$ with $\Sigma(\vartheta) = J(\vartheta)\Lambda(\vartheta)J(\vartheta)^\top$ as the optimal limit distribution of estimators of $\kappa(\vartheta)$.

Proposition 5.13.1 implies that $\tilde{\ell}_n^\bullet(\cdot, \vartheta)$ and $n^{1/2}(\kappa^{(n)} - \kappa_(\vartheta)) - K(\vartheta)\Lambda(\vartheta)\tilde{\ell}^\bullet(\cdot, \vartheta)$ are asymptotically independent. Hence the optimal limit distribution for estimator sequences $(\kappa(n), n \in \mathbb{N}$, that are regular with respect to all directions, is $N(0, \Sigma(\vartheta))$.

Applied with $p = k$ and $\kappa(\vartheta) = \vartheta$, this leads to $J = I$, the $k \times k$ unit matrix, i.e. to $\Sigma(\vartheta) = \Lambda(\vartheta)$.

For readers who dislike the condition of regular convergence on which the Convolution Theorem is based, we offer an alternative which is based on the continuity of $\Sigma(\vartheta)$:

Theorem 5.13.3 *Assume that $P_\vartheta^n \circ n^{1/2}(\kappa^{(n)} - \kappa(\vartheta))$, $n \in \mathbb{N}$, converges to a limit distribution Q_ϑ which is continuous at ϑ_0. If $\Sigma(\vartheta)$ is continuous at ϑ_0, then*

$$\int \ell(u)N(0, \Sigma(\vartheta_0))(du) \le \int \ell(u)Q_{\vartheta_0}(du)$$

for every subconvex loss function ℓ.

For a proof see (5.14.2).

Some scholars are misguided by formulations like "What is the best estimator sequence for ϑ_1 in the presence of nuisance parameters $\vartheta_2, \dots, \vartheta_p$?". Such formulations ignore an essential aspect, namely: The quality of the estimator $\vartheta_1^{(n)}$ for ϑ_1 is limited by the condition that $(\vartheta_1^{(n)})_{n\in\mathbb{N}}$ must be regular with respect to all parameters. Whether estimators of nuisance parameters are required does not matter.

According to our intuition, it should be possible to estimate ϑ_1 more accurately if we know the nuisance parameters $\vartheta_2, \ldots, \vartheta_p$. This idea occurs occasionally in the literature. In connection with the Cramér-Rao Theorem, Cramér (1946b) [p. 94] mentions this phenomenon explicitly. Kati (1983) [p. 302] objects against Bar-Lev and Reiser (1983) [p. 300] for estimating two parameters of a Gamma-distribution simultaneously. A somewhat twisted statement in Stuart and Ord (1991) [p. 676, and p. 646, Exercise 17.20] asserts that, for any component, the minimal attainable variance is "no less than if a single parameter were being estimated". Surprising is the remark in Witting and Müller-Funk (1995) [p. 426] on this problem. Given estimators $(\vartheta_1^{(n)}, \ldots, \vartheta_k^{(n)})$ for $(\vartheta_1, \ldots, \vartheta_k)$ and $p < k$, they say "auf die Frage der asymptotischen Effizienz von $[(\vartheta_1^{(n)}, \ldots, \vartheta_p^{(n)})$ als Schätzer für $(\vartheta_1, \ldots, \vartheta_p)]$ soll wegen des Auftretens der "Nebenparameter" $\vartheta_{p+1}, \ldots, \vartheta_k$ hier nicht weiter eingegangen werden."

To emphasize this point, we consider the functional $\kappa(\vartheta) = \vartheta_1$. Specialized to this case, Proposition 5.13.1 says that $n^{1/2}(\vartheta_1^{(n)} - \vartheta_1) - \sum_{j=1}^k \Lambda_{1j}(\vartheta)\tilde{\ell}^{(j)}(\cdot, \vartheta)$ is stochastically independent of $\tilde{\ell}^{\bullet}(\cdot, \vartheta)$. Hence $n^{1/2}(\vartheta_1^{(n)} - \vartheta_1)$ is stochastically independent of every $\tilde{\ell}^{(i)}(\cdot, \vartheta)$, $i \neq 1$ (which was considered by Parke 1986, p. 356, a "surprising fact").

Proposition 5.13.1 implies that $n^{1/2}(\vartheta^{(n)} - \hat{\vartheta}^{(n)})$ and $n^{1/2}(\hat{\vartheta}^{(n)} - \vartheta)$ are stochastically independent if $\hat{\vartheta}^{(n)}$, $n \in \mathbb{N}$, is asymptotically optimal. This asymptotic stochastic independence occurs in a weaker form in R.A. Fisher (1925, p. 706): $n^{1/2}(\hat{\vartheta}^{(n)} - \vartheta)$ and $n^{1/2}(\vartheta^{(n)} - \hat{\vartheta}^{(n)})$ are symptotically uncorrelated. (Hint: If $\hat{\vartheta}^{(n)}$ has minimal asymptotic variance, the asymptotic variance of $(1 - \alpha)\hat{\vartheta}^{(n)} + \alpha\vartheta^{(n)}$ cannot be smaller.)

Discussion of the Convolution Theorem

The limit distributions in the Convolution Theorem are of the form $Q = \overline{Q} * R$ with $\overline{Q}$ and R $\mathfrak{Q}$ in some class $\mathfrak{Q}$ of probability measures on $\mathbb{B}^k$.

Obviously, $\overline{Q}$ is unique up to shifts only: If $Q = \overline{Q} * R$, then $Q = \overline{Q}_c * R_{-c}$ for every $c \in \mathbb{R}^k$ (where P_c is a shifted version of P). If $\mathfrak{Q}$ is closed under shifts, then $\overline{Q}$ is unique up to a shift (i.e., if $\overline{Q}_1$ and $\overline{Q}_2$ have the property specified above, then they differ by a shift only. (For a proof see Pfanzagl 2000a, p. 3, Proposition 3.1.)

Some authors consider the assertion that a given limit distribution, say Q, can be represented as a convolution product like $Q = \overline{Q} * R$ as a final result (see e.g. Roussas 1972, or Janssen and Ostrovski 2013). These authors ignore the fact that a relation like $Q = \overline{Q} * R$ admits an operational interpretation in terms of concentration on certain sets in special cases such as $\overline{Q} = N(0, \Sigma)$ only. They accept without discussion that $\overline{Q}$ is "better" than $\overline{Q} * R$ and that, therefore, an estimator sequence with limit distribution Q is asymptotically better than an estimator sequence with limit distribution $\overline{Q} * R$, as the latter one is subject to a "disturbance", expressed by R. First doubts about the weight of this idea arise from the fact that the representation of a probability measure as a convolution product is not unique. Is it convincing to say that every $\overline{Q}_c$ is better than $\overline{Q} * R$? For a discussion of this question see Section 2.8.

The concept of *multivariate optimality* based on the *peak-order* is an outgrowth of Anderson's Theorem and is therefore restricted to models for which the optimal limit distribution is normal. Prior to the Convolution Theorem there was no convincing idea of a concept of multivariate optimality. Starting from the univariate situation where the concentration on symmetric intervals was universally accepted as an expression of the accuracy of limit distributions, it was not clear what the adequate generalization from $\mathbb{R}^1$ to $\mathbb{R}^k$ could be: symmetric rectangles (symmetric in which sense) or balls? Kaufman, before he settled on arbitrary symmetric convex sets, thought of the concentration on ellipsoids $\{u \in \mathbb{R}^k : u^\top \Lambda(\vartheta) u \leq r\}$ (see his Abstract from 1965).

In order to create an operationally significant concept of multivariate concentration on $\mathbb{B}^k$ one needs a family of sets $\mathscr{B}_p \subset \mathbb{B}^p$ which can be used to define "Q_0 is more concentrated than Q_1 if $Q_0(B) \geq Q_1(B)$ for every $B \in \mathscr{B}_p$".

By Anderson's Theorem, such an assertion holds for $\mathscr{B}_p = \mathscr{C}_p$ if Q_1 is a convolution product with a factor Q_0 which is symmetric and unimodal. It appears that there is no family $\mathscr{B}_p$ of interesting sets larger than $\mathscr{C}_p$ (except for the stripes).

The optimality concepts for probability measures on $\mathbb{B}^p$ and $\mathbb{B}^q$ are linked in a meaningful way if for $q < p$ the family $\mathscr{B}_q$ consists of all sets with cylinders in $\mathscr{B}_p$, i.e., if $\mathscr{B}_q$ consists of all sets $B \in \mathbb{B}^q$ such that $\{u \in \mathbb{R}^p : Ku \in B\} \in \mathscr{B}_p$ for some $q \times p$ matrix K. In this case,

$$Q_0(B) \geq Q_1(B) \quad \text{for } B \in \mathscr{B}_p$$

implies

$$Q_0 \circ (u \to Ku)(B) \geq Q_1 \circ (u \to Ku)(B) \quad \text{for } B \in \mathscr{B}_q. \tag{5.13.13}$$

Relation (5.13.13) is in particular fulfilled if $\mathscr{B}_p = \mathscr{C}_p$ and $\mathscr{B}_q = \mathscr{C}_q$.

Functions of Functionals

What has been said about the quality of estimators of functions of the parameter can be generalized to functions, say $G_i(\kappa_1(P), \ldots, \kappa_p(P))$, $i = 1, \ldots, q$, of functionals $(\kappa_1(P), \ldots, \kappa_p(P))$. These reflections are needed to develop an operational concept of multivariate optimality.

Throughout the following we assume that the function $G : \mathbb{R}^p \to \mathbb{R}^q$ is continuously differentiable at $\kappa(P)$ with Jacobian

$$J(\kappa(P)) = (G_i^{(j)}(\kappa(P)))_{i=1,\ldots,q;\, j=1,\ldots,p}.$$

Then

$$n^{1/2}(G(\kappa^{(n)}) - G(\kappa(P))) = J(\kappa(P)) n^{1/2}(\kappa^{(n)} - \kappa(P)) + o(n^0, P^n). \tag{5.13.14}$$

This makes it easy to link the asymptotic concentration of $G(\kappa^{(n)})$ in certain subsets of $\mathbb{B}^q$, say $\mathscr{B}_q$, to the asymptotic concentration of $\kappa^{(n)}$ in corresponding subsets of $\mathbb{B}^p$.

Convexity is a natural requirement for the sets in $\mathscr{B}_p$ and $\mathscr{B}_q$. According to the Convolution Theorem, concentrations of regularly attainable limit distributions are comparable on convex sets that are symmetric about the origin. This is a property inherited from $\kappa^{(n)}$ by $G(\kappa^{(n)})$.

A particular consequence: If $\hat{\kappa}^{(n)}$ is better than $\kappa^{(n)}$ on $\mathscr{C}_p$, then $G(\hat{\kappa}^{(n)})$ is better than $G(\kappa^{(n)})$ on $\mathscr{C}_q$. This does, however, not imply that *optimality* of $\kappa^{(n)}$ implies *optimality* of $(G(\kappa^{(n)})$ among *all* regular estimator sequences of $G(\kappa(P)$. For this, the Convolution Theorem is indispensable.

Under suitable regularity conditions, the canonical gradient of $G \circ \kappa$ is of the form $J(\kappa(P))\kappa^*(\cdot, P)$. Hence the covariance matrix of the optimal limit distribution of regular estimator sequences of $G(\kappa(P))$ is $J(\kappa(P))\Sigma(P)J(\kappa(P))^\top$, where $\Sigma(P) = \int \kappa^*(\cdot, P)\kappa^*(\cdot, P)^\top dP$ is the covariance matrix of the optimal limit distribution of regular estimator sequences of κ.

It now follows from (5.13.14) that the estimator sequence $G(\kappa^{(n)})$ is optimal on $\mathscr{C}_q$ among all regular estimator sequences if $\kappa^{(n)}$ is optimal on $\mathscr{C}_p$. In particular: If $\kappa^{(n)}$ is optimal on $\mathscr{C}_p$, then each component $G_i(\kappa^{(n)})$ is optimal on $\mathscr{C}_1$, the family of all intervals symmetric about 0. Hence a multidimensionally optimal estimator sequence consists of optimal components.

Conversely, if $\kappa_1^{(n)}, \dots, \kappa_p^{(n)}$ are p optimal estimator sequences, then each of these admits the stochastic expansion

$$n^{1/2}(\kappa_i^{(n)} - \kappa_i(P)) = \tilde{\kappa}_i^*(\cdot, P) + o(n^0, P^n).$$

Therefore, $(\kappa_1^{(n)}, \dots, \kappa_p^{(n)})$ estimates $(\kappa_1(P), \dots, \kappa_p(P))$ and has the limit distribution $N(0, \Sigma(P))$, which is optimal according to the Convolution Theorem.

To summarize: In a comparison between estimator sequences based on the asymptotic concentration on convex and symmetric sets, we find that *functions of better estimator sequences are again better. In LAN-models, functions of optimal estimator sequences are again optimal. In particular, optimal multivariate estimator sequences consist of optimal components*.

The following example shows that the phrase "componentwise optimality implies joint optimality" has to be interpreted with some care.

Example There exist probability measures $Q_0|\mathbb{B}^2$ with the following properties.

(i) For $i = 1, 2$, the marginals Q_{0i} are of minimal spread.
(ii) There exists $Q|\mathbb{B}^2$ such that Q_i has the same median, say m_i, as Q_{0i}, so that $Q_i(I) \le Q_{0i}(I)$ for every interval I containing m_i.
(iii) There are intervals $I_1 \ni m_1$ and $I_2 \ni m_2$ such that

$$Q(I_1 \times I_2) > Q_0(I_1 \times I_2). \tag{5.13.15}$$

Hence component-wise optimality does not imply optimality on a relevant subset of $\mathbb{B}^2$, even if the comparison is restricted to probability measures with comparable marginals.

Since Q is in this example a convolution product with the factor Q_0, relation (5.13.15) also disproves the idea that "convolution always spreads out mass" (see Section 2.8).

Here is such an example. Let $q(x) := 1_{(0,\infty)}(x)e^{-x}$, $x \in \mathbb{R}$, be the density of the exponential distribution. Let $Q_0|\mathbb{B}^2$ be the probability measure with λ^2-density $(x_1, x_2) \to q(x_1)q(x_2)$. Let $R|\mathbb{B}$ be a probability measure with support $(-\log 2, \infty)$. (This is just for convenience.) Let $Q|\mathbb{B}^2$ be the probability measure with λ^2-density

$$(x_1, x_2) \to \int q(x_1 + y)q(x_2 + y)R(dy).$$

Essential for the following computations is the relation

$$\int_{t+y} q(x)dx = e^{-(t+y)} \quad \text{if} \quad t \geq \log 2 \text{ and } y > -\log 2.$$

For $i = 1, 2$, the probability measure Q_{0i} has the median $\log 2$, and Q_i has the same median if $\int e^{-y}R(dy) = 1$. Moreover,

$$Q((t', t'') \times (t', t'')) = \int Q_0((t' + y, t'' + y) \times (t' + y, t'' + y))R(dy)$$
$$= \int (e^{-t'} - e^{-t''})^2 \int e^{-2y} R(dy) \quad \text{if } \log 2 \leq t' < t''.$$

Since $\int e^{-2y}R(dy) > \int e^{-y}R(dy) = 1$ (unless $R\{0\} = 1$), this implies

$$Q((t', t'') \times (t', t'')) > Q_0((t', t'') \times (t', t'')),$$

with strict inequality if R is chosen appropriately. Since the (strict) inequality holds for every $t' \geq \log 2$, it also holds for $t' < \log 2$ close to $\log 2$, in which case $(t', t'') \times (t', t'')$ contains the medians $(\log 2, \log 2)$. □

The following example shows that "componentwise better" does not always imply "jointly better".

Example There are $Q_0 = N(0, \Sigma_0)$ and $Q = N(0, \Sigma)|\mathbb{B}^2$ such that

$$Q_0\{|u_i| \leq t\} > Q\{|u_i| \leq t\} \quad \text{for } i = 1, 2 \text{ and every } t > 0$$

but

$$Q_0\{|u_i| \leq t \text{ for } i = 1, 2\} < Q\{|u_i| \leq t \text{ for } i = 1, 2\} \quad \text{for some } t > 0.$$

(See Pfanzagl 1994, pp. 291/2, Example 8.5.7.) □

Example There is a (symmetric) probability measure $Q|\mathbb{B}^2$ such that

$$N(0, I)\{|a^\top u| \le t\} \ge Q\{|a^\top u| \le t\} \quad \text{for } a \in \mathbb{R}^2 \text{ and } t > 0$$

but

$$N(0, I)\{u^\top u \le t\} < Q\{u^\top u \le t\} \quad \text{for } 0 < t < 1.$$

(See Pfanzagl 1994, p. 84, Example 2.4.3.) If we interpret N and Q as limit distributions of estimator sequences $(\vartheta_1^{(n)}, \vartheta_2^{(n)})_{n\in\mathbb{N}}$ and $(\hat{\vartheta}_1^{(n)}, \hat{\vartheta}_2^{(n)})_{n\in\mathbb{N}}$, then we obtain that every $\kappa(\hat{\vartheta}_1^{(n)}, \hat{\vartheta}_2^{(n)})$ is better than $\kappa(\vartheta_1^{(n)}, \vartheta_2^{(n)})$, though $(\vartheta_1^{(n)}, \vartheta_2^{(n)})$ is more concentrated than $(\hat{\vartheta}_1^{(n)}, \hat{\vartheta}_2^{(n)})$ on small circles around $(\vartheta_1, \vartheta_2)$. □

We conclude this section by a result on the relation between multivariate concentration and the concentration on its marginals, the inequality

$$N(0, \Sigma)([-t_1, t_1] \times \cdots \times [-t_k, t_k]) \ge \prod_{i=1}^{k} N(0, \sigma_{ii})[-t_i, t_i] \quad \text{for all } t_i > 0,\ i = 1, \ldots, k. \tag{5.13.16}$$

This relation was proved by Dunn (1958) under rather restrictive conditions. After an abortive attempt by Scott (1967), it was proved by Sidak (1967, p. 628, Corollary 1), using Anderson's theorem. According to Jogdeo (1970) [p. 408, Theorem 3] equality in (5.13.16) for *some* $t_1, \ldots, t_k$ implies that Σ is a diagonal matrix.

5.14 The Extended Role of Intrinsic Bounds

Let $\mathfrak{P}$ be a family of probability measures $P|\mathscr{A}$ and $\kappa : \mathfrak{P} \to \mathbb{R}^p$ a functional. Given an estimator $\kappa^{(n)}$ for κ, let

$$Q_P^{(n)} := P^n \circ c_n(\kappa^{(n)} - \kappa(P)).$$

We assume that $Q_P^{(n)} \Rightarrow Q_P$ for $P \in \mathfrak{P}$.

Sections 5.9, 5.11 and 5.13 present bounds for the concentration of Q_P if the convergence of $Q_P^{(n)}$, $n \in \mathbb{N}$, to Q_P is—in some sense—distinguished. Depending on the conceptual framework, this may be "regular convergence" or "locally uniform median unbiasedness". For $P \in \mathfrak{P}$ let $\overline{Q}_P$ be optimal among these distinguished limit distributions; i.e. $Q_P \le \overline{Q}_P$ for $P \in \mathfrak{P}$; with the interpretation that $\int \ell d Q_P \ge \int \ell d\overline{Q}_P$ for certain loss functions ℓ. Under suitable regularity conditions on $\mathfrak{P}$, this inequality holds for subconvex loss functions.

Our intention is to show that $\overline{Q}_P$, defined as a bound for "distinguished" limit distributions, is valid for a larger class of limit distributions. The basic idea: Depending on the general framework, any limit distribution Q_P is distinguished for P in a dense subset $\mathfrak{P}_1 \subset \mathfrak{P}$, so that

$$Q_P \le \overline{Q}_P \quad \text{for } P \in \mathfrak{P}_1. \tag{5.14.1}$$

If both, Q_P and $\overline{Q}_P$, depend separately on P, relation (5.14.1) holds for every $P \in \mathfrak{P}$. Under these conditions, an estimator sequence with continuous limit distribution cannot be superefficient. This generalizes an argument occurring in Bahadur (1964, Sect. 3, pp. 1550/1).

Warning: If $Q_P^{(n)}$, $n \in \mathbb{N}$, converges to Q_P (locally) uniformly, then the inequality $Q_P \le \overline{Q}_P$ establishes $\overline{Q}_P$ as a bound for $Q_P^{(n)}$ for large n. This is not the case any more if this inequality is, by a trick, extended to limit distributions Q_P which are not attained (locally) uniformly.

It remains to be shown how the general program indicated above can be carried through.

(i) *Parametric families.* The Convolution Theorem applied to parametric families $\{P_\vartheta : \vartheta \in \Theta\}$, $\Theta \subset \mathbb{R}^k$, asserts that a regular estimator sequence $\kappa^{(n)}$, $n \in \mathbb{N}$, is optimal iff

$$n^{1/2}(\kappa^{(n)} - \kappa(\vartheta)) = \tilde{\kappa}^*(\cdot, \vartheta) + o(n^0, P_\vartheta^n),$$

where $\kappa^*(\cdot, \vartheta)$ is the canonical gradient of κ with respect to the tangent space $\{a^\top \ell^\bullet(\cdot, \vartheta) : a \in \mathbb{R}^k\}$. It is easy to see that the proof of the Convolution Theorem (yielding the optimality of $Q_\vartheta = N(0, \Lambda(\vartheta))$ among all regularly attainable limit distributions) remains valid if the convergence of $Q_\vartheta^{(n)}$ to Q_ϑ holds just for a subsequence $\mathbb{N}_0$ and for $Q^{(n)}_{\vartheta + n^{-1/2}a} \Rightarrow Q_\vartheta$ with a in a dense subset of $\mathbb{R}^k$. According to Bahadur's Lemma , these assumptions are fulfilled for ϑ in a dense subset $\Theta_0 \subset \Theta$. Hence for $\ell \in \mathscr{L}_s$,

$$\int \ell(u) N(0, \Lambda(\vartheta))(du) \le \int \ell(u) Q_\vartheta(du) \quad \text{for } \vartheta \in \Theta_0. \tag{5.14.2}$$

If Λ is continuous on Θ then no continuous limit distribution can be better than $N(0, \Sigma(\vartheta))$. The idea to use continuity for extending an a.e.-relation to a relation which holds everywhere occurs already in Bahadur (1964). It was, in fact, one motivation for introducing what is now called Bahadur's Lemma.

(ii) *Nonparametric families.* Let $\mathfrak{P}$ be a general family with tangent set $T(P, \mathfrak{P})$, and $\kappa : \mathfrak{P} \to \mathbb{R}^p$ a differentiable functional with canonical gradient $\kappa^*(\cdot, P) \in T(P, \mathfrak{P})^p$. By the Convolution Theorem the optimal limit distribution of regular estimator sequences of κ is $\overline{Q}_P = N(0, \Sigma_*(P))$ with

$$\Sigma_*(P) := \int \kappa^*(\cdot, P) \kappa^*(\cdot, P)^\top dP.$$

More precisely,

$$Q_P \le N(0, \Sigma_*(P))$$

if $Q_P^{(n)}$, $n \in \mathbb{N}$, converges to Q_P regularly in the parametric subfamily spanned by the tangent set $\{\kappa_1^*(\cdot, P), \ldots, \kappa_p^*(\cdot, P)\}$.

For the following we need a metric ρ on $\mathfrak{P}$ such that $P \to \int \ell dQ_P^{(n)}$ is continuous for $\ell \in \mathscr{L}_u$. By the Addendum to Lemma 5.14.1, applied for $X = \mathfrak{P}$ with $f_n(P) = \int \ell dQ_P^{(n)}$ and $f_0(P) = \int \ell dQ_P$, the convergence of $\int \ell dQ_P^{(n)}$, $n \in \mathbb{N}$, to $\int \ell dQ_P$ is locally uniform (hence also regular) except for a set $\mathfrak{P}_1$ of first category in $(\mathfrak{P}, \rho)$. If ρ is the sup-metric d and κ is continuous, this holds by Proposition 5.3.2 for $\ell \in \mathscr{L}_u$. The class $\mathscr{L}_u$ contains a countable weak convergence determining class, for example by appropriately truncating the convergence determining class $x \to \exp[itx]$, $t \in \mathbb{Q}^k$. We obtain $Q_P \leq N(0, \Sigma_*(P))$ for $P \in \mathfrak{P} - \mathfrak{P}_1$.

In the case of a parametric family, the exceptional set, being of λ^k-measure 0, could legitimately be considered "small". This does not necessarily apply to the set $\mathfrak{P}_1$. However: If $(\mathfrak{P}, \rho)$ is complete, $\mathfrak{P} - \mathfrak{P}_1$ is dense in $\mathfrak{P}$ by Lemma 5.14.2.

Instances of a metric ρ with the desired properties can be found in Pfanzagl (2002a) [Sect. 7]. Here are two examples of general families with a natural metric ρ such that no estimator sequence can be superefficient on some open subset of $(\mathfrak{P}, \rho)$.

Example 1. Let $\mathfrak{P}$ be the family of all probability measures $P|\mathbb{B}$ with Lebesgue density such that $\int |x| P(dx) < \infty$. Let $\kappa(P) := \int x P(dx)$. Then $\mathfrak{P}$, endowed with the metric $\rho(P, Q) := \int (1 + |x|)|p(x) - q(x)|dx$, is complete, and κ is continuous.

Example 2. Let $\mathfrak{P}$ be the family of all probability measures $P|\mathbb{B}$ with a positive, bounded and continuous Lebesgue density. Let $\kappa(P)$ be the median of P. Then $\mathfrak{P}$, endowed with the metric $\rho(P, Q) := \|p - q\|_1 + \|p - q\|_\infty$, is complete and κ is continuous.

(ii) Here is a suggestion how the inequality $Q_P \leq \overline{Q}_P$ can be extended to limit distributions Q_P which are continuous in all parametric subfamilies. Let $\Pi_a := Q_{P_a}$. This notation is to distinguish between $\overline{Q}_P$ (the optimal limit distribution in the family $\mathfrak{P}$), applied with $P = P_a$, and $\hat{\Pi}_a$, the optimal limit distribution in the family $a \to Q_{P_a}$. If $P \to Q_P$ is continuous in all parametric subfamilies, $a \to \Pi_a$ is continuous. Since P_a is a path with $P_a = P_0$ for $a = 0$, we have $\Pi_0 = Q_{P_0}$.

Let $\overline{\Pi}_a$ denote the optimal limit distribution within the path P_a, $a \in A$. Since $a \to \Pi_a$ is continuous, we obtain from the results obtained in (i) that

$$\Pi_a \leq \overline{\Pi}_a \quad \text{for } a \in A. \tag{5.14.3}$$

The essential point is to choose the subfamily such that $\overline{\Pi}_a$ is continuous.

If the path $P_a, a \in A$, converges to P_0 from the least favourable direction, we have $\overline{\Pi}_0 = \overline{Q}_{P_0}$. According to (5.14.3), applied for $a = 0$,

$$Q_{P_0} = \Pi_0 \leq \overline{\Pi}_0 = \overline{Q}_{P_0}.$$

Hence the inequality $Q_P \leq \overline{Q}_{P_0}$ holds for all limit distributions Q_P which are continuous on all parametric subfamilies.

The application of this idea requires to define a least favourable parametric subfamily with the desired continuity properties. A natural choice is the density $x \to p_0(x)(1 + \vartheta\kappa^*(x, P_0))$, but these probability measures will usually not be elements of $\mathfrak{P}$; the choice of $\mathfrak{P}_0$ requires much care.

As an example we mention the estimation of the β-quantile $\kappa_\beta(P)$ of P in the family $\mathfrak{P}$ of all distributions on $\mathbb{B}$ with a positive, symmetric and differentiable Lebesgue density. Presuming (w.l.g) $\beta \ge 1/2$, the optimal limit distribution of regular estimator sequences is $N(0, \sigma_\beta^2(P))$ with (see Sect. 5.6)

$$\sigma_\beta^2(P) := \left(\int \frac{p'(x)^2}{p(x)} dx\right)^{-1} + (\beta - \frac{1}{2})(1 - \beta)/(p(\kappa_\beta(P)))^2.$$

Regular estimator sequences attaining the optimal limit distribution do exist. The optimal limit distribution depends continuously on P along any sufficiently regular one-parameter subfamily. Hence, no limit distribution sharing this continuity property can be superefficient at some $P \in \mathfrak{P}$.

(Warning: The reader who tries to find a clear-cut description of this procedure in Pfanzagl (2002a) will be disappointed. There are various examples where this is carried through, but the ingredients of this procedure need to be collected from Lemma 4.1, Theorem 6.1, etc. See also Pfanzagl 1999a, p. 74, Theorem.)

Auxiliary Results

In the following we show that under various conditions on a sequence of functions $f_n : X \to \mathbb{R}$, "convergence everywhere" implies "continuous convergence somewhere". We first present results for a general set X, and then stronger results for a Euclidean X.

Lemma 5.14.1 *Let $(X, \mathscr{U})$ be a Hausdorff space. For $n \in \mathbb{N}$ let $f_n : X \to \mathbb{R}$ be upper semicontinuous and $f_0 : \mathscr{X} \to \mathbb{R}$ lower semicontinuous. If*

$$\liminf_{n\to\infty} f_n(x) \ge f_0(x) \quad for \quad x \in X,$$

then there exists a set $X_1 \subset X$ which is of first category in X such that for every $x_0 \in X - X_1$ and for every sequence $x_n \to x_0$,

$$\liminf_{n\to\infty} f_n(x_n) \ge f_0(x_0). \tag{5.14.4}$$

Addendum. *For continuous functions $f_n : X \to Y$, the following is true. If*

$$\lim_{n\to\infty} f_n(x) = f_0(x) \quad \text{for} \quad x \in X,$$

then there exists a set $X_1 \subset X$ of first category in X such that for every $x_0 \in X - X_1$ and every sequence $x_n \to x_0$,

$$\lim_{n\to\infty} f_n(x_n) = f_0(x_0), \tag{5.14.5}$$

i.e. f_n *converges on* $X - X_1$ *continuously to* f_0*, and* f_0 *is continuous on* $X - X_1$*. Therefore,* $f_n \to f_0$ *locally uniformly on* $X - X_1$*.*

This holds for Y an arbitrary metric space. Observe that continuity of f_0 is not required. If $Y = \mathbb{R}$ and f_0 is continuous, relation (5.14.5) follows from (5.14.4) and the corresponding assertion for lower/upper semicontinuous functions.

Lemma 5.14.1 is a generalized version of a theorem which goes back to (Osgood 1897, p. 173, Corollary). For comments and a convenient proof see Pfanzagl (2003, p. 108, Proposition 5.2, and 2002a, pp. 94/5, Lemma 9.1 for the Addendum.)

For statistical applications one needs to know more about the exceptional set X_1 of "first category". In particular: Can X_1 be justifiably considered as "small", or: Is the set of continuous convergence a large subset of X_0? In fact, the set $X_0 - X_1$ of continuous convergence may be even empty. Therefore, the following Lemma is important.

Lemma 5.14.2 *Let* (X, ρ) *be a metric space which is complete or locally compact. Let* $X_0 \subset X$ *be an open subset. If* X_1 *is of first category in* (X_0, ρ)*, then* $X_0 - X_1$ *is dense in* X_0*.*

Proof A convenient proof for the case of a complete metric space can be found in Pfanzagl (2002a) [p. 96, Lemma 9.2]. In order to establish the assertion for locally compact metric spaces, observe that for any compact $C \subset X_0$ with $C^\circ \neq \emptyset$, the set $X_1 \cap C$ is of first category in C if X_1 is of first category in X_0. Since (C, ρ) is complete, $C - X_1$ is dense in C. The assertion follows by applying this with C a compact neighbourhood of an arbitrary element of X. (See also Kelley 1955, p. 200, Theorem 34.) □

If X is Euclidean, the following result is more useful for statistical applications.

Lemma 5.14.3 *Let* X_0 *be a subset of* $\mathbb{R}^k$*. Let* $f_0, f_n : \mathscr{X}_0 \to \mathbb{R}$ *be measurable functions such that*

$$\liminf_{n\to\infty} f_n(x) \geq f_0(x) \quad \textit{for} \quad x \in X_0.$$

Then for every sequence $(y_n)_{n\in\mathbb{N}} \to 0$ *there exists an infinite subsequence* $\mathbb{N}_0$ *and a* λ^k*-null set* X_1 *such that*

$$\liminf_{n\in\mathbb{N}_0} f_n(x + y_n) \geq f_0(x) \quad \textit{for} \quad x \in X_0 - X_1.$$

Addendum. *Together with the analogous version with* lim inf *and* $\geq$ *replaced by* lim sup *and* $\leq$*, this yields that* $\lim_{n\to\infty} f_n(x) = f_0(x)$ *for* $x \in X_0$ *implies*

$$\lim_{n\in\mathbb{N}_0} f_n(x + y_n) = f_0(x) \quad \text{for} \quad x \in X_0 - X_1,$$

with X_1 *a* λ^k*-null set and* $\mathbb{N}_0 \subset \mathbb{N}$ *an infinite subsequence.*

This Addendum is the well known Lemma of Bahadur (1964) [p. 1549, Lemma 4]. See also Droste and Wefelmeyer (1984) [p. 140, Proposition 3.7]. For a proof of Lemma 5.14.3 see Pfanzagl (2003) [p. 107, Proposition 5.1].

In contrast to Lemmas 5.14.1, 5.14.3 does not necessarily require the functions f_n to be continuous. On the other hand, Bahadur's Lemma asserts a version of local uniformity which is weaker than continuous convergence in two respects. It asserts $f_n(x_n) \to f_0(x)$ (i) not for all sequences $x_n \to x$, but only for $x_n = x + y_n$, for a countable family of sequences $(y_n)_{n\in\mathbb{N}} \to 0$, and (ii) only for some subsequence $\mathbb{N}_0 \subset \mathbb{N}$. Yet, this is all one needs for applications in statistical theory: The proof of the Convolution Theorem requires $Q^{(n)}_{\vartheta_n} \Rightarrow Q_{\vartheta_0}$ just for those sequences $\vartheta_n = \vartheta_0 + c_n^{-1}a$ with $a \in \mathbb{Q}$.

Though the occurrence of some subsequence $\mathbb{N}_0$ in Bahadur's Lemma does not impair its usefulness in statistical applications, Bahadur (1964) [p. 1550] raised the question whether $\lim_{n\in\mathbb{N}_0} f_n(x+y_n) = f_0(x)$ for λ^k-a.a. $x \in \mathbb{R}^k$ for some subsequence $\mathbb{N}_0$ could be strengthened to $\lim_{n\to\infty} f_n(x+y_n) = f_0(x)$ for λ^k-a.a. $x \in \mathbb{R}^k$. An example by Rényi and Erdős (see Schmetterer 1966, pp. 304/5) exhibits a sequence of functions with $f_n(x) \to 0$ for λ-a.a. $x \in (0,1)$, but $\limsup_{n\to\infty} f_n(x + n^{-1/2}) = 1$ for every $x \in (0,1)$. (Be aware of a misprint on p. 306). Since the functions f_n in this example are discontinuous, this does not exclude a sharper result for continuous functions.

In the case of a parametric family $\mathfrak{P} := \{P_\vartheta : \vartheta \in \Theta\}$, $\Theta \subset \mathbb{R}^k$, the natural application is with Lemma 5.14.3. As an alternative one might entertain the use of Lemma 5.14.1, applied with $X = \Theta$ and $f_n(\vartheta) = \int \ell(c_n(\vartheta^{(n)} - \vartheta))dP^n_\vartheta$ (rather than $X = \{P_\vartheta : \vartheta \in \Theta\}$). If $\vartheta \to P_\vartheta$ is continuous with respect to the sup-distance on $\mathfrak{P}$, the functions f_n are continuous. (The proof is about the same as that of Lemma 5.8.3). Hence the Addendum to Lemma 5.14.1 implies continuous convergence on Θ except for a set Θ_1 of first category. Lemma 5.14.3 yields the weaker (but equally useful) regular convergence along some subsequence, except for a set Θ_2 which is of Lebesgue-measure zero. Obviously, $\Theta_2 \subset \Theta_1$, i.e., the set of continuous convergence is smaller than the set of regular convergence. If one is satisfied with the weaker assertion, then one can do without Bahadur's Lemma , provided Θ is complete. Yet it adds much to the interpretation of this result if we know that the set of regular convergence is not just dense in Θ but equal to Θ up to a set of Lebesgue-measure zero. (Recall, in this connection, that a set of first category in $\mathbb{R}^k$ may be of positive Lebesgue-measure.)

If $X = \mathbb{R}^k$ and the functions $f_n, n = 0, 1, \ldots$, are continuous, both versions of the lemmas apply. To make use of the fact that the exceptional set X_1 in Lemma 5.14.1 is of first category, one might assume that X_0 is complete or locally compact. Then one could conclude that (see Lemma 5.14.2) convergence of $f_n(x)$ to $f_0(x)$ for every $x \in X_0$ implies locally uniform convergence on a dense subset of X_0.

Lemma 5.14.3 implies some sort of "qualified" regularity only, i.e., the convergence of $f_n(x + c_n^{-1}a)$, $n \in \mathbb{N}_0$, to $f_0(x)$ along a subsequence $\mathbb{N}_0$ for λ-a.a. $x \in X_0$ and all a in a countable subset of $\mathbb{R}^k$. For certain proofs this qualified regularity suffices. In such cases, one will prefer a result for λ-a.a. $x \in X_0$ over a result for all x in countable subset of X_0.

Some Additional Results

Let $\{(P_\vartheta^{(n)})_{n\in\mathbb{N}} : \vartheta \in \Theta\}$ be a family of sequences of probability measures with $\Theta \subset \mathbb{R}^k$, and let $\kappa : \Theta \to \mathbb{R}^p$ with $p \le k$ be a differentiable functional. Let

$$Q_\vartheta^{(n)} := P_\vartheta^{(n)} \circ c_n(\kappa^{(n)} - \kappa(\vartheta)).$$

As a consequence of Lemma 5.8.3, for ℓ is bounded and continuous,

$$\vartheta \to \int \ell d Q_\vartheta^{(n)} \quad \text{is continuous for every } n \in \mathbb{N}.$$

Asymptotic assertions about the function sequences

$$a \to \int \ell d Q^{(n)}_{\vartheta + c_n^{-1} a}, \quad n \in \mathbb{N},$$

seem not to have been considered so far.

Theorem 5.14.4 *Let $\{(P_\vartheta^{(n)})_{n\in\mathbb{N}} : \vartheta \in \Theta\}$, $\Theta \subset \mathbb{R}^k$, be a family such that*

$$\limsup_{n\to\infty} \|a - a_0\|^{-1} d(P^{(n)}_{\vartheta + c_n^{-1} a}, P^{(n)}_{\vartheta + c_n^{-1} a_0}) < \infty \tag{5.14.6}$$

for two vectors $a, a_0 \in \mathbb{R}^k$, and let $\kappa : \Theta \to \mathbb{R}^p$ with $p \le k$ be a differentiable function. Then the functions

$$a \to \int \ell d Q^{(n)}_{\vartheta + c_n^{-1} a}, \quad n \in \mathbb{N},$$

are equicontinuous for $a \in \mathbb{R}^k$ if ℓ is bounded and uniformly continuous.

Proof Since

$$\begin{aligned}
&\left| \int \ell d Q^{(n)}_{\vartheta + c_n^{-1} a} - \int \ell d Q^{(n)}_{\vartheta + c_n^{-1} a_0} \right| \\
&\le \left| \int \ell(c_n(\kappa^{(n)} - \kappa(\vartheta + c_n^{-1} a))) d P^{(n)}_{\vartheta + c_n^{-1} a} \right. \\
&\qquad \left. - \int \ell(c_n(\kappa^{(n)} - \kappa(\vartheta + c_n^{-1} a))) d P^{(n)}_{\vartheta + c_n^{-1} a_0} \right| \\
&+ \int \left| \ell(c_n(\kappa^{(n)} - \kappa(\vartheta + c_n^{-1} a))) - \ell(c_n(\kappa^{(n)} - \kappa(\vartheta + c_n^{-1} a_0))) \right| d P^{(n)}_{\vartheta + c_n^{-1} a_0} \\
&\le d(P^{(n)}_{\vartheta + c_n^{-1} a}, P^{(n)}_{\vartheta + c_n^{-1} a_0}) + \sup_{u \in \mathbb{R}^k} \left| \ell(u + c_n(\kappa(\vartheta + c_n^{-1} a) - \kappa(\vartheta + c_n^{-1} a_0))) - \ell(u) \right|,
\end{aligned}$$

the assertion follows from (5.14.6) and the differentiability of κ. □

Proposition 5.14.5 *Under condition (5.14.6), and for compact A,*

$$\limsup_{n\to\infty}\sup_{a\in A}\int \ell\, dQ^{(n)}_{\vartheta+c_n^{-1}a}=\sup_{a\in A}\limsup_{n\to\infty}\int \ell\, dQ^{(n)}_{\vartheta+c_n^{-1}a} \tag{5.14.7}$$

if the loss function $\ell:\mathbb{R}^k\to[0,1]$ *is continuous and* $\{u\in\mathbb{R}^k:\ell(u)\le t\}$ *is bounded for every* $t\in[0,1)$.

Proof Since the functions $a\to\int \ell\, dQ^{(n)}_{\vartheta+c_n^{-1}a}$ are equicontinuous, $a_n\to a$ implies $\int \ell\, dQ^{(n)}_{\vartheta+c_n^{-1}a_n}-\int \ell\, dQ^{(n)}_{\vartheta+c_n^{-1}a}\to 0$. Hence relation (5.14.7) follows from Lemma 5.14.6, applied with $f_n(a)=\int \ell\, dQ^{(n)}_{\vartheta+c_n^{-1}a}$. □

Lemma 5.14.6 *Let* (X,ρ) *be a metric space. For* $n\in\mathbb{N}$ *let* $f_n:X\to\mathbb{R}$ *be such that for every* $x\in X$, *the relation* $\lim_{n\to\infty}\rho(x_n,x)=0$ *implies*

$$\lim_{n\to\infty}(f_n(x_n)-f_n(x))=0.$$

Then for A compact,

$$\limsup_{n\to\infty}\sup_{x\in A}f_n(x)=\sup_{x\in A}\limsup_{n\to\infty}f_n(x). \tag{5.14.8}$$

Addendum 1. We have

$$\lim_{r\to\infty}\limsup_{n\to\infty}\sup_{\|x\|\le r}f_n(x)=\sup_{x\in X}\limsup_{n\to\infty}f_n(x).$$

(From (5.14.7), applied with $A=\{x\in\mathbb{R}^p:\|x\|\le r\}$.)

Addendum 2. *If* $\lim_{n\to\infty}f_n(x)=0$ *for every* $x\in X$, *then for A compact,*

$$\lim_{n\to\infty}\sup_{x\in A}|f_n(x)|=0.$$

(From (5.14.8) applied with $|f_n|$ in place of f_n.)

Proof For $n\in\mathbb{N}$ let $x_n\in A$ be such that $\sup_{x\in A}f_n(x)\le f_n(x_n)+n^{-1}$. Since A is compact, there exists $x_0\in A$ and $\mathbb{N}_0\subset\mathbb{N}$ such that $\lim_{n\in\mathbb{N}_0}x_n=x_0$. Hence $\lim_{n\in\mathbb{N}_0}(f_n(x_n)-f_n(x_0))=0$, which implies

$$\limsup_{n\to\infty}\sup_{xd\in A}f_n(x)\le\limsup_{n\in\mathbb{N}_0}f_n(x_0)\le\limsup_{n\to\infty}f_n(x_0)\le\sup_{x\in A}\limsup_{n\to\infty}f_n(x).$$

Since the converse inequality is obvious, this proves (5.14.8). □

5.15 Local Asymptotic Minimaxity

Chernoff (1956) [p. 12, Theorem 1] presents for regular one-parameter families and i.i.d. observations a result which seems to be the forefather of various asymptotic "minimax" theorems. Rewritten in our notations, with $\vartheta^{(n)}$ an estimator of ϑ and

$$Q_\vartheta^{(n)} = P_\vartheta^n \circ n^{1/2}(\vartheta^{(n)} - \vartheta),$$

and assuming that $\Lambda(\vartheta) := 1/I(\vartheta)$ is continuous at ϑ_0, Chernoff's Theorem (attributed by Hájek 1972, p. 177, to an unpublished paper by Stein and Rubin) asserts that for *every* estimator sequence $\vartheta^{(n)}$,

$$\lim_{r\to\infty} \liminf_{n\to\infty} \sup_{|\vartheta-\vartheta_0|\le n^{-1/2}r} \int \min\{u^2, r^2\} Q_\vartheta^{(n)}(du) \ge \Lambda(\vartheta_0).$$

Chernoff's Theorem is supplemented by a result by Huber (1966) [Theorem], proved under the usual regularity conditions: If for *some* estimator sequence,

$$\limsup_{n\to\infty} \sup_{|\vartheta-\vartheta_0|\le n^{-1/2}r} \int \min\{u^2, r^2\} Q_\vartheta^{(n)}(du) \le \Lambda(\vartheta_0) \text{ for every } r > 0,$$

then

$$Q_{\vartheta_0}^{(n)} \Rightarrow N(0, \Lambda(\vartheta_0)) = \overline{Q}_{\vartheta_0}.$$

The interrelation between the length of the interval $|\vartheta - \vartheta_0| \le n^{-1/2}r$ over which the sup is taken, and the loss function $u \to \min\{u^2, r^2\}$, makes the interpretation of such results difficult; the restriction to the quadratic loss function impairs their relevance.

Motivated by the results of Chernoff (1956) and Huber (1966), Hájek (1972) felt the need of supplementing the Convolution Theorem for regular estimator sequences converging regularly to a limit distribution by a result for arbitrary estimator sequences. This led him to the so-called "local asymptotic Minimax Theorem": For regular estimator sequences, the asymptotic risk $\int \ell dQ_\vartheta^{(n)}$ is constant in shrinking neighborhoods $|\vartheta - \vartheta_0| \le n^{-1/2}r$ for every r. The minimax theorem says that even for non-regular estimators, the maximal risk in such neighborhoods, or more precisely

$$\lim_{r\to\infty} \liminf_{n\to\infty} \sup_{|\vartheta-\vartheta_0|\le n^{-1/2}r} \int \ell dQ_\vartheta^{(n)},$$

cannot be lower than this constant. Chernoff already seemed to be ill at ease with the interpretation of his theorem. "For an arbitrary estimate [$\Lambda(\vartheta_0)$] is "essentially" asymptotically a lower bound for the asymptotic variance..." After more than 50 years this construct is still lacking an operationally significant interpretation. For Le

Cam, Hájek's Minimax Theorem is another chance to invent an abstract version (see e.g. Le Cam 1979, p. 124, Theorem 1).

The following refers to a general family $\{(P_\vartheta^{(n)})_{n\in\mathbb{N}} : \vartheta \in \Theta\}$, $\Theta \subset \mathbb{R}^k$ fulfilling an LAN condition

$$\log(dP^{(n)}_{\vartheta+c_n a}/dP^{(n)}_\vartheta) = a^\top \Delta_n - \frac{1}{2}a^\top L(\vartheta)a + o(n^0, P^{(n)}_\vartheta).$$

We write $\Lambda(\vartheta) = L(\vartheta)^{-1}$ for the the "optimal" asymptotic covariance matrix for estimators of ϑ.

Hájek's Minimax Theorem *For one-parameter families fulfilling LAN, the following is true for loss functions $\ell \in \mathscr{L}_s$.*

(i) For any estimator sequence $\vartheta^{(n)}$,

$$\lim_{r\to\infty}\liminf_{n\to\infty}\sup_{|\vartheta-\vartheta_0|\le c_n r}\int \ell dQ^{(n)}_\vartheta \ge \int \ell dN(0,\sigma^2_*(\vartheta_0)). \tag{5.15.1}$$

(ii) If

$$\lim_{r\to\infty}\lim_{n\to\infty}\sup_{|\vartheta-\vartheta_0|\le c_n r}\int \ell dQ^{(n)}_\vartheta = \int \ell dN(0,\Lambda(\vartheta_0)) \tag{5.15.2}$$

for some non-constant loss function $\ell \in \mathscr{L}_s$, then

$$c_n^{-1}(\vartheta^{(n)} - \vartheta_0) - \Lambda(\vartheta_0)\Delta_n(\cdot,\vartheta_0) \to 0 \quad (P^{(n)}_{\vartheta_0}). \tag{5.15.3}$$

Hájek (1972) [p. 186, Theorem 4.1.]

Remark Relation (5.15.1) occurs in Hájek's Theorem 4.1 with

$$\lim_{\delta\to 0}\liminf_{n\to\infty}\sup_{|\vartheta-\vartheta_0|<\delta}$$

rather than

$$\lim_{r\to\infty}\liminf_{n\to\infty}\sup_{|\vartheta-\vartheta_0|\le c_n r},$$

but the latter version is indicated in his Remark 1, p. 189. (Beware of a misprint in Hájek's relation 4.17.) Hájek's remark does not explicitly refer to part(ii), but Ibragimov and Has'minskii 1981, p. 168, claim (without proof) that part (ii) holds for neighbourhoods $|\vartheta - \vartheta_0| \le c_n r$ as well.) Fabian and Hannan (1982) discuss the distinction between fixed and shrinking neighbourhoods in detail. In Sect. 3, pp. 463/4, they point out that estimator sequences fulfilling relation (5.15.2) with $\sup_{|\vartheta-\vartheta_0|<\delta}$ may not exist.

Hájek's Minimax Theorem is restricted to the case $\Theta \subset \mathbb{R}$, with a brief remark (p. 188, Theorem 4.2) concerning the bound for estimators of ϑ_1 if $\vartheta = (\vartheta_1, \ldots, \vartheta_k)$. The full k-dimensional version of part (i), referring to loss functions which are symmetric, subconvex, and which increase with $\|u\| \to \infty$ not too quickly, is due to Ibragimov and Has'minskii (1981) [p. 162, Theorem 12.1, with the Russian original dating from 1979].

For readers who are amazed at finding on p. 162 in Ibragimov and Has'minskii the assertion "... is possible if and only if ...": The "if" is an extra of the translator. The Russian original says correctly: "... is possible only if ...".

This is not the only slip in the translation of this theorem. The theorem refers to the class $W_{e,2}$ of all loss functions growing more slowly than $u \to \exp[\varepsilon u^2]$ for every $\varepsilon > 0$. In the translation, this class appears as $W_{\varepsilon,2}$, a slip which is not without the danger of confusion, since ε corresponds in the model of Ibragimov and Has'minskii to n^{-1}, and the asymptotic assertion refers explicitly to $\varepsilon \to 0$. Hence the reader might be confused by an asymptotic assertion for $\varepsilon \to 0$ which holds for every loss function in the class $W_{\varepsilon,2}$.

In their Theorem 12.1, the authors present the version of (5.15.1) with

$$\lim_{\delta \to 0} \liminf_{n\to\infty} \sup_{|\vartheta - \vartheta_0| < \delta},$$

but they mention on p. 168, Remark 12.2, the version with

$$\lim_{r\to\infty} \liminf_{n\to\infty} \sup_{|\vartheta - \vartheta_0| \le c_n^{-1/2} r}.$$

A detailed proof of the latter version may be found in Witting and Müller-Funk (1995) [p. 457, Satz 6.229].

Observe that $\lim_{r\to\infty}$ in relation (5.15.1) is essential. For the Hodges estimator with $\alpha = 1/2$, we have

$$\lim_{n\to\infty} \sup_{|\vartheta| \le c_n r} \int \ell \, d Q_\vartheta^{(n)} < \int \ell \, d N(0,1) \quad \text{for} \quad \ell(u) = u^2$$

as long as $r < \sqrt{3}$. (See Lehmann and Casella 1998, p. 440, Example 2.5 and p. 442, Example 2.7.)

When Hájek wrote his paper on the Minimax Theorem, $N(0, \Lambda(\vartheta_0))$ was known as a bound for *regular* estimator sequences. When he tried in 1972 to characterize $N(0, \Lambda(\vartheta_0))$ as a bound for *arbitrary* estimator sequences (Hájek 1972), he obviously was not familiar with the details of Le Cam's fundamental from (1953). In Theorem 14, page 327, Le Cam had shown, for $\Theta \subset \mathbb{R}$, bounded and continuous loss functions ℓ, and under Cramér-type regularity conditions, that

$$\limsup_{n\to\infty} \int \ell \, d Q_{\vartheta_0}^{(n)} < \int \ell \, d \overline{Q}_{\vartheta_0}$$

implies

$$\limsup_{n\to\infty} \int \ell d Q_{\vartheta_n}^{(n)} > \int \ell d\overline{Q}_{\vartheta_0} \quad \text{for some sequence } \vartheta_n \to \vartheta_0.$$

This implies that

$$\limsup_{n\to\infty} \sup_{|\vartheta-\vartheta_0|\le\delta} \int \ell d Q_{\vartheta}^{(n)} \ge \int \ell d\overline{Q}_{\vartheta_0} \quad \text{for } \delta > 0.$$

Since this relation holds for every subsequence, it follows that

$$\liminf_{n\to\infty} \sup_{|\vartheta-\vartheta_0|\le\delta} \int \ell d Q_{\vartheta}^{(n)} \ge \int \ell d\overline{Q}_{\vartheta_0} \quad \text{for } \delta > 0,$$

which is the original version of Hájek's relation (5.15.1). Hence in part (i) of his Minimax Theorem, Hájek had just rediscovered earlier results of Le Cam. Part (ii) contains something new.

With the bound $N(0, \Lambda(\vartheta_0))$ determined by (5.15.1), it is natural to define "optimality" (or "local asymptotic minimaxity") by equality in (5.15.1). One could object that Hájek's definition (5.15.1) presumes the existence of

$$\lim_{n\to\infty} \sup_{|\vartheta-\vartheta_0|\le c_n r} \int \ell d Q_{\vartheta}^{(n)}$$

for every $r > 0$. This suggests replacing the definition of optimality by the apparently stronger but more intuitive condition

$$\limsup_{n\to\infty} \sup_{|\vartheta-\vartheta_0|\le c_n r} \int \ell d Q_{\vartheta}^{(n)} \le \int \ell d N(0, \Lambda(\vartheta_0)) \quad \text{for every } r > 0. \tag{5.15.4}$$

With

$$F_n(r) := \sup_{|\vartheta-\vartheta_0|\le c_n r} \int \ell d Q_{\vartheta}^{(n)} - \int \ell d\overline{Q}_{\vartheta_0}, \tag{5.15.5}$$

Le Cam's result may be rewritten as

$$\liminf_{n\to\infty} F_n(0) < 0 \quad \text{implies} \quad \lim_{r\to\infty} \limsup_{n\to\infty} F_n(r) > 0.$$

Le Cam writes this relation with neighbourhoods $|\vartheta - \vartheta_0| < \delta$, but his proof is valid for $|\vartheta - \vartheta_0| < c_n r$ as well. Relation (5.15.4), rewritten as $\limsup_{n\to\infty} F_n(r) \le 0$, for every $r > 0$, therefore implies $\liminf_{n\to\infty} F_n(0) \ge 0$. Hence

$$0 \le \liminf_{n\to\infty} F_n(0) \le \liminf_{n\to\infty} F_n(r) \le \limsup_{n\to\infty} F_n(r) \le 0,$$

so that

$$\lim_{n\to\infty} F_n(r) = 0 \quad \text{for every } r > 0,$$

or, written explicitly,

$$\lim_{n\to\infty} \sup_{|\vartheta-\vartheta_0|\le c_n r} \int \ell d Q_\vartheta^{(n)} = \int \ell d N(0, \Lambda(\vartheta_0)),$$

which is Hájek's definition (5.15.3) of optimality.

According to Hájek's Minimax Theorem (ii), optimal estimator sequences have the stochastic expansion (5.15.3).

Levit (1974) [pp. 333/4 Theorem 1.1 and p. 336, Theorem 1.3], presents nonparametric versions of Hájek's Minimax Theorem (i) and (ii). (See also Koshevnik and Levit (1976) [p. 745, Theorem 2.]

A proof of Hájek (i) can be found in Ibragimov and Has'minskii (1981) [p. 162, Theorem II 12.1], Millar (1983) [p. 45, Definition 2.4], and Rüschendorf (1988) [p. 209, Satz 6.11]. Minimaxity in its relation to superefficiency is discussed in Pfanzagl (1994) [pp. 298–303].

Some authors ignore Hájek's main result. Witting and Müller-Funk (1995) [p. 457] present in Satz 6.229 on "Hájek's Minimax Theorem", proved for continuous (!) loss functions, a second part which should not be mistaken for Hájek (ii). It just asserts the existence of estimator sequences attaining the bound given in (5.15.1). A path breaking book like Bickel et al. (1993) with important results on asymptotic optimality of regular estimator sequences contains a reference to Hájek (1972) in several places but it never mentions what the subject of this paper is, optimality in the minimax sense.

According to Hájek's Minimax Theorem, the stochastic expansion (5.15.3) is necessary for optimality in the minimax sense. Neither Hájek nor any of the authors mentioned above show that their stochastic expansion is also sufficient. Sufficiency follows if the estimator sequence $\vartheta^{(n)}$ is regular in the sense with a replaced by bounded sequences a_n. Apparently, Hájek and his followers have been unaware of the fact that this kind of regularity holds automatically for estimator sequences with the asymptotic expansion (5.15.3). (More generally: For estimator sequences with the asymptotic expansion pertaining to optimal regular estimator sequences. See Proposition 5.5.1.) This is almost trivial for the particular case in question. The relation

$$c_n^{-1}(\vartheta^{(n)} - \vartheta_0) - \Lambda(\vartheta_0)\Delta_n(\cdot, \vartheta_0) \to 0 \quad (P_\vartheta^{(n)})$$

implies

$$\begin{aligned} c_n^{-1}(\vartheta^{(n)} - (\vartheta_0 + c_n a_n)) &= c_n^{-1}(\vartheta^{(n)} - \vartheta_0) - a_n \\ &= \Lambda(\vartheta_0)\Delta_n(\cdot, \vartheta_0) - a_n + o(n^0, P_\vartheta^{(n)}). \end{aligned}$$

If the LAN-condition holds with a_n in place of a, we have

$$P^{(n)}_{\vartheta+n^{-1/2}a_n} \circ (\Lambda(\vartheta_0)\Delta_n(\cdot, \vartheta_0) - a_n) \Rightarrow N(0, \Lambda(\vartheta_0)),$$

hence also

$$P^{(n)}_{\vartheta_0+c_n a_n} \circ c_n^{-1}(\vartheta^{(n)} - (\vartheta_0 + c_n a_n)) \Rightarrow N(0, \Lambda(\cdot, \vartheta_0)),$$

which may be rewritten as

$$\lim_{n\to\infty} \sup_{|a|\le r} \int \ell(c_n^{-1}(\vartheta^{(n)} - (\vartheta_0 + n^{-1/2}a))dP^{(n)}_{\vartheta_0+c_n a} = \int \ell dN(0, \Lambda(\vartheta_0))$$

for every $r > 0$, a relation stronger than (5.15.2).

The only paper claiming sufficiency of (5.15.3) is Fabian and Hannan (1982). On p. 467, Theorem 6.3, they proved the sufficiency for LAN-families with a replaced by a bounded sequence $(a_n)_{n\in\mathbb{N}}$. (Note: If the authors speak of "regular" estimator sequences, they mean estimator sequences with a stochastic expansion.)

Being poorly organized, this paper is not easy to read. (In fact the authors had trouble getting it published.) This perhaps explains why it is ignored throughout. Though it contains a section on adaptivity, it is not listed among the nearly 500 references in the book by Bickel et al. (1993) on "efficient and adaptive estimation".

As shown above, a condition weaker than (5.15.2) suffices for the asymptotic expansion (5.15.3). The fact that (5.15.3) implies (5.15.2) does not exclude the possibility of conditions nicer than (5.15.2) that imply (5.15.3). By Theorem 8.3.2 in Strasser (1985) [p. 439], specialized to the present framework, the stochastic expansion (5.15.3) follows from the relation

$$\limsup_{n\to\infty} \int \ell dQ^{(n)}_{\vartheta+c_n^{-1}a} \le \int \ell dN(0, \Lambda(\vartheta)) \quad \text{for every } a \in \mathbb{R}.$$

For the origin of this result, Strasser gives unspecified references to Hájek (1972) and Le Cam (1953), without emphasizing the improvement from "$\sup_{|a|\le r}$ for all r" to "for all a". (The reader who tries to discover traces of this result in Le Cam (1953), will fail. A rather abstract version of Hájek (ii) appears in Le Cam (1972).)

Observe that the implication from (5.15.4) to (5.15.3) does not extend to families with $\Theta \subset \mathbb{R}^k$, $k \ge 3$. Stein's estimator (1956), see Sect. 5.14, Example 4, below, fulfills at $\vartheta_0 = 0$ relation (5.15.4), hence also (5.15.2), for the loss function $\ell(u) = \|u\|^2$, but not (5.15.3).

For $\Theta \subset \mathbb{R}^k$ with $k > 1$, Hájek does not mention that a k-dimensional estimator sequence $\vartheta^{(n)} = (\vartheta_1^{(n)}, \dots, \vartheta_k^{(n)})$ which admits a stochastic expansion

$$c_n^{-1}(\vartheta^{(n)} - \vartheta_0) - \Lambda(\vartheta_0)\Delta_n(\cdot, \vartheta_0) \to 0 \quad (P^{(n)}_\vartheta)$$

has the joint limit distribution $N(0, \Lambda(\vartheta_0))$. For a discussion of "component-wise optimality" versus "joint optimality" see Sect. 5.13.

Since the result of Hájek (ii) does not extend from $\Theta \subset \mathbb{R}$ to arbitrary dimensions k (think of Stein's shrinking estimators), the adequate generalization is to one-dimensional functionals defined on $\Theta \subset \mathbb{R}^k$. A result of this kind appears in Strasser (1997) [p. 372, Theorem (iii)]. Specialized to the present framework of parametric LAN-families with $\Theta \subset \mathbb{R}^k$ and continuously differentiable functionals $\kappa : \Theta \to \mathbb{R}$ with gradient $K(\vartheta)$ at ϑ, Strasser's result reads as follows. Let $\mathscr{L}_s$ denote the set of subconvex and symmetric functions $L : [0, 1] \to \mathbb{R}^k$.

If for some estimator sequence $\kappa^{(n)}$, $n \in \mathbb{N}$,

$$\limsup_{n\to\infty} \int \ell(c_n(\kappa^{(n)} - \kappa(\vartheta + c_n^{-1}a)))dP^{(n)}_{\vartheta+c_n^{-1}a} \tag{5.15.6}$$
$$\leq \int \ell dN(0, K(\vartheta)\Lambda(\vartheta)K(\vartheta)^\top) \quad \text{for } a \in \mathbb{R}^k \text{ and every } \ell \in \mathscr{L}_s,$$

then

$$c_n(\kappa^{(n)} - \kappa(\vartheta)) - K(\vartheta)\Lambda(\vartheta)\Delta_n(\cdot, \vartheta) \to 0 \quad (P^{(n)}_\vartheta). \tag{5.15.7}$$

According to Strasser, the result for $\Theta \subset \mathbb{R}^k$ is an easy consequence of the results for $\Theta \subset \mathbb{R}$: "... the complete assertion of part (iii) for $k > 1$ follows since $\mathscr{L}_s$ contains sufficiently many loss functions which depend only on one single component of κ". For statisticians with a less penetrating intuition it might be easier to obtain this result for $\Theta \subset \mathbb{R}^k$ by an application of the result for one-parameter subfamilies to the least favourable one-parameter subfamily . With this argument it becomes clear that (5.15.6) for one particular $\ell \in \mathscr{L}$ suffices to imply (5.15.7), and that condition (5.15.6) may be confined to one particular $a \in \mathbb{R}^k$, the "least favourable direction".

The Role of the Supremum in the Minimax Theorem

The occurrence of $\sup_{|a|\leq r}$ in Hájek's Minimax Theorem makes its interpretation difficult. It occurs twice, in (5.15.1) and (5.15.3).

For $\Theta \subset \mathbb{R}^k$,

$$\lim_{r\to\infty} \liminf_{n\to\infty} \sup_{|a|\leq r} \int \ell dQ^{(n)}_{\vartheta+c_n^{-1}a} \geq \int \ell d\overline{Q}_\vartheta \quad \text{for every } \ell \in \mathscr{L}_s \tag{5.15.8}$$

and for $\Theta \subset \mathbb{R}$ (in the modified version (5.15.3)),

$$\limsup_{n\to\infty} \sup_{|a|\leq r} \int \ell dQ^{(n)}_{\vartheta+c_n^{-1}a} \leq \int \ell d\overline{Q}_\vartheta \quad \text{for some } \ell \in \mathscr{L}_s \text{ and every } r > 0 \tag{5.15.9}$$

implies

$$c_n(\vartheta^{(n)} - \vartheta) - \Lambda(\vartheta)\Delta_n(\cdot, \vartheta) \to 0 \quad (P^{(n)}_\vartheta). \tag{5.15.10}$$

As already mentioned above, a result of Strasser shows that (5.15.10) follows from a weaker version of (5.15.9) in which $\sup_{r\to\infty}\limsup_{n\to\infty}\sup_{|a|\le r}$ is replaced by $\sup_{a\in\mathbb{R}}\limsup_{n\to\infty}$.

For the particular case of loss functions which are bounded and *continuous*, we have (apply Proposition 5.14.5 for $A=\{a\in\mathbb{R}^k : |a|\le r\}$)

$$\lim_{r\to\infty}\limsup_{n\to\infty}\sup_{\|a\|\le r}\int \ell\, dQ^{(n)}_{\vartheta+c_n^{-1}a}=\sup_{a\in\mathbb{R}^k}\limsup_{n\to\infty}\int \ell\, dQ^{(n)}_{\vartheta+c_n^{-1}a}.$$

With the help of this relation, one obtains from Hájek's results (5.15.6) and (5.15.7) or (5.15.8) that

$$\sup_{a\in\mathbb{R}}\limsup_{n\to\infty}\int \ell\, dQ^{(n)}_{\vartheta+c_n^{-1}a}\ge\int \ell\, d\overline{Q}_\vartheta$$

and, more importantly, that

$$\sup_{a\in\mathbb{R}}\limsup_{n\to\infty}\int \ell\, dQ^{(n)}_{\vartheta+c_n^{-1}a}\le\int \ell\, d\overline{Q}_\vartheta$$

implies (5.15.10).

Corresponding results are true if $Q^{(n)}_\vartheta=P^{(n)}_\vartheta\circ c_n(\vartheta^{(n)}-\vartheta)$ is replaced by $Q^{(n)}_\vartheta=P^{(n)}_\vartheta\circ c_n(\kappa^{(n)}-\kappa(\vartheta))$.

Asymptotic Optimality is Independent of the Loss Function

The Convolution Theorem provides an asymptotic bound for the concentration of an important class of estimator sequences: Those converging regularly to some limit distribution. If one takes the concept of a loss function seriously, one might ask whether there are estimator sequences adjusted to the "true" loss function that are asymptotically better than the best regular estimator sequences. The Convolution Theorem, after all, assumes regular convergence of $\int \ell\, dQ^{(n)}_\vartheta$, $n\in\mathbb{N}$, to $\int \ell\, dQ_\vartheta$ for *some* Q_ϑ and *every* $\ell\in\mathscr{L}$.

In order to bring in a certain regularity if the evaluation is restricted to some ℓ_0, one might require the existence of a function $\tilde{\ell}_0:\Theta\to[0,\infty)$ which is approached by $\int \ell_0\, dQ^{(n)}_\vartheta$ in the sense that

$$\lim_{\delta\downarrow 0}\lim_{n\to\infty}\sup_{|\vartheta-\vartheta_0|\le\delta}\left|\int \ell_0 d Q^{(n)}_\vartheta-\bar{\ell}_0(\vartheta)\right|=0. \tag{5.15.11}$$

Recall that continuity of $\vartheta\to\int \ell_0 d Q^{(n)}_\vartheta$ implies continuity of $\vartheta\to\bar{\ell}_0(\vartheta)$ by Lemma 5.3.9. Together with (5.15.1), relation (5.15.11) implies that

$$\bar{\ell}_0(\vartheta_0)\ge\int \ell_0 d\overline{Q}_{\vartheta_0}.$$

Hence any optimal regular estimator sequence is also optimal in the larger class of all estimator sequences for which $\int \ell_0 dQ_\vartheta^{(n)}$, $n \in \mathbb{N}$, converges locally uniformly to a limit which may serve as an approximation to $\int \ell_0 dQ_\vartheta^{(n)}$. In other words: Asymptotic optimality does not depend on the loss function; a conjecture indicated already in Millar (1983) [p. 146].

Some Auxiliary Results

In the discussion of the Minimax Theorem, the following lemmas for nondecreasing functions $F_n : [0, \infty) \to \mathbb{R}$ have been applied with F_n defined by (5.15.5). In the proofs of these lemmas, we use repeatedly that any subsequence $\mathbb{N}_1$ a subsequence $\mathbb{N}_0$ such that $\lim_{n\in\mathbb{N}_0} F_n(r)$ exists for every $r \geq 0$.

Lemma 5.15.1 *The assertions (i) and (ii) are equivalent.*
(i)

$$\lim_{r\to\infty} \liminf_{n\to\infty} F_n(r) \geq 0 \tag{5.15.12}$$

and

$$\limsup_{n\to\infty} F_n(r) \leq 0 \quad \textit{for every } r \geq 0. \tag{5.15.13}$$

(ii) If, for some subsequence $\mathbb{N}_0$, $\lim_{n\in\mathbb{N}_0} F_n(r)$ *exists for every* $r \geq 0$, *then*

$$\lim_{r\to\infty} \lim_{n\in\mathbb{N}_0} F_n(r) = 0. \tag{5.15.14}$$

Observe that (5.15.12) is based on relation (5.15.1), i.e., part (i) of Hájek's Minimax Theorem, which is not restricted to one-parameter families.

Proof (i) *implies* (ii). Let $\mathbb{N}_0$ be a subsequence such that $\lim_{n\in\mathbb{N}_0} F_n(r)$ exists for every $r \geq 0$. Since

$$\liminf_{n\to\infty} F_n(r) \leq \lim_{n\in\mathbb{N}_0} F_n(r) \leq \limsup_{n\to\infty} F_n(r),$$

we obtain from (5.15.12) and (5.15.13) that

$$\lim_{r\to\infty} \lim_{n\in\mathbb{N}_0} F_n(r) \geq 0 \quad \text{and} \quad \lim_{n\in\mathbb{N}_0} F_n(r) \leq 0 \quad \text{for} \quad r \geq 0,$$

hence

$$\lim_{r\to\infty} \lim_{n\in\mathbb{N}_0} F_n(r) = 0.$$

(ii) *implies* (i). Since F_n is nondecreasing, any subsequence $\mathbb{N}_1$ contains a subsequence $\mathbb{N}_0$ such that $\lim_{n\in\mathbb{N}_0} F_n(r)$ exists for every $r \geq 0$. By assumption, this implies (5.15.14).

Equation (5.15.12): If $\lim_{r\to\infty} \liminf_{n\to\infty} F_n(r) < 0$, there exists a sequence $(r_n)_{n\in\mathbb{N}_1} \to \infty$ such that $\lim_{n\in\mathbb{N}_1} F_n(r_n) < 0$. Since there exists a subsequence $\mathbb{N}_0 \subset \mathbb{N}_1$ such that $\lim_{r\to\infty} \lim_{n\in\mathbb{N}_0} F_n(r) = 0$, the relation $\lim_{n\in\mathbb{N}_0} F_n(r) \leq \lim_{n\in\mathbb{N}_0} F_n (r_n) < 0$ for $r \geq 0$ leads to a contradiction.

Equation (5.15.13): If $\limsup_{n\to\infty} F_n(r_0) > 0$ for some $r_0 \geq 0$, there exists a subsequence $\mathbb{N}_1$ such that $\lim_{n\in\mathbb{N}_1} F_n(r_0) > 0$. This is in contradiction to the existence of a subsequence $\mathbb{N}_0 \subset \mathbb{N}_1$, such that $\lim_{n\in\mathbb{N}_0} F_n(r)$ exists for every $r > 0$ whence $\lim_{r\to\infty} \lim_{n\in\mathbb{N}_0} F_n(r) = 0$. □

Lemma 5.15.2 *The implications (i), (ii) and (iii) are equivalent.*

(i) For every subsequence $\mathbb{N}_0$,

$$\limsup_{n\in\mathbb{N}_0} F_n(0) < 0 \quad \textit{implies} \quad \lim_{r\to\infty} \limsup_{n\in\mathbb{N}_0} F_n(r) > 0.$$

(ii) For every subsequence $\mathbb{N}_0$,

$$\liminf_{n\in\mathbb{N}_0} F_n(0) < 0 \quad \textit{implies} \quad \lim_{r\to\infty} \limsup_{n\in\mathbb{N}_0} F_n(r) > 0.$$

(iii) For every subsequence $\mathbb{N}_0$,

$$\limsup_{n\in\mathbb{N}_0} F_n(0) < 0 \quad \textit{implies} \quad \lim_{r\to\infty} \liminf_{n\in\mathbb{N}_0} F_n(r) > 0.$$

Addendum. *If any of these implications holds true, then*

$$\lim_{r\to\infty} \liminf_{n\to\infty} F_n(r) \geq 0.$$

Proof (i) *implies* (ii). If $\liminf_{n\in\mathbb{N}_0} F_n(0) < 0$, there exists $\mathbb{N}_1 \subset \mathbb{N}_0$ such that

$$\lim_{n\in\mathbb{N}_1} F_n(0) = \liminf_{n\in\mathbb{N}_0} F_n(0) < 0.$$

By (i) this implies $\lim_{r\to\infty} \limsup_{n\in\mathbb{N}_1} F_n(r) > 0$, whence

$$\lim_{r\to\infty} \limsup_{n\in\mathbb{N}_0} F_n(r) > 0.$$

(ii) *implies* (iii). Assume that

$$\lim_{r\to\infty} \liminf_{n\in\mathbb{N}_0} F_n(r) \leq 0.$$

There exists a subsequence $\mathbb{N}_1 \subset \mathbb{N}_0$ such that

$$\lim_{r\to\infty} \limsup_{n\in\mathbb{N}_1} F_n(r) = \lim_{r\to\infty} \liminf_{n\in\mathbb{N}_0} F_n(r),$$

hence

$$\lim_{r\to\infty} \limsup_{n\in\mathbb{N}_1} F_n(r) \le 0.$$

By (ii) this implies $\liminf_{n\in\mathbb{N}_1} F_n(0) \ge 0$, whence

$$\limsup_{n\in\mathbb{N}_0} F_n(0) \ge 0.$$

The implication from (iii) to (i) is obvious.

To prove the addendum, assume that $\lim_{r\to\infty} \liminf_{n\to\infty} F_n(r) < 0$. There exists $\mathbb{N}_0$ such that

$$\lim_{r\to\infty} \limsup_{n\in\mathbb{N}_0} F_n(r) = \lim_{r\to\infty} \liminf_{n\to\infty} F_n(r) < 0.$$

However, $\lim_{r\to\infty} \limsup_{n\in\mathbb{N}_0} F_n(r) < 0$ for some subsequence $\mathbb{N}_0$ is impossible if (i) is true. □

5.16 Superefficiency

Let $\{(P_\vartheta^{(n)})_{n\in\mathbb{N}} : \vartheta \in \Theta\}$, $\Theta \subset \mathbb{R}^k$ fulfill LAN, and let $\kappa : \Theta \to \mathbb{R}^p$ be a functional with Jacobian K. The normal distribution $\overline{Q}_\vartheta = N(0, \Sigma_*(\vartheta))$, with $\Sigma_*(\vartheta) = K(\vartheta)\Lambda(\vartheta)K(\vartheta)^\top$, comes out as asymptotically optimal in various aspects: If $P_\vartheta^{(n)} \circ c_n(\kappa^{(n)} - \kappa(\vartheta))$, $n \in \mathbb{N}$, converges regularly to a limit distribution Q_ϑ, then Q_ϑ is a convolution product with the factor $N(0, \Sigma_*(\vartheta))$, whence $\int \ell dQ_\vartheta \ge \int \ell dN(0, \Sigma_*(\vartheta))$ for every $\ell \in \mathscr{L}_s$. If $\kappa^{(n)} : X \to \mathbb{R}$, $n \in \mathbb{N}$, is required to be asymptotically median unbiased (see Sect. 5.11), then

$$\liminf_{n\to\infty} \int \ell dQ_\vartheta^{(n)} \ge \int \ell dN(0, \sigma_*^2(\vartheta)) \quad \text{for } \ell \in \mathscr{L}.$$

In Hájek's Minimax-Theorem, $\int \ell dN(0, \Sigma_*(\vartheta))$ occurs as a lower bound for

$$\lim_{r\to\infty} \liminf_{n\to\infty} \sup_{|a|\le r} \int \ell dQ^{(n)}_{\vartheta+c_n^{-1}a}.$$

All these results are based on some kind of local uniformity condition. Even if a condition like regular convergence to a limit distribution may be justified from the operational point of view (as an outgrowth of locally uniform convergence), it is of interest whether a bound like $N(0, \Sigma_*(\vartheta))$ keeps its role under less restrictive conditions, say as a bound for limit distributions which are attained for every $\vartheta \in \Theta$ (but not necessarily in a locally uniform sense). This problem is dealt with in Sect. 5.14. The present section is on the phenomenon of "superefficiency".

Definition 5.16.1 The family $\mathfrak{P}_0 \subset \mathfrak{P}$ is for the estimator sequence $(\kappa^{(n)})_{n\in\mathbb{N}}$ and the loss function ℓ a *set of superefficiency* if there exists a subsequence $\mathbb{N}_0$ such that

$$\limsup_{n\in\mathbb{N}_0} \int \ell(c_n(\kappa^{(n)} - \kappa(P)))dP^{(n)} < \int \ell d\overline{Q}_P \quad \text{for } P \in \mathfrak{P}_0. \tag{5.16.1}$$

The "optimal" limit distribution $\overline{Q}_P$ to which the definition of superefficiency refers is characterized by two properties:

(i) It is attainable, i.e., there exists a "regular" estimator sequence $(\hat{\kappa}^{(n)})_{n\in\mathbb{N}}$ such that

$$P^{(n)} \circ c_n(\hat{\kappa}^{(n)} - \kappa(P)) \Rightarrow \overline{Q}_P.$$

(ii) $\overline{Q}_P$ is optimal in the sense that for every "regular" estimator sequence $(\kappa^{(n)})_{n\in\mathbb{N}}$

$$\lim_{n\to\infty} \int \ell(c_n(\kappa^{(n)} - \kappa(P))dP^{(n)} \geq \int \ell d\overline{Q}_P \quad \text{for every } \ell \in \mathscr{L}_0.$$

For parametric LAN-families we have

$$\overline{Q}_{P_\vartheta} = N(0, \Sigma_*(\vartheta)) \quad \text{with} \quad \Sigma_*(\vartheta) = K(\vartheta)\Lambda(\vartheta)K(\vartheta)^\top.$$

See Sect. 5.13.

The question is whether there are estimator sequences $\kappa^{(n)}$ such that the risk $\int \ell(c_n(\kappa^{(n)} - \kappa(P)))dP^{(n)}$ remains, for some subsequence and some loss function ℓ, smaller than $\int \ell d\overline{Q}_P$ for P in a substantial subset of $\mathfrak{P}$.

The common definition requires (5.16.1) with $\mathbb{N}$ in place of $\mathbb{N}_0$. The present definition takes into account that the statistician will be satisfied with superefficiency along some infinite subsequence from which the sample size could be chosen.

Some authors (see e.g. Ibragimov and Has'minskii 1981, p. 170) use lim inf rather than lim sup in the definition of superefficiency. This seems to be questionable. Smallness of $\liminf_{n\to\infty} \int \ell(c_n(\kappa^{(n)} - \kappa(P)))dP^n$ for every P is of no relevance if the sample sizes corresponding to small values of $\int \ell(c_n(\kappa^{(n)} - \kappa(P)))dP^{(n)}$ depend on P. This is convincingly demonstrated by the following example, due to van der Vaart (1997, p. 407).

Example 1. (i) Let $u_n \in (0, 1)$, $n \in \mathbb{N}$, be such that $n^{1/2}|u_n - u|$, $n \in \mathbb{N}$, has, for every $u \in (0, 1)$, the accumulation point zero. If $u_{2^m+k} := k2^{-m}$ for $k = 1, \dots, 2^m$, then $|u_n - u| < n^{-1}$ for $n = 2^m + k_m$, if $k_m \in \{1, \dots, 2^m\}$ is chosen such that $(k_m - 1)2^{-m} < u \leq k_m 2^{-m}$.

(ii) Let $\{P_\vartheta : \vartheta \in (0, 1)\}$ be a parametric family. Then the estimator sequence $\vartheta^{(n)}(\mathbf{x}_n) := u_n$ (which is independent of $\mathbf{x}_n$) fulfills

$$\liminf_{n\to\infty} \int \ell(n^{1/2}(\vartheta^{(n)} - \vartheta))dP_\vartheta^n = 0$$

for every $\vartheta \in (0,1)$ if ℓ with $\ell(0) = 0$ is bounded and continuous. □

Recall that the evaluation of estimator sequences in k-parameter families (or, more generally, of estimator sequences for k-dimensional functionals on a general family) is restricted to loss functions which are subconvex and symmetric, provided the optimal distribution $\overline{Q}_P|\mathbb{B}^k$ is normal. Accordingly, the concept of "superefficiency" must be based on such loss functions. The exception is a one-dimensional parametric family (or a one-dimensional functional on a general family) if the comparison is restricted to asymptotically median unbiased estimator sequences. In this case, a definition based on an arbitrary (i.e., not necessarily symmetric) loss function makes sense.

Applied to estimator sequences with limit distribution $Q_P := \lim_{n\to\infty} Q_P^{(n)}$, superefficiency at P with respect to the loss function ℓ just means that $\int \ell dQ_P < \int \ell d\overline{Q}_P$.

It took surprisingly long until such a simple example demonstrated that there is no bound for the concentration of limit distributions without some asymptotic uniformity of the estimator sequence. The first example of a superefficient estimator sequence is due to Hodges (see Sect. 5.9). Using the idea underlying the example of Hodges, it is easy to construct estimator sequences superefficient on a given countable subset of Θ. Le Cam (1953, p. 291, Example 4) suggested how to construct an estimator sequence superefficient on an uncountable subset of $\Theta \subset \mathbb{R}$. His suggestion has never been worked out in detail.

Examples of Superefficiency

The first example of a superefficient estimator sequence is due to Hodges (see Le Cam 1953, p. 280).

Example 1. For the family $\{(N(\vartheta,1)^n)_{n\in\mathbb{N}} : \vartheta \in \mathbb{R}\}$, the estimator

$$\vartheta^{(n)}(\mathbf{x}_n) := \begin{cases} \overline{x}_n \\ \alpha\overline{x}_n \end{cases} \text{if } |\overline{x}| \begin{matrix} > \\ \leq \end{matrix} n^{-1/4},$$

with $\alpha \in (0,1)$, is superefficient at $\vartheta = 0$. We have

$$N(\vartheta,1)^n \circ n^{1/2}(\vartheta^{(n)} - \vartheta) \Rightarrow \begin{cases} N(0,\alpha^2) \\ N(0,1) \end{cases} \text{for } \vartheta \begin{matrix} = \\ \neq \end{matrix} 0.$$

Example 2. In the following we construct for $\alpha \in (0,1)$ an estimator sequence for ϑ in $\{(N(\vartheta,1)^n)_{n\in\mathbb{N}} : \vartheta \in \mathbb{R}\}$ which converges for the subsequence $\{\mathbb{N}_0 := 2^{m^2} : m \in \mathbb{N}\}$ on an uncountable subset of $\mathbb{R}$ to the limit distribution $N(0,\alpha^2)$, and to $N(0,1)$ elsewhere. By means of a more subtle construction one can probably obtain an estimator sequence which is superefficient on the whole sequence.

(i) To prepare the definition of a superefficient estimator we define a sequence of sets $(S_m)_{m\in\mathbb{N}}$ as follows. S_m consists of 2^m intervals $I_{\delta_1,\ldots,\delta_m}$, $(\delta_1,\ldots,\delta_m) \in \{0,1\}^m$, which are defined by the following inductive procedure:

$$I_0 := [-1, -1/2], \quad I_1 := [1/2, 1].$$

If, for $m > 1$, $I_{\delta_1,\ldots,\delta_{m-1}} = [a_{\delta_1,\ldots,\delta_{m-1}}, b_{\delta_1,\ldots,\delta_{m-1}}]$, we define

$$I_{\delta_1,\ldots,\delta_{m-1},0} := [a_{\delta_1,\ldots,\delta_{m-1}}, a_{\delta_1,\ldots,\delta_{m-1}} + 2^{-m(m+1)/2}],$$
$$I_{\delta_1,\ldots,\delta_{m-1},1} := [b_{\delta_1,\ldots,\delta_{m-1}}, -2^{-m(m+1)/2}, b_{\delta_1,\ldots,\delta_{m-1}}].$$

The length of $I_{\delta_1,\ldots,\delta_m}$ is $2^{-m(m+1)/2}$, and the minimum distance between the two intervals $I_{\delta_1,\ldots,\delta_m}$ and $I_{\delta'_1,\ldots,\delta'_m}$ is

$$\Delta_m = 2^{-(m-1)m/2} - 2 \cdot 2^{-m(m+1)/2} = 2^{-m(m-1)/2}(1 - 2^{-m+1}).$$

Finally, let $n_m := 2^{m^2}$.

(ii) We define a sequence of estimators $(\vartheta^{(n_m)})_{m\in\mathbb{N}}$ as follows: If $d(\overline{x}_{n_m}, S_m) \geq \frac{1}{2}\Delta_m$, we define $\vartheta^{(n_m)}(\mathbf{x}_n) = \overline{x}_{n_m}$. If $d(\overline{x}_{n_m}, S_m) < \frac{1}{2}\Delta_m$, then there exists a uniquely determined m-tuple $(\delta_1, \ldots, \delta_m)$ such that $d(\overline{x}_{n_m}, I_{\delta_1,\ldots,\delta_m}) < \frac{1}{2}\Delta_m$. In this case we define $\vartheta^{(n_m)}(\mathbf{x}_n) := (1-\alpha)z_{\delta_1,\ldots,\delta_m} + \alpha\overline{x}_{n_m}$, where $z_{\delta_1,\ldots,\delta_m} \in I_{\delta_1,\ldots,\delta_m}$ is arbitrarily fixed and $\alpha \in (0, 1)$.

(iii) Obviously, $\vartheta^{(n_m)}$ is measurable. We shall show that

$$N(\vartheta, 1)^{n_m} \circ n_m^{1/2}(\vartheta^{(n_m)} - \vartheta) \Rightarrow \begin{cases} N(0, \alpha^2) \\ N(0, 1) \end{cases} \quad \text{for } \vartheta \begin{Bmatrix} \in \\ \notin \end{Bmatrix} \bigcap_{m=1}^{\infty} S_m.$$

(iv) If $\vartheta \notin \cap_{m=1}^{\infty} S_m$, there exists $m_0 \in \mathbb{N}$ such that $\vartheta \notin S_{m_0}$, whence $d(\vartheta, S_{m_0}) > 0$. Let m_1 be such that $\frac{1}{2}\Delta_{m_1} < d(\vartheta, S_{m_0})$. As $(\overline{x}_{n_m})_{m\in\mathbb{N}} \to \vartheta$ $N\vartheta, 1)^{\mathbb{N}}$-a.e., there exists $m(\mathbf{x}_n) \geq \max\{m_0, m_1\}$ (depending on ϑ) such that $d(\bar{x}_{n_m}, \vartheta) \leq d(\vartheta, S_{m_0}) - \frac{1}{2}\Delta_{m_1}$ for all $m \geq m(\mathbf{x}_n)$. Hence $m \geq m(\mathbf{x}_n)$ implies

$$d(\overline{x}_{n_m}, S_m) \geq d(\overline{x}_{n_m}, S_{m_0}) \geq d(\vartheta, S_{m_0}) - d(\overline{x}_{n_m}, \vartheta) \geq \frac{1}{2}\Delta_{m_1} \geq \frac{1}{2}\Delta_m.$$

Hence we have $\vartheta^{(n_m)}(\mathbf{x}_n) = \overline{x}_{n_m}$ for all $m \geq m(\mathbf{x}_n)$ and $N(\vartheta, 1)^{\mathbb{N}}$-a.a. $\mathbf{x}_n \in \mathbb{R}^{\mathbb{N}}$. This, however, implies that $N_{(\vartheta,1)}^{n_m} \circ n_m^{1/2}(\vartheta^{(n_m)} - \vartheta), m \in \mathbb{N}$, has the same limit distribution as $N_{(\vartheta,1)}^{n_m} \circ n_m^{1/2}(\bar{x}_{n_m} - \vartheta)$, $m \in \mathbb{N}$, namely $N(0, 1)$.

(v) If $\vartheta \in \cap_{m=1}^{\infty} S_m$, there exists $(\delta_i)_{i\in\mathbb{N}} \in \{0, 1\}^{\mathbb{N}}$ such that $\vartheta \in I_{\delta_1,\ldots,\delta_m}$ for all $m \in \mathbb{N}$. Then

$$\{\mathbf{x}_n \in \mathbb{R}^{\mathbb{N}} : \vartheta^{(n_m)}(\mathbf{x}_n) = (1-\alpha)z_{\delta_1,\ldots,\delta_m} + \alpha\overline{x}_{n_m}\}$$
$$= \{\mathbf{x}_n \in \mathbb{R}^{\mathbb{N}} : a_{\delta_1,\ldots,\delta_m} - \frac{1}{2}\Delta_m < \overline{x}_{n_m} < b_{\delta_1,\ldots,\delta_m} + \frac{1}{2}\Delta_m\}$$
$$\supset \{\mathbf{x}_n \in \mathbb{R}^{\mathbb{N}} : -\frac{1}{2}\Delta_m n_m^{1/2} < n_m^{1/2}(\overline{x}_{n_m} - \vartheta) < \frac{1}{2}\Delta_m n_m^{1/2}\},$$

since $a_{\delta_1,\ldots,\delta_m} < \vartheta < b_{\delta_1,\ldots,\delta_m}$. As $\frac{1}{2}\Delta_m n_m^{1/2} = 2^{m/2-1}(1-2^{-m+1})$, we have

$$\lim_{m\to\infty} N(\vartheta,1)^{\mathbb{N}}\{\mathbf{x}_n \in \mathbb{R}^{\mathbb{N}} : \vartheta^{n_m}(\mathbf{x}_n) = (1-\alpha)z_{\delta_1,\ldots,\delta_m} + \alpha\overline{x}_{n_m}\} = 1.$$

Therefore, $N^{n_m}_{(\vartheta,1)} \circ n_m^{1/2}(\vartheta^{(n_m)} - \vartheta)$, $m \in \mathbb{N}$, has the same limit distribution as $N^{n_m}_{(\vartheta,1)} \circ n_m^{1/2}((1-\alpha)z_{\delta_1,\ldots,\delta_m} + \alpha\bar{x}_{n_m} - \vartheta)$, $m \in \mathbb{N}$, namely $N(0,\alpha^2)$.

Furthermore, $\vartheta \in I_{\delta_1,\ldots,\delta_m}$ implies

$$\frac{|((1-\alpha)z_{\delta_1,\ldots,\delta_m} + \alpha\vartheta) - \vartheta|}{\alpha n_m^{-1/2}} = \frac{1-\alpha}{\alpha}|z_{\delta_1,\ldots,\delta_m} - \vartheta| n_m^{1/2}$$
$$\le \frac{1-\alpha}{\alpha} 2^{-m(m+1)/2} 2^{m^2/2} = \frac{1-\alpha}{\alpha} 2^{-m/2} \downarrow 0.$$

Therefore,

$$N^{n_m}_{(\vartheta,1)} \circ n_m^{1/2}(\vartheta^{(n_m)} - \vartheta) \Rightarrow N(0,\alpha^2).$$

(vi) For every sequence $(\delta_i)_{i\in\mathbb{N}} \in \{0,1\}^{\mathbb{N}}$ there exists a point $r_{(\delta_i)_{i\in\mathbb{N}}}$ in the set $\bigcap_{m=1}^{\infty} I_{\delta_1,\ldots,\delta_m}$. (Since $(I_{\delta_1,\ldots,\delta_m})_{m\in\mathbb{N}}$ is a decreasing sequence of compact sets with diameter converging to zero, this intersection consists of exactly one point.) As $\{0,1\}^{\mathbb{N}}$ is uncountable, $\bigcap_{m=1}^{\infty} S_m$ is uncountable. Being a set of superefficiency, $\bigcap_{m=1}^{\infty} S_m$ is necessarily of Lebesgue measure zero. This also follows directly from $\lambda(\bigcap_{m=1}^{\infty} S_m) \le \lambda(S_m) = 2^m 2^{-m(m+1)/2} = 2^{-m(m-1)/2}$ for $m \in \mathbb{N}$. □

Whereas it requires some efforts (as in Example 2) to construct an estimator sequence which is superefficient on an uncountable subset of $\mathbb{R}$, superefficient estimator sequences on uncountable subsets of $\mathbb{R}^k$, $k > 1$, are straightforward.

Example 3. For $\Theta = \mathbb{R}^2$ let $P_{(\vartheta_1,\vartheta_2)} := N(\vartheta_1,1) \times N(\vartheta_2,1)$. The problem is to estimate $\kappa(P_{(\vartheta_1,\vartheta_2)}) = \vartheta_1$. The optimal limit distribution of regular estimator sequences is $N(0,1)$. The estimator sequence

$$\vartheta^{(n)}((x_{1\nu},x_{2\nu})_{\nu=1,\ldots,n}) := \begin{cases} \overline{x}_{1n} \\ (\overline{x}_{1n} + \overline{x}_{2n})/2 \end{cases} \text{if } |\overline{x}_{1n} - \overline{x}_{2n}| \begin{matrix} > \\ \le \end{matrix} n^{-1/4}$$

is superefficient on $\{(\vartheta_1,\vartheta_2) \in \mathbb{R}^2 : \vartheta_1 = \vartheta_2\}$, with limit distribution $N(0,1/2)$.

The Set of Superefficiency is Small

Le Cam (1953) was the first to show that a set of superefficiency for a k-parameter family is necessarily of λ^k-measure zero. For a correct interpretation of Le Cam's result, observe that his definition of superefficiency (p. 283, Definitions 3 and 4) is more restrictive. To present his definition in the present framework, we presume that the M.L. sequence occurring explicitly in Le Cam's definition converges to the optimal limit distribution $\overline{Q}_\vartheta := N(0,\sigma_*^2(\vartheta))$. Moreover, we use the now common concept of a loss function rather than Le Cam's gain functions. With Anderson's Theorem not yet available, Le Cam had to invent a concept for loss functions suitable for expressing the optimality of a normal limit distribution. The decisive property (see p. 312) is that $z \to \int \ell(u+z)\exp[-u^\top \Sigma u]du$ attains, for every Σ, its maximum at $z=0$. This is a concept adjusted to the particular framework in which optimal limit distributions are normal; it is not derived from a general concept of optimality which is significant in a more general framework. It is merely an accident that the operationally meaningful subconvex and symmetric loss functions share the property required by Le Cam.

With the necessary modifications, Le Cam's definition of superefficiency reads as follows: The estimator sequence with distribution $Q_\vartheta^{(n)}$, $n \in \mathbb{N}$, is *superefficient* on Θ_0 if

$$\limsup_{n\to\infty} \int \ell dQ_\vartheta^{(n)} \le \int \ell d\overline{Q}_\vartheta \quad \text{for every} \quad \vartheta \in \Theta$$

and

$$\limsup_{n\to\infty} \int \ell dQ_\vartheta^{(n)} < \int \ell d\overline{Q}_\vartheta \quad \text{for some} \quad \vartheta \in \Theta_0.$$

Since the interesting results on superefficiency are to the negative (in the sense that "sets of superefficiency are necessarily small") they are stronger if based on Definition 5.16.1 which requires nothing about the performance of the estimator sequences for ϑ outside Θ_0.

In his Corollary 8:1, p. 314, Le Cam asserts for $\Theta \subset \mathbb{R}^k$ under Cramér-type regularity conditions that the set of superefficiency is necessarily of Lebesgue measure 0. His result is based on Bayesian arguments, which were in the air at this time. Wolfowitz, too (1953a, p. 116), gave an (informal) Bayesian proof for the fact that superefficiency is possible on a set of Lebesgue measure zero only, but he assumes that the estimator sequence converges to a normal limit distribution. An elegant proof under the same assumption was later given by Bahadur (1964) in a paper which is remarkable in various respects.

The best result concerning sets of superefficiency for k-parameter LAN-families is implicitly contained in Strasser (1978a). His Proposition, p. 37, adjusted to the present framework, reads as follows.

For any probability measure $\Pi|\mathbb{B}^k \ll \lambda^k|\mathbb{B}^k$ and any subconvex loss function $\ell \in \mathscr{L}_s$,

$$\liminf_{n\to\infty} \int\int \ell dQ_\vartheta^{(n)} \Pi(d\vartheta) \ge \int\int \ell dN(0, \Sigma_*(\vartheta))\Pi(d\vartheta). \tag{5.16.2}$$

The emphasis of Strasser's paper is on the Convolution Theorem. He concludes from (5.16.2) that $Q_\vartheta^{(n)} \Rightarrow Q_\vartheta$ for every $\vartheta \in \Theta$ implies

$$\lim_{n\to\infty} \int \ell dQ_\vartheta^{(n)} \ge \int \ell dN(0, \Sigma_*(\vartheta)) \quad \text{for } \lambda^k - a.a.\, \vartheta \in \Theta,$$

but he ignores an important consequence of (5.16.2): By Fatou's Lemma,

$$\lambda^k\left\{\vartheta \in \Theta : \limsup_{n\to\infty} \int \ell dQ_\vartheta^{(n)} < \int \ell dN(0, \Sigma_*(\vartheta))\right\} = 0 \text{ for every } \ell \in \mathscr{L}_s.$$

An intelligible version of Le Cam's original proof can be found in van der Vaart (1997, pp. 398–401).

For parametric families, "smallness" of a set can be expressed as being "of Lebesgue measure 0". For general families, a set of superefficiency may be shown to be of "first category" (see Pfanzagl (2003), p. 97, Theorem 2.1). Conditions under which sets of first category can be considered as "small" are discussed in Sect. 5.14.

Does Superefficiency Exclude Local Uniformity?

After having shown that sets of superefficiency are necessarily of Lebesgue measure zero, the next step for Le Cam would naturally have been to find operationally meaningful conditions on the estimator sequence which preclude the irritating phenomenon of superefficiency. Le Cam approached this problem indirectly by exhibiting the irregularity of superefficient estimator sequences for one-dimensional parameters. He proved (Le Cam 1953, p. 327, Theorem 14) under Cramér-type conditions the following.

Lemma 5.16.2 *Let $\Theta \subset \mathbb{R}$. For every loss function ℓ which is sufficiently smooth, bounded and symmetric about 0,*

$$\limsup_{n\to\infty} \int \ell dQ_{\vartheta_0}^{(n)} < \int \ell d\overline{Q}_{\vartheta_0} \tag{5.16.3}$$

implies

$$\limsup_{n\to\infty} \int \ell dQ_{\vartheta_n}^{(n)} > \int \ell d\overline{Q}_{\vartheta_0} \quad \text{for some sequence} \quad \vartheta_n \to \vartheta_0. \tag{5.16.4}$$

Based on this result, it would have been easy to show that uniformly convergent estimator sequences of real parameters cannot be superefficient; more precisely:

Proposition 5.16.3 *If Q_{ϑ_0} is a limit distribution such that*

$$\lim_{n\to\infty} \int \ell dQ_{\vartheta_n}^{(n)} = \int \ell dQ_{\vartheta_0} \quad \text{for every sequence} \quad \vartheta_n \to \vartheta_0, \tag{5.16.5}$$

then

$$\int \ell dQ_{\vartheta_0} \geq \int \ell d\overline{Q}_{\vartheta_0}. \tag{5.16.6}$$

Proof If

$$\int \ell dQ_{\vartheta_0} < \int \ell d\overline{Q}_{\vartheta_0}, \tag{5.16.7}$$

then relation (5.16.5), applied with $\vartheta_n \equiv \vartheta_0$, implies (5.16.3), which, in turn, implies (5.16.4), i.e.

$$\int \ell dQ_{\vartheta_0} = \lim_{n\to\infty} \int \ell dQ_{\vartheta_n}^{(n)} > \int \ell d\overline{Q}_{\vartheta_0},$$

which is in contradiction to (5.16.7). □

It is clear from Le Cam's proof that Lemma 5.16.2 holds true already with $\vartheta_n = \vartheta_0 + n^{-1/2}a$. Hence regularity in the sense of (5.16.5) with $\vartheta_n = \vartheta_0 + n^{-1/2}a$ suffices to exclude superefficiency. Since Le Cam missed this opportunity, such results did not appear until 1963 (C.R. Rao, Wolfowitz; see Sect. 5.8).

The implication from (5.16.3) to (5.16.4) was established by Le Cam for $\Theta \subset \mathbb{R}$, and he remarks that its extension to arbitrary dimensions of Θ poses certain problems. This is confirmed by Stein's shrinkage estimator.

Example 4. We consider for $k \geq 3$ the family $\{N(\vartheta, I_k) : \vartheta \in \mathbb{R}^k\}$, where I_k is the unit-matrix in $\mathbb{R}^k$. The problem is to estimate $\vartheta = (\vartheta_1, \ldots, \vartheta_k)$. The optimal limit distribution for regular estimator sequences is $N(0, I_k)$, which is attained locally uniformly by the sample mean $\bar{\mathbf{x}}_n := (\overline{x}_{1n}, \ldots, \overline{x}_{kn})$ for every $n \in \mathbb{N}$. Evaluated by the loss function $\ell(u) = \|u\|^2$, we have

$$\int \ell(n^{1/2}(\bar{\mathbf{x}}_n - \vartheta))N(\vartheta, I_k)^n(d\mathbf{x}_n) = k = \int \ell dN(0, I_k) \quad \text{for } \vartheta \in \mathbb{R}^k \text{ and } n \in \mathbb{N}.$$

For Stein's estimator

$$\vartheta^{(n)}(\mathbf{x}_n) := \left(1 - \frac{k-2}{n\|\bar{\mathbf{x}}_n\|^2}\right)\bar{\mathbf{x}}_n$$

we have

$$\int \ell(n^{1/2}(\vartheta^{(n)} - \vartheta))dN(\vartheta, I_k)^n = \int \ell dN(0, I_k) - n^{-1}\int \|\bar{\mathbf{x}}_n\|^{-2}N(\vartheta, I_k)^n(d\mathbf{x}_n).$$

Since the presentation in Stein (1956) is not very transparent, compare Ibragimov and Has'minskii (1981) [p. 27] or Lehmann and Casella (1998, p. 355, Theorem 5.5.1) for details. According to Casella and Hwang (1982) [p. 306, Lemma 1],

$$\frac{1}{(k-2) + n\|\vartheta\|^2} < n^{-1}\int \|\bar{\mathbf{x}}_n\|^{-2}N(\vartheta, I_k)^n(d\mathbf{x}_n) \leq \frac{1}{k-2}\cdot\frac{1}{1 + n\|\vartheta\|^2/k},$$

which implies for $\vartheta \in \mathbb{R}^k$ and $n \in \mathbb{N}$,

$$\int \ell(n^{1/2}(\vartheta^{(n)} - \vartheta))dN^n_{(\vartheta, I_k)} < \int \ell dN(0, I_k) - \frac{1}{(k-2) + n\|\vartheta\|^2}.$$

Hence Stein's estimator, evaluated by the loss function $\ell(u) = \|u\|^2$, is superefficient at $\vartheta = 0$, and nowhere inefficient. Stein was interested in the performance of $\vartheta^{(n)}$ for finite sample sizes. This might explain why he overlooked the relevance his invention has in connection with Le Cam's Theorem 14. □

With Stein's example restricted to the dimension $k \geq 3$, the problem arises whether Le Cam's Theorem 14 is valid for $k = 2$. Proposition 6 in Le Cam 1974, p. 187, seems to give an affirmative answer to this question, but I was unable to follow the proof.

The reader who is aware of the problem raised by Le Cam's Theorem 14, will be disappointed by the treatment Le Cam gives to this problem in (1986, p. 144). He presents a hitherto unknown version of Stein's estimator which reads in our notations as

$$\Big(1 - \frac{k-2}{1 + \|\bar{\mathbf{x}}_n\|^2}\Big)\bar{\mathbf{x}}_n.$$

Omitting all details, he gives his opinion about the limit distribution of this estimator sequence, but he avoids to discuss what is of interest here, namely the performance of these estimators near $\vartheta = 0$, the point of superefficiency.

According to Sect. 5.15, regular estimator sequences are optimal with respect to all loss functions in $\mathscr{L}_s$ if they are optimal with respect to one of these. This implies, in particular, that the components of an optimal multidimensional estimator sequence are optimal themselves. In contrast to that, superefficiency may be tied to a particular loss function.

If an estimator sequence $(\vartheta_1^{(n)}, \ldots, \vartheta_k^{(n)})$ is superefficient for $(\vartheta_1, \ldots, \vartheta_k)$ with respect to some loss function ℓ_0, then at least one of the components $\vartheta_i^{(n)}$, $i = 1, \ldots, k$, fails to converge regularly to the optimal limit distribution for ϑ_i.

Remark That Le Cam's paper is mainly known for Hodges' example and not for its many deep results is, perhaps, due to the fact that it is not easy to read. (It contains 14 theorems, 4 corollaries and addenda, and 8 lemmas, all of which are somehow interrelated.) Its mathematical deficiencies were criticized by Wolfowitz (1965) [p. 249]; Hájek (1972, p. 177) politely says that this paper "contains some omissions". Le Cam (1974) [p. 254] explains why his paper is "rather incorrect". As a consequence of the intricate presentation, it escaped notice that Le Cam's results do not require convergence to a limit distribution. Miraculously, Le Cam himself published his superefficiency result once more for estimator sequences with limit distribution (1958, p. 33). Bahadur (1964) suggests an easier way to this result. Though he explicitly refers to Le Cam's papers of 1953 and 1958 (Le Cam 1953, 1958), he overlooked the fact that the paper of 1953 is on arbitrary estimator sequences. The same misunderstanding occurs in many textbooks. (See e.g. Stuart and Ord 1991, pp. 660/1. Witting and Müller-Funk 1995, p. 200, Satz 6.33 and p. 417, Satz 6.199; Lehmann and Casella 1998, p. 440.)

Remark on a Concept of Global Superefficiency

Brown et al. (1997) [p. 2612] suggest a new concept of superefficiency. To illustrate the difficulties with the interpretation of this concept, we consider the following problem: Let $\{P_\vartheta : \vartheta \in \Theta\}$, $\Theta = (\vartheta_0, \infty)$, be a family of probability measures on some measurable space $(X, \mathscr{A})$. For any estimator $\vartheta^{(n)}$ let

$$R_\vartheta^{(n)}(\vartheta^{(n)}) := n \int (\vartheta^{(n)} - \vartheta)^2 dP_\vartheta^n.$$

According to Brown et al.'s new concept, the estimator sequence $(\hat{\vartheta}^{(n)})_{n\in\mathbb{N}}$ is (asymptotically]) *superefficient* at ϑ if

$$\limsup_{n\to\infty} R_\vartheta^{(n)}(\hat{\vartheta}^{(n)}) < \limsup_{n\to\infty} \inf_{\vartheta^{(n)}} \sup\{R_\tau^{(n)}(\vartheta^{(n)}) : \tau \in \Theta, \tau \le \vartheta\}.$$

Let

$$r_\vartheta^{(n)} := \inf_{\vartheta^{(n)}} R_\vartheta^{(n)}(\vartheta^{(n)}) \quad \text{and} \quad r(\vartheta) := \limsup_{n\to\infty} r_\vartheta^{(n)}.$$

Since

$$\inf_{\vartheta^{(n)}} \sup\{R_\tau^{(n)}(\vartheta^{(n)}) : \tau \in \Theta,\ \tau \le \vartheta\} \ge \sup\{r_\tau^{(n)} : \tau \in \Theta, \tau \le \vartheta\},$$

the estimator sequence $(\hat{\vartheta}^{(n)})_{n\in\mathbb{N}}$ is superefficient at ϑ if

$$\limsup_{n\to\infty} R_\vartheta^{(n)}(\hat{\vartheta}^{(n)}) < \sup\{r(\tau) : \tau \in \Theta, \tau \le \vartheta\}.$$

If $(\hat{\vartheta}^{(n)})_{n\in\mathbb{N}}$ is asymptotically efficient with respect to the quadratic loss function, then $\lim_{n\to\infty} R_\vartheta^{(n)}(\hat{\vartheta}^{(n)}) = r(\vartheta)$. Hence any asymptotically efficient estimator sequence is superefficient at ϑ if $r(\vartheta) < \sup\{r(\tau) : \tau \in \Theta, \tau \le \vartheta\}$. Therefore, any estimator sequence which is asymptotically efficient for every $\vartheta \in \Theta$ is by definition automatically superefficient on Θ if the function r is decreasing. An example of this kind is the family $\{N(\vartheta, \vartheta^{-1}) : \vartheta \in (1, \infty)\}$, where $r(\vartheta) = 2\vartheta^2/(1 + 2\vartheta^3)$. It is straightforward to modify these considerations in such a way that estimator sequences which are asymptotically inefficient for every $\vartheta \in \Theta$ are asymptotically superefficient on a large subset of Θ.

5.17 Rates of Convergence

For LAN-families and one-dimensional differentiable functionals, it is comparatively easy to find a concept of asymptotic optimality: There is an optimal rate $c_n = n^{1/2}$ (or $c_n = n^{1/2}(\log n)^a$ in "almost regular" models), and there exists a normal distribution with variance $\sigma^2(P)$, determined by the local structure of $\mathfrak{P}$ and κ at P, such that no "regularly attainable" limit distribution Q_P can be more concentrated than $N(0, \sigma^2(P))$ on symmetric intervals containing 0.

This favourable situation is typical for parametric families. It occurs in some nonparametric models, too. As an example we mention the estimation of the if $\mathfrak{P}$ is the family of all symmetric $P|\mathbb{B}$ with a sufficiently regular density. Typical for nonparametric families is the existence of an estimator sequence such that $c_n(\kappa^{(n)} -$

$\kappa(P))$ remains stochastically bounded for a rate $(c_n)_{n\in\mathbb{N}}$ which tends to infinity more slowly than $n^{1/2}$; even if $P^n \circ c_n(\kappa^{(n)} - \kappa(P))$ converges to a limit distribution, the optimality of this limit distribution remains open.

For the purpose of illustration we consider the estimation of a density at a given point. This is also the problem where the question of optimal rates took its origin. The natural framework: a family of probability measures on $\mathbb{B}$, with a Lebesgue density p fulfilling a certain smoothness condition. The early papers in this area were just concerned with the construction of "good" estimators. Histogram estimators were a natural choice: For a fixed value of ξ, $p(\xi)$ is estimated by

$$p^{(n)}(\xi, \mathbf{x}_n) := n^{-1} \sum_{\nu=1}^{n} h_n^{-1} 1_{[\xi - h_n/2, \xi + h_n/2]}(x_\nu).$$

That the estimator sequence $p^{(n)}(\xi, \cdot)$ is consistent for $p(\xi)$ under mild conditions on p (continuity at ξ) provided $h_n \to 0$ is usually attributed to Fix and Hodges (1951) [p. 244, Lemma 3]. It occurs, from what is heard, already in Glivenko's "Course in Probability Theory" (1939). In a paper unknown to Fix and Hodges, Smirnov (1950, p. 191, Theorem 3) had already obtained certain results on the rate at which

$$\max\{|p^{(n)}(\xi, \cdot) - p(\xi)|/p(\xi)^{1/2} : \xi \in [a, b]\}$$

converges to zero, depending on the choice of h_n.

In a fundamental paper, Rosenblatt (1956) [p. 835] shows for the sequence of histogram estimators with $h_n = O(n^{-1/5})$ that

$$n^{4/5} \int (p^{(n)}(\xi, \mathbf{x}_n) - p(\xi))^2 P^n(d\mathbf{x}_n)$$

converges for every $\xi \in \mathbb{R}$ to a finite and positive number if p admits a continuous second derivative at ξ. In Sect. 4 of this paper, he shows that the same result holds for arbitrary kernel estimators, defined by $n^{-1} \sum_{\nu=1}^{n} h_n^{-1} K(h_n^{-1}(\xi - x_\nu))$ with $K \geq 0$ and $\int K(\xi) d\xi = 1$. On p. 837, Rosenblatt raises the question whether estimators with the rate n (in place of $n^{4/5}$) exist.

If p admits a continuous r-th derivative at ξ, Parzen (1962) [p. 1074, relation (4.16)] shows that for kernel estimators with bandwidth

$$h_n = n^{1/(2r+1)} C_r(K) p(\xi) |p^{(r)}(\xi)|^{-2/(2r+1)}$$

one has

$$\lim_{n\to\infty} n^{2r/(2r+1)} \int (p^{(n)}(\xi, \mathbf{x}_n) - p(\xi))^2 P^n(d\mathbf{x}_n)$$
$$= C_r(K) p(\xi)^{2r/(2r+1)} |p^{(r)}(\xi)|^{2/(2r+1)}.$$

Parzen's result refers to a particular class of kernels, and he neglects the question of how to choose a sequence h_n not depending on p.

It is not the purpose of the present section to deal with the theory of kernel estimators in more detail. What is essential for our problem is the fact that other techniques of density estimation like orthogonal expansions (Čentsov 1962), Fourier series (Kronmal and Tarter 1968) or polynomial algorithms (Wahba 1971) lead under intuitively comparable regularity conditions on the densities to the same rates of convergence.

We just mention one more paper which was at that time almost entirely ignored by scholars working on density estimation. Prakasa Rao (1969) [Theorem 6.3, p. 35] obtained for the family of all distributions on $\mathbb{B} \cap (0, \infty)$ with a nonincreasing Lebesgue density the following result for the distribution of the ML-sequence

$$P^n \circ n^{1/3}(p^{(n)}(\xi, \cdot) - p(\xi)) \Rightarrow Q,$$

where $Q|\mathbb{B}$ is a certain (nonnormal) distribution, symmetric about 0, with moments of all orders. Prakasa Rao's interest was just in the distribution of the maximum likelihood estimator. He was not concerned with the question whether the maximum likelihood estimator, known to be asymptotically optimal in regular parametric families, retains this property also here. If Prakasa Rao says that the ML-sequence is "suboptimal", he just means that the rate is $n^{1/3}$, in contrast to the usual $n^{1/2}$. He had, perhaps, missed Parzen's paper (1962) according to which the same rate $n^{1/3}$ is achieved by kernel estimators under comparable smoothness conditions (densities with one derivative).

Since different techniques of density estimation lead under comparable regularity conditions to the same rate of convergence (which cannot be improved by refinements of the specific techniques), this suggests that this common rate is the best possible one, that there exists a bound for the rate of convergence, determined by the properties of the density, which cannot be surpassed by any method of density estimation. This motivated Farrell's paper (1972). His Theorem 1.2, p. 173, specialized to densities of dimension 1 and written in our notations, asserts that

$$\limsup_{n\to\infty} c_n n^{-r/(2r+1)} = \infty$$

implies

$$\limsup_{n\to\infty} \sup_{P \in \mathfrak{P}_r} c_n^2 \int (p^{(n)}(\xi, \mathbf{x}_n) - p(\xi))^2 P^n(d\mathbf{x}_n) = \infty$$

for every estimator sequence $p^{(n)}(\xi, \cdot)$. In this relation, $\mathfrak{P}_r$ is (somewhat simplified) the family of all probability measures in a Lipschitz neighbourhood of a given measure P_0 admitting a density with a bounded r-th derivative.

At the time Farrell wrote his paper, it was clear from the study of parametric families that a meaningful concept of asymptotic optimality had to be built upon a condition of (locally) uniform convergence. Farrell requires uniformity on $\mathfrak{P}_r$

without giving second thoughts to this question. To establish the rate $c_n = n^{r/(2r+1)}$ as optimal, it was therefore necessary to show an estimator sequence $\hat{p}^{(n)}(\xi, \cdot), n \in \mathbb{N}$, such that

$$\limsup_{n\to\infty} \sup_{P\in\mathfrak{P}_r} n^{2r/(2r+1)} \int (\hat{p}_n(\xi, \mathbf{x}_n) - p(\xi))^2 P^n(d\mathbf{x}_n) < \infty.$$

According to Farrell's Lemma 1.4, p. 173, this holds true for kernel estimators.

In the proof of his Theorem 1.2, Farrell uses (see p. 174) what corresponds to "least favourable paths" $(p_n)_{n\in\mathbb{N}}$, converging to p such that $p_n(\xi) - p_0(\xi)$ is large, and P_n is close to P_0. Farrell's complicated construction on pp. 174–177 shows that the invention of a least favourable path may be a nontrivial task. A better arranged version of Farrell's proof can be found in Wahba (1975) [pp. 27–29].

Farrell (1972) was also aware of a problem which did not find due attention until twenty years later: The problem of rate adaptivity. His generally neglected Theorem 1.3, p. 173, is a somewhat vague expression of the fact that estimator sequences attaining the optimal rate $n^{r/(2r+1)}$ uniformly on $\mathfrak{P}_r$ cannot, at the same time, attain the better rate $n^{\bar{r}/(2\bar{r}+1)}$ locally uniformly on the smaller family $\mathfrak{P}_{\bar{r}}$ if $\bar{r} > r$.

Farrell's paper (1972) has a forerunner in his paper (1967), giving a lower bound for the quadratic risk of sequential estimator sequences. In this paper, Farrell suspects (see pp. 471/2) that for densities with a continuous second derivative,

$$\liminf_{n\to\infty} \inf_{p^{(n)}(\xi,\cdot)} \sup_{P\in\mathfrak{P}} n^{2/3} \int (p^{(n)}(\xi, \mathbf{x}_n) - p(\xi))^2 P^n(d\mathbf{x}_n) > 0$$

(On p. 472, line 1, this statement occurs without the factor $n^{2/3}$. It makes sense only if one assumes that this factor became victim of a misprint.) As remarked by Kiefer (1982) [p. 424], Farrell's proof is not entirely correct since he uses a least favourable subfamily which is not in $\mathfrak{P}$.

The best result now available is Theorem 5.1 in Ibragimov and Has'minskii (1981) [p. 237] saying that for every $\xi \in \mathbb{R}$

$$\liminf_{n\to\infty} \inf_{p^{(n)}(\xi,\cdot)} \sup_{P\in\mathfrak{P}_{r,L}} \int \ell\big(n^{r/(2r+1)}(p^{(n)}(\xi, \mathbf{x}_n) - p(\xi))\big) P^n(d\mathbf{x}_n) > 0 \qquad (5.17.1)$$

for arbitrary symmetric loss functions ℓ.

This is the counterpart to their Theorem 4.2, p. 236, asserting the existence of estimator sequences attaining the rate $n^{r/(2r+1)}$:

$$\limsup_{n\to\infty} \sup_{P\in\mathfrak{P}_{r,L}} \sup_{\xi\in\mathbb{R}} \int \ell\big(n^{r/(2r+1)}(p^{(n)}(\xi, \mathbf{x}_n) - p(\xi))\big) P^n(d\mathbf{x}_n) < \infty \qquad (5.17.2)$$

for symmetric loss functions ℓ increasing not too quickly ($u \to \ell(u)\exp[-\varepsilon u^2]$ bounded). In these relations, $\mathfrak{P}_{r,L}$ is the class of all probability measures on $\mathbb{B}$ the densities of which have $r - 1$ derivatives with $|p^{(r-1)}(x) - p^{(r-1)}(y)| \leq L|x - y|$.

Prior to Farrell (1972) there was a paper by Weiss and Wolfowitz (1967a) asserting the existence of an estimator sequence $p^{(n)}(\xi,\cdot)$, $n \in \mathbb{N}$, such that $P^n \circ n^{2/5}(p^{(n)}(\xi,\cdot) - p(\xi))$ is asymptotically maximally concentrated in intervals symmetric about zero (see p. 331, relation (2.22)). Here $p^{(n)}(\xi,\cdot)$ is some kind of maximum probability estimator, and the assertion refers to densities admitting a certain Taylor expansion of order 2. For the proof the authors refer to Theorem 3.1 in Weiss and Wolfowitz (1966) [p. 65], which, however, is on the estimation of a real parameter; the optimality assertion in that theorem holds for one particular symmetric interval (involved in the construction of the estimator) and (of course) for estimator sequences fulfilling a certain local uniformity condition.

What Is an Optimal Rate?

To discuss the idea of an "optimal rate " in a more general (i.i.d.) context, let now $\mathfrak{P}$ be an arbitrary family of probability measures P on a measurable space $(X, \mathscr{A})$, let $\kappa : \mathfrak{P} \to \mathbb{R}$ be a functional and $\kappa^{(n)} : X^n \to \mathbb{R}$ an estimator sequence. To deal with (local) uniformity, we introduce a sequence $\mathfrak{P}_n \subset \mathfrak{P}$ which could mean $\mathfrak{P}_n = \mathfrak{P}$ or $\mathfrak{P}_n \downarrow \{P_0\}$.

Definition 5.17.1 The estimator sequence $(\kappa^{(n)})_{n\in\mathbb{N}}$ *attains the rate* $(c_n)_{n\in\mathbb{N}}$ *uniformly* on $(\mathfrak{P}_n)_{n\in\mathbb{N}}$ if

$$\lim_{n\to\infty} \sup_{P\in\mathfrak{P}_n} P^n\{c_n|\kappa^{(n)} - \kappa(P)| > u_n\} = 0 \quad \text{for every} \quad u_n \to \infty. \tag{5.17.3}$$

The rate $(c_n)_{n\in\mathbb{N}}$ is attained iff $c_n(\kappa^{(n)} - \kappa(P))$ is stochastically bounded, uniformly on $\mathfrak{P}_n$. This is in particular the case if $P^n \circ c_n(\kappa^{(n)} - \kappa(P))$ converges to a limit distribution, uniformly on $\mathfrak{P}_n$.

We define an order relation between rates by

$$(c'_n)_{n\in\mathbb{N}} \preceq (c_n)_{n\in\mathbb{N}} \quad \text{if} \quad \limsup_{n\to\infty} \frac{c'_n}{c_n} < \infty.$$

If $(c_n)_{n\in\mathbb{N}}$ is attainable, then, by Definition 5.17.1, any rate $(c'_n)_{n\in\mathbb{N}} \preceq (c_n)_{n\in\mathbb{N}}$ is attainable, too. The interest, therefore, is in attainable rates which are as large as possible. How can we determine whether an attainable rate is the best possible one?

For this purpose, we introduce the concept of a "rate bound". Roughly speaking, $(c_n)_{n\in\mathbb{N}}$ is a rate bound if no better rate is attainable. A second thought suggests a more restrictive definition: An estimator sequence attaining a better rate for infinitely many $n \in \mathbb{N}$ would certainly be considered as an improvement. This suggests the following definition.

Definition 5.17.2 $(c_n)_{n\in\mathbb{N}}$ is a *rate bound for uniform convergence* on $(\mathfrak{P}_n)_{n\in\mathbb{N}}$ if the following is true. If the rate $(c'_n)_{n\in\mathbb{N}}$ is attained along an infinite subsequence $\mathbb{N}_0$, i.e., if

$$\lim_{n\in\mathbb{N}_0} \sup_{P\in\mathfrak{P}_n} P^n\{c'_n|\kappa^{(n)} - \kappa(P)| > u_n\} = 0 \tag{5.17.4}$$

for some estimator sequence $(\kappa^{(n)})_{n\in\mathbb{N}}$ and every $u_n \to \infty$, then

$$\limsup_{n\in\mathbb{N}_0} c_n'/c_n < \infty.$$

Hence an attainable rate is optimal if it is, at the same time, a rate bound.

Definition 5.17.2 is not so easy to handle if it comes to proving that a given rate $(c_n)_{n\in\mathbb{N}}$ is a rate bound. Here is another, equivalent, definition expressing the idea of a rate bound.

Definition 5.17.3 $(c_n)_{n\in\mathbb{N}}$ is a *rate bound for uniform convergence* on $(\mathfrak{P}_n)_{n\in\mathbb{N}}$ if for every estimator sequence $(\kappa^{(n)})_{n\in\mathbb{N}}$,

$$\liminf_{n\to\infty} \sup_{P\in\mathfrak{P}_n} P^n\{c_n|\kappa^{(n)} - \kappa(P)| > u_n\} > 0 \quad \text{for every} \quad u_n \to 0. \tag{5.17.5}$$

The following Lemma implies the equivalence of Definitions 5.17.2 and 5.17.3.

Lemma 5.17.4 *For any sequence of nonincreasing functions* $H_n : [0,\infty) \to [0,\infty)$, *the following assertions are equivalent.*

(i) For every $\mathbb{N}_0 \subset \mathbb{N}$,

$$\lim_{n\in\mathbb{N}_0} H_n(\delta_n u_n) = 0 \ \textit{for} \ (u_n)_{n\in\mathbb{N}_0} \to \infty \ \textit{implies} \ \liminf_{n\in\mathbb{N}_0} \delta_n > 0.$$

(ii) $\liminf_{n\to\infty} H_n(u_n) > 0$ *for* $(u_n)_{n\in\mathbb{N}} \to 0$.

Proof (i) If for some infinite subsequence $\mathbb{N}_0 \subset \mathbb{N}$ there exists $(\overline{\delta}_n)_{n\in\mathbb{N}}$ fulfilling $\liminf_{n\in\mathbb{N}_0} \overline{\delta}_n = 0$ and

$$\lim_{n\in\mathbb{N}_0} H_n(\overline{\delta}_n u_n) = 0 \ \text{ for every } (u_n)_{n\in\mathbb{N}} \to \infty,$$

then

$$\lim_{n\in\mathbb{N}_0} H_n(\overline{\delta}_n^{1/2}) = \lim_{n\in\mathbb{N}_0} H_n(\overline{\delta}_n \overline{\delta}_n^{-1/2}) = 0.$$

Hence (ii) is violated for $u_n = \overline{\delta}_n^{1/2}$.

(ii) If $(\overline{u}_n)_{n\in\mathbb{N}} \to 0$ and $\liminf_{n\to\infty} H_n(\overline{u}_n) = 0$ then $\lim_{n\in\mathbb{N}_0} H_n(\overline{u}_n) = 0$ for some $\mathbb{N}_0 \subset \mathbb{N}$. Since $H_n(\overline{u}_n u_n) \le H_n(\overline{u}_n)$ eventually if $u_n \to \infty$, relation (i) is violated for $\delta_n = \overline{u}_n$. □

If $(c_n)_{n\in\mathbb{N}}$ is a rate bound, then $(c_n')_{n\in\mathbb{N}}$ with $(c_n')_{n\in\mathbb{N}} \succeq (c_n)_{n\in\mathbb{N}}$ is a rate bound, too. As a side result: If every $(c_n')_{n\in\mathbb{N}} \succeq (c_n)_{n\in\mathbb{N}}$ is a rate bound, then $(c_n)_{n\in\mathbb{N}}$ is a rate bound itself.

If $(c_n)_{n\in\mathbb{N}}$ is attainable, and $(c_n')_{n\in\mathbb{N}}$ is a rate bound, then $(c_n)_{n\in\mathbb{N}} \preceq (c_n')_{n\in\mathbb{N}}$. This implies $(c_n)_{n\in\mathbb{N}} \approx (c_n')_{n\in\mathbb{N}}$ if both sequences are attainable rate bounds. Hence an optimal rate is unique up to equivalence.

This can be seen as follows. Assume that $\limsup c_n/c'_n = \infty$, and let $\mathbb{N}_0$ be such that $\lim_{n\in\mathbb{N}_0} c'_n/c_n = \infty$. Since $(c_n)_{n\in\mathbb{N}}$ is attainable, (5.17.3) implies

$$\lim_{n\in\mathbb{N}_0} \sup_{P\in\mathfrak{P}_n} P^n\{c_n|\kappa_0^{(n)} - \kappa(P)| > (c_n/c'_n)^{1/2}\} = 0 \quad \text{for some } (\kappa_0^{(n)})_{n\in\mathbb{N}}.$$

Since $(c'_n)_{n\in\mathbb{N}}$ is a rate bound, (5.17.5) implies

$$\liminf_{n\in\mathbb{N}_0} \sup_{P\in\mathfrak{P}_n} P^n\{c'_n|\kappa^{(n)} - \kappa(P)| > (c'_n/c_n)^{1/2}\} > 0$$

for every $(\kappa^{(n)})_{n\in\mathbb{N}}$, hence in particular for $(\kappa_0^{(n)})_{n\in\mathbb{N}}$. Since

$$\{c'_n|\kappa_0^{(n)} - \kappa(P)| > (c'_n/c_n)^{1/2}\} = \{c_n|\kappa_0^{(n)} - \kappa(P)| > (c_n/c'_n)^{1/2}\},$$

this is impossible. Hence $(c_n)_{n\in\mathbb{N}} \preceq (c'_n)_{n\in\mathbb{N}}$. □

As an alternative to Definition 5.17.2 we mention a less stringent concept of optimality which can be justified on methodological grounds: That $(c_n)_{n\in\mathbb{N}}$ is optimal if it cannot be improved for a.a. $n \in \mathbb{N}$. That means: If

$$\lim_{n\to\infty} \sup_{P\in\mathfrak{P}_n} P^n\{c'_n|\kappa^{(n)} - \kappa(P)| > u_n\} = 0 \tag{5.17.6}$$

for some estimator sequence $(\kappa^{(n)})_{n\in\mathbb{N}}$ and every $u_n \to \infty$, then

$$\liminf_{n\in\mathbb{N}} c'_n/c_n < \infty.$$

This is a definition equivalent to

$$\limsup_{n\to\infty} \sup_{P\in\mathfrak{P}_n} P^n\{c_n|\kappa^{(n)} - \kappa(P)| > u_n\} > 0 \text{ for } \quad u_n \to 0. \tag{5.17.7}$$

Though the definitions (5.17.4) and (5.17.6) as well as (5.17.5) and (5.17.7) are conceptionally distinct, the difference will usually be irrelevant. If (5.17.6) holds true, then in all practical cases the stronger condition (5.17.4) will be fulfilled, too. That means: If $(c_n)_{n\in\mathbb{N}}$ cannot be improved for a.a. $n \in \mathbb{N}$, then, usually, it cannot be improved along an infinite subsequence.

The concept of an *attainable rate* defined by (5.17.3) is generally accepted. (See Farrell 1972, p. 172, relation (1.4); Stone 1980, p. 1348, relation (1.3); Kiefer 1982, p. 420; Hall and Welsh 1984, Sect. 3, pp. 1083–1084.) With $c_n = n^{1/2}$, relation (5.17.3) occurs as $\sqrt{n}$-consistency in Bickel et al. (1993) [p. 18, Definition 2]. Akahira and Takeuchi (1995, p. 77, Definition 3.5.1) call the property defined by (5.17.3) "consistency of order $(c_n)_{n\in\mathbb{N}}$".

There is less agreement on the concept of a *rate bound* (often occurring only implicitly in the proof that a certain rate is optimal). Since (5.17.5) is equivalent to (5.17.4), this is the weakest condition on the sequence $(c_n)_{n\in\mathbb{N}}$ which guarantees that an estimator sequence attaining a better rate for infinitely many $n \in \mathbb{N}$ is impossible.

The literature offers a plethora of intuitively plausible conditions for a rate bound. As an example of such a condition we mention the following:

$$\liminf_{n\to\infty} \sup_{P\in\mathfrak{P}_n} P^n\{c_n|\kappa^{(n)} - \kappa(P)| > u\} > 0 \text{ for every } (\kappa^{(n)})_{n\in\mathbb{N}} \text{ and every } u > 0. \tag{5.17.8}$$

This condition, which implies (5.17.5), occurs in (Stone (1980), p. 1348) and in Hall (1989) [p. 50, (3.3); see also p. 51, Example 3.1]. In addition to (5.17.8), Stone requires one more condition, (1.2) (see also Kiefer 1982, p. 420), which is equivalent to

$$\lim_{n\to\infty} \sup_{P\in\mathfrak{P}_n} P^n\{c_n|\kappa^{(n)} - \kappa(P)| > u_n\} = 1 \text{ for every } (\kappa^{(n)})_{n\in\mathbb{N}} \text{ and } u_n \to 0. \tag{5.17.9}$$

Needless to say that (5.17.9), too, implies (5.17.5).

In (1983, p. 393) Stone requires—in a different context—a condition which corresponds to

$$\lim_{n\to\infty} \sup_{P\in\mathfrak{P}_n} P^n\{c_n|\kappa^{(n)} - \kappa(P)| > u\} = 1 \quad \text{for some} \quad u > 0.$$

In this paper, Stone had a deviating definition of attainability (p. 394, relation (2)), namely

$$\lim_{n\to\infty} \sup_{P\in\mathfrak{P}_n} P^n\{c_n|\kappa^{(n)} - \kappa(P)| > u\} = 0 \quad \text{for some} \quad u > 0.$$

That a competent author like Stone is at variance with himself illustrates how vague the idea of an optimal rate is unless intuition is guided by methodological principles. His argument (p. 394) that "these definitions...were formulated this way mainly because they could be verified in the present context" is not compelling. Stone abstains from relating his concept(s) of a rate bound to the concept Farrell had used ten years ago.

Farrell (1972), p. 173 (see also Hall and Welsh 1984, p. 1080, and Carroll and Hall 1988, p. 1185) considers the rate $(c_n)_{n\in\mathbb{N}}$ as optimal if

$$\lim_{n\to\infty} \sup_{P\in\mathfrak{P}_n} P^n\{|\kappa^{(n)} - \kappa(P)| > a_n\} = 0 \text{ for every } (\kappa^{(n)})_{n\in\mathbb{N}} \text{ implies } \lim_{n\to\infty} a_n c_n = \infty,$$

a condition which, on second thought, turns out to be equivalent to (5.17.8) with lim sup in place of lim inf.

The question of optimal rates did not find much attention in textbooks. Prakasa Rao, in his monograph *Nonparametric Functional Estimation* (1983), presents no general thoughts concerning optimal rates or optimal limit distributions. In the particular context of estimating the value of a density, he presents in Theorem 2.0.2, p. 31, the conditions of Farrell (1972) for "best possible rates", and the conditions of Stone (1980) in Example 3, p. 153, for "optimal rates" without comments on the mutual relationship.

Rates Based on Loss Functions

Some scholars hold the opinion that for a crude concept of optimality like that of an optimal rate (as opposed to an optimal limit distribution) a detailed consideration of $\sup_{P\in\mathfrak{P}_n} P^n\{c_n|\kappa^{(n)} - \kappa(P)| > u\}$ as a function of u is to no avail. This suggests to base the concept of an optimal rate on a global measure like "risk". Let ℓ be a symmetric loss function. The following definitions are rather plausible:

The estimator sequence $(\kappa^{(n)})_{n\in\mathbb{N}}$ *attains* the rate $(c_n)_{n\in\mathbb{N}}$ *uniformly on* $(\mathfrak{P}_n)_{n\in\mathbb{N}}$ if

$$\limsup_{n\to\infty} \sup_{P\in\mathfrak{P}_n} \int \ell(c_n(\kappa^{(n)} - \kappa(P)))dP^n < \infty. \tag{5.17.10}$$

The attainable rate $(c_n)_{n\in\mathbb{N}}$ is *optimal uniformly on* $(\mathfrak{P}_n)_{n\in\mathbb{N}}$ if

$$\liminf_{n\to\infty} \sup_{P\in\mathfrak{P}_n} \int \ell(c_n(\kappa^{(n)} - \kappa(P)))dP^n > 0 \quad \text{for every} \quad (\kappa^{(n)})_{n\in\mathbb{N}}. \tag{5.17.11}$$

Based on a global measure, relation (5.17.11) can be given in a slightly stronger version, as

$$\liminf_{n\to\infty} \inf_{\kappa^{(n)}} \sup_{P\in\mathfrak{P}_n} \int \ell(c_n(\kappa^{(n)} - \kappa(P)))dP^n > 0 \quad \text{for every} \quad (\kappa^{(n)})_{n\in\mathbb{N}},$$

the proof of which is not more difficult than the proof of (5.17.11), which suffices to establish the optimality of an attainable rate.

In order to express that $(c_n)_{n\in\mathbb{N}}$ is attainable, relation (5.17.10) with $\ell(u) = u^2$ was used from the beginning. In Theorem 1.2, p. 173, Farrell (1972) (see also Wahba 1975, p. 16) defines an attainable rate $(c_n)_{n\in\mathbb{N}}$ as *optimal* if

$$\limsup_{n\to\infty} \sup_{P\in\mathfrak{P}_n} a_n^{-2} \int (\kappa^{(n)} - \kappa(P))^2 dP^n < \infty \quad \text{implies} \quad \liminf_{n\to\infty} c_n a_n > 0.$$

This is a rather circumstantial way of saying that (5.17.11) holds true for $\ell(u) = u^2$.

Recall that relations (5.17.1) and (5.17.2), proved by Ibragimov and Has'minskii (1981) [pp. 236/7] to establish $n^{r/(2r+1)}$ as an optimal rate for density estimators, imply (5.17.10) and (5.17.11).

In the following we discuss the connection between rate concepts based on $u \to P^n\{c_n|\kappa^{(n)} - \kappa(P)| > u\}$, and rate concepts based on $\int \ell(c_n(\kappa^{(n)} - \kappa(P)))dP^n$. The connection between these concepts is conveyed by the following lemma, applied with $Q_P^{(n)} := P^n \circ c_n(\kappa^{(n)} - \kappa(P))$.

Lemma 5.17.5 *Let $\mathscr{L}_0$ be the class of all symmetric loss functions $\ell : \mathbb{R} \to [0, \infty]$ which are continuous, and positive on $(0, \infty]$. Given probability measures $Q_P^{(n)}$ on $\mathbb{B} \cap [0, \infty)$, we consider the following statements.*

$$\liminf_{n\to\infty} \sup_{P\in\mathfrak{P}_n} Q_P^{(n)}[u_n, \infty) > 0 \quad \textit{for every} \quad u_n \to 0, \tag{5.17.12}$$

$$\liminf_{n\to\infty} \sup_{P\in\mathfrak{P}_n} \int \ell(v) Q_P^{(n)}(dv) > 0. \tag{5.17.13}$$

Then the following is true.

(i) Relation (5.17.13) for some bounded $\ell \in \mathscr{L}_0$ implies (5.17.12).

(ii) Relation (5.17.12) implies (5.17.13) for every $\ell \in \mathscr{L}_0$.

The same assertions hold true if lim inf *is replaced with* lim sup *in both (5.17.12) and (5.17.13).*

Moreover, $\limsup_{n\to\infty} \sup_{P\in\mathfrak{P}_n} \int \ell(v) Q_P^{(n)}(dv) < \infty$ *for an unbounded loss function implies*

$$\lim_{n\to\infty} \sup_{P\in\mathfrak{P}_n} Q_P^{(n)}(u_n, \infty) = 0 \quad \textit{for every } u_n \to \infty.$$

Observe that nothing is assumed about $\mathfrak{P}_n$. Hence the assertions hold in particular with $\mathfrak{P}_n = \{P_0\}$ for $n \in \mathbb{N}$.

Proof Since ℓ is nondecreasing, we have

$$\ell(u) 1_{[u,\infty)}(v) \le \ell(v) \quad \text{for} \quad u, v > 0.$$

Since ℓ is bounded by 1,

$$\ell(v) \le \ell(u) + 1_{[0,\infty)}(v) \quad \text{for} \quad u, v > 0.$$

Hence

$$\ell(u) Q_P^{(n)}[u, \infty) \le \int \ell(v) Q_P^{(n)}(dv) \le \ell(u) + Q_P^{(n)}[u, \infty) \quad \text{for} \quad u \ge 0. \tag{5.17.14}$$

(i) Since $u_n \to 0$ implies $\ell(u_n) \to 0$, we obtain from the right-hand side of (5.17.14), applied with $u = u_n$, that (5.17.12) follows from (5.17.13).

(ii) Assume there exists a bounded $\ell \in \mathscr{L}_0$ such that

$$n \to \infty \sup_{P\in\mathfrak{P}_n} \int \ell(v) Q_P^{(n)}(dv) = 0.$$

Then we obtain from the left-hand side of (5.17.14) that

$$\liminf_{n\to\infty} \sup_{P\in\mathfrak{P}_n} Q_P^{(n)}[u, \infty) = 0 \quad \text{for} \quad u > 0.$$

This, in turn, implies the existence of a sequence $u_n \to 0$ such that

$$\liminf_{n\to\infty} \sup_{P\in\mathfrak{P}_n} Q_P^{(n)}[u_n, \infty) = 0,$$

in contradiction to (5.17.12). (Hint: For every m there exists $n(m) > n(m-1)$ such that $\sup_{P\in\mathfrak{P}_{n(m)}} Q_P^{(n(m))}[1/m, \infty) < 1/m$.) □

Rate Bounds Without Local Uniformity

In Sect. 5.16 it was shown that limit distributions which are optimal subject to the condition of locally uniform convergence, cannot be surpassed on open subsets by any limit distribution. In the following we shall establish an analogous result for optimal rates. More precisely, we shall show the following: If $(c_n)_{n\in\mathbb{N}}$ is, for every P in an open subset $\mathfrak{P}_0$, a rate bound under the condition of local uniformity, then it is (under suitable regularity conditions on κ and $\mathfrak{P}$) also a rate bound (without a local uniformity condition) for P in a dense subset of $\mathfrak{P}_0$. In other words: There is no estimator sequence attaining a better rate for every P in an open subset of $\mathfrak{P}$.

The proof is based on Lemma 5.14.1. This lemma applies with X replaced by $\mathfrak{P}$, endowed with a suitable metric ρ. If $P \to \kappa(P)$ is ρ-continuous, and if ρ is at least as strong as the sup-metric, then the function $P \to \int \ell(c_n(\kappa^{(n)} - \kappa(P)))dP^n$ is continuous for every $n \in \mathbb{N}$, provided ℓ is bounded and continuous. (The proof is about the same as that of Corollary 5.1 in Pfanzagl 2003, p. 109.)

Assume now that $(c_n)_{n\in\mathbb{N}}$ is, for every P_0 in an open subset $\mathfrak{P}_0 \subset \mathfrak{P}$, a rate bound (in the sense of (5.17.7) for locally uniform convergence. Using the characterization by means of loss functions given in Lemma 5.17.5, this can be expressed as follows: For every $(\kappa^{(n)})_{n\in\mathbb{N}}$ and every $\ell \in \mathscr{L}_0$,

$$\limsup_{n\to\infty} \sup_{P\in\mathfrak{P}_n(P_0)} \int \ell(c_n(\kappa^{(n)} - \kappa(P)))dP^n > 0. \tag{5.17.15}$$

We shall show that this implies

$$\limsup_{n\to\infty} \int \ell(c_n(\kappa^{(n)} - \kappa(P)))dP^n > 0$$

for P in some dense subset of $\mathfrak{P}_0$. (Observe that this subset depends on $(\kappa^{(n)})_{n\in\mathbb{N}}$.)

Proof Assume that for some $(\kappa^{(n)})_{n\in\mathbb{N}}$ and some $\ell \in \mathscr{L}_0$ the relation

$$\lim_{n\to\infty} \int \ell(c_n(\kappa^{(n)} - \kappa(P)))dP^n = 0$$

holds for every P in some open subset $\mathfrak{P}_1 \subset \mathfrak{P}_0$. Since $P \to \int \ell(c_n(\kappa^{(n)} - \kappa(P)))dP^n$ is continuous for every $n \in \mathbb{N}$, there exists (by Lemmas 5.14.1 and 5.14.2) a dense subset $\mathfrak{P}_2 \subset \mathfrak{P}_1$ such that, for every $P_0 \in \mathfrak{P}_2$, the relation $\lim_{n\to\infty} \rho(P_0, P_n) = 0$ implies

$$\lim_{n\to\infty}\int \ell(c_n(\kappa^{(n)}-\kappa(P_n)))dP_n^n=0.$$

If $(\mathfrak{P}_n(P_0))_{n\in\mathbb{N}}$ shrinks to P_0 with respect to the metric ρ, this implies

$$\lim_{n\to\infty}\sup_{P\in\mathfrak{P}_n(P_0)}\int \ell(c_n(\kappa^{(n)}-\kappa(P)))dP^n=0,$$

in contradiction to (5.17.15). □

For examples where the conditions of these assertions (such as the existence of a metric ρ for which $\mathfrak{P}$ is complete) see Pfanzagl (2002a) .

For families with a Euclidean parameter, the optimal rate usually is $n^{1/2}$ (or $n^{1/2}(\log n)^a$), and estimator sequences converging at this rate to a limit distribution are the custom. Hence assertions on the pointwise validity of locally uniform rate bounds are, in this case, of minor interests. Just for sake of completeness we mention that, as a consequence of Bahadur's Lemma, locally uniform rate bounds are valid pointwise for all Θ except for a set of Lebesgue measure 0.

Optimal Rates and Optimal Limit Distributions

In most cases, there are several estimator sequences converging to the estimand with the same, optimal, rate. As an example, we mention the estimation of the center of symmetry of an (unknown) distribution on $\mathbb{B}$. Examples of estimator sequences with the optimal rate $n^{1/2}$ are the Bickel and Hodges estimator (1967, Sect. 3), the estimator of Rao et al. (1975) [Theorem 4, p. 866] based on the Kolmogorov distance, or the (adaptive) estimator by van Eeden (1970). Under natural conditions (see Sect. 5.5), the center of symmetry is a differentiable functional, so that the LAN-theory yields the asymptotic concentration bound, which is attained by various estimator sequences (see Sect. 5.13).

There are many more nonparametric models of this kind: a differentiable functional and an LAN-condition based on a rate $n^{1/2}L(n)$ (usually with $L(n)=(\log n)^a$), which distinguishes certain estimator sequences as asymptotically maximally concentrated.

Typical for nonparametric models are, however, optimal rates slower than $n^{1/2}$. [This is shorthand for a rate $n^aL(n)$ with $a<1/2$.] Even if estimator sequences attaining this optimal rate converge to a limit distribution, there is no asymptotic bound for the quality of these estimator sequences. This is not a lacuna in the theory; such bounds are impossible.

First we mention a result which illustrates the principal difficulties relating to limit distributions attained with an optimal rate slower than $n^{1/2}$. Then we turn to more specific conditions which determine the optimal rate, and which, at the same time, exclude the existence of estimator sequences converging locally uniformly to some limit distribution with the optimal rate, if the latter is slower than $n^{1/2}$.

If there exists an estimator sequence converging with a rate $n^a L(n)$, $a \in [0, 1/2)$ locally uniformly to a limit distribution with expectation 0 and finite variance, then there exists an estimator sequence converging with a better rate locally uniformly to a normal limit distribution with mean 0. (See Pfanzagl 1999b, p. 759, Theorem 2.1 (ii).)

For a precise formulation of this result we need the concept of locally uniform convergence, holding uniformly on some subset $\mathfrak{P}_0 \subset \mathfrak{P}$. (The formulation of the Theorem cited above is somewhat careless in this respect.) Given for every $P_0 \in \mathfrak{P}$ a nondecreasing sequence $\mathfrak{P}_n(P_0)$, $n \in \mathbb{N}$, we say that $Q_P^{(n)}$, $n \in \mathbb{N}$, converges locally uniformly to Q_P, uniformly on $\mathfrak{P}_0$, if $\sup_{P \in \mathfrak{P}_n(P_0)} D(Q_P^{(n)}, Q_P)$, $n \in \mathbb{N}$, converges to 0, uniformly for $P_0 \in \mathfrak{P}_0$. (This is, in particular, the case if $\sup_{P \in \mathfrak{P}_0} D(Q_P^{(n)}, Q_P)$, $n \in \mathbb{N}$, converges to 0.)

If there exists an estimator sequence $\kappa^{(n)}$ such that $P^n \circ c_n(\kappa^{(n)} - \kappa(P))$, $n \in \mathbb{N}$, converges in this sense with $c_n = n^a L(n)$, $a \in [0, 1/2)$, to a limit distribution Q_P, then there is an estimator sequence $\hat{\kappa}^{(n)}$ such that $P^n \circ \hat{c}_n(\hat{\kappa}^{(n)} - \kappa(P))$, converges in this sense to $N(0, \sigma^2(P))$, with a rate$(\hat{c}_n)_{n \in \mathbb{N}}$ such that $\lim_{n \to \infty} \hat{c}_n / c_n = \infty$. This holds under the assumption that Q_P has expectation 0, that u^2 is Q_P-integrable uniformly for $P \in \mathfrak{P}_0$, and $\sigma^2(P) = \int u^2 Q_P(du)$ is bounded and bounded away from 0 on $\mathfrak{P}_0$. A special, yet more concise version is this: If $P^n \circ c_n(\kappa^{(n)} - \kappa(P))$, $n \in \mathbb{N}$, converges with $c_n = n^a L(n)$, $a \in [0, 1/2)$ to a $N(0, \sigma^2(P))$, then there exists an estimator sequence $(\hat{\kappa}^{(n)})_{n \in \mathbb{N}}$ such that $P^n \circ \hat{c}_n(\hat{\kappa}^{(n)} - \kappa(P))$, $n \in \mathbb{N}$, converges to the same normal limit distribution with a rate $(\hat{c}_n)_{n \in \mathbb{N}}$ better than $(c_n)_{n \in \mathbb{N}}$, provided $\sigma^2(P)$ is bounded and bounded away from 0 for $P \in \mathfrak{P}_0$.

The conclusion: In situations which are typical for nonparametric models, it is impossible to define the concept of a maximally concentrated limit distribution in the usual sense, based on a local uniformity.

This result does not exclude the existence of estimator sequences which converge locally uniformly to a limit distribution with a rate slower than the optimal one, and they do not exclude the existence of estimator sequences converging to a limit distribution with this rate for every $P \in \mathfrak{P}$.

Observe the connection with what has been said above: If it is possible to find a metric ρ "stronger" than d, such that $\mathfrak{P}$ is complete and κ continuous, then convergence of $P^n \circ c_n(\kappa^{(n)} - \kappa(P))$ to a limit distribution Q_P for every P in an open subset $\mathfrak{P}_0$ implies locally uniform convergence on a dense subset of $\mathfrak{P}_0$. However, this local uniformity is with respect to ρ, and locally ρ-uniform convergence at some P_0 does not imply uniform convergence on $\{P \in \mathfrak{P} : d(P_0^n, P^n) \leq \alpha\}$.

How to Determine the Optimal Rate

The following theorem shows explicitly how optimal rate bounds can be determined, and that estimator sequences converging with this optimal rate locally uniformly to some limit distribution do not exist if this optimal rate is slower than $n^{1/2}$.

Theorem 5.17.6 *Assume that for $(c_n)_{n \in \mathbb{N}}$ there exists a sequence $P_n \in \mathfrak{P}$, $n \in \mathbb{N}$, such that*

$$\limsup_{n \to \infty} d(P_0^n, P_n^n) < 1 \tag{5.17.16}$$

and

$$\liminf_{n\to\infty} c_n|\kappa(P_n) - \kappa(P_0)| > 0. \tag{5.17.17}$$

Then the following is true.

(i) *There is an* $\alpha < 1$ *such that* $(c_n)_{n\in\mathbb{N}}$ *is a rate bound for uniform convergence on*

$$\mathfrak{P}_{n,\alpha}(P_0) := \{P \in \mathfrak{P} : d(P_0^n, P^n) \le \alpha\}, \text{ for some } \alpha < 1.$$

(ii) *If*

$$\lim_{m\to\infty} m^{-1/2} \limsup_{n\to\infty} c_{mn}/c_n = 0, \tag{5.17.18}$$

then there exists no estimator sequence converging with this rate uniformly on $\mathfrak{P}_{n,\alpha}(P_0)$ *for some* $\alpha \in (0, 1)$ *to a non-degenerate limit distribution.*

Remark Condition (5.17.18) is in particular fulfilled if $c_n = n^a L(n)$ with $a \in [0, 1/2)$. Recall that all regularly varying sequences $(c_n)_{n\in\mathbb{N}}$ are of the type $c_n = n^a L(n)$.

Proof (i) By assumptions (5.17.16) and (5.17.17) there is $\alpha < 1$ and $\lambda > 0$ such that for $n \in \mathbb{N}$,

$$d(P_0^n, P_n) < \alpha \tag{5.17.19}$$

and

$$c_n|\kappa(P_n) - \kappa(P_0)| > \lambda. \tag{5.17.20}$$

W.l.g. we assume that

$$c_n(\kappa(P_n) - \kappa(P_0)) > \lambda \quad \text{for} \quad n \in \mathbb{N}.$$

These relations imply

$$\begin{aligned} &P_0^n\{c_n(\kappa^{(n)} - \kappa(P_0)) > u + \lambda\} \\ &> -\alpha + P_n^n\{c_n(\kappa^{(n)} - \kappa(P_0)) > u + \lambda\} \\ &\ge -\alpha + P_n^n\{c_n(\kappa^{(n)} - \kappa(P_n)) > u\}. \end{aligned} \tag{5.17.21}$$

(ii) Relation (5.17.21), applied with $u = -\lambda/2$, yields

$$P_0^n\{c_n(\kappa^{(n)} - \kappa(P_0)) > \lambda/2\} + P_n^n\{c_n(\kappa^{(n)} - \kappa(P_n)) \le -\lambda/2\} > 1 - \alpha.$$

Since $P_n \in \mathfrak{P}_{n,\alpha}(P_0)$, this implies

$$\sup_{P\in\mathfrak{P}_{n,\alpha}(P_0)} P^n\{c_n|\kappa^{(n)} - \kappa(P)| \ge \lambda/2\} > \frac{1}{2}(1 - \alpha) > 0.$$

Hence, by Definition 5.17.3, $(c_n)_{n\in\mathbb{N}}$ is a rate bound for uniform convergence on $(\mathfrak{P}_{n,\hat{\alpha}})_{n\in\mathbb{N}}$.

(iii) Assume now that $P^n \circ c_n(\kappa^{(n)} - \kappa(P))$, $n \in \mathbb{N}$, converges weakly to $Q_{P_0}|\mathbb{B}$, uniformly for $P \in \mathfrak{P}_{n,\beta}$ for some $\beta > 0$. By definition of uniform weak convergence, there exists a dense subset $B \subset \mathbb{R}$ such that

$$\lim_{n\to\infty} \sup_{P\in\mathfrak{P}_{n,\beta}} |P^n\{c_n(\kappa^{(n)} - \kappa(P)) > u\} - Q_{P_0}(u, \infty)| = 0 \text{ for every } u \in B. \tag{5.17.22}$$

Relation (5.17.24), applied with P replaced by P_0^n and Q replaced by P_{mn}^n (with m fixed), yields

$$\begin{aligned} d(P_0^n, P_{mn}^n) &\le m^{-1/2}\sqrt{2/(1-\alpha)}d(P_0^{mn}, P_{mn}^{mn})^{1/2} \\ &\le m^{-1/2}\sqrt{2\alpha/(1-\alpha)}. \end{aligned} \tag{5.17.23}$$

With P_{mn} in place of P_n, relation (5.17.19) therefore holds with

$$\hat{\alpha} = m^{-1/2}\sqrt{2\alpha/(1-\alpha)}.$$

From (5.17.20),

$$c_n|\kappa(P_{mn}) - \kappa(P_0)| \ge \frac{c_n}{c_{mn}}\lambda.$$

Let

$$\varepsilon_m := m^{-1/2} \limsup_{n\to\infty} c_{mn}/c_n.$$

By assumption (5.17.18), $\lim_{m\to\infty} \varepsilon_m = 0$. For every $m \in \mathbb{N}$, there is n_m such that

$$c_{mn}/c_n < 2 \limsup_{n\to\infty} c_{mn}/c_n \quad \text{for} \quad n \ge n_m,$$

hence

$$c_{mn}/c_n \le 2m^{1/2}\varepsilon_m \text{ for } n \ge n_\varepsilon.$$

This implies

$$c_n|\kappa(P_{mn}) - \kappa(P_0)| \ge \frac{c_n}{c_{mn}}\lambda \ge \frac{\lambda}{2}m^{-1/2}\varepsilon_m^{-1}.$$

Hence, with P_{mn} in place of P_n, relation (5.17.20) holds for $n \ge n_m$ with $\hat{\lambda} = \frac{\lambda}{2}m^{-1/2}\varepsilon_m^{-1}$.

Relation (5.17.21), with P_{mn} in place of P_n and $\hat{\alpha}, \hat{\lambda}$ in place of α, λ, respectively, becomes

$$P_0^n\{c_n(\kappa^{(n)} - \kappa(P_0)) > u + \frac{\lambda}{2}m^{-1/2}\varepsilon_m^{-1}\}$$

$$\geq -m^{-1/2}\sqrt{2\alpha/(1-\alpha)} + P^n_{mn}\{c_n(\kappa^{(n)} - \kappa(P_{mn})) > u\} \quad \text{for } n \geq n_m.$$

Since $P_{mn} \in \mathfrak{P}_{n,\beta}$ for every $\beta > 0$ (by (5.17.23), relation (5.17.22) implies

$$Q_{P_0}(u + \frac{\lambda}{2}m^{-1/2}\varepsilon_m^{-1}, \infty) \geq -m^{-1/2}\sqrt{2\alpha/(1-\alpha)} + Q_{P_0}(u, \infty),$$

i.e.

$$Q_{P_0}(u, u + \frac{\lambda}{2}m^{-1/2}\varepsilon_m^{-1}] \leq m^{-1/2}\sqrt{2\alpha(1-\alpha)} \quad \text{for } u \in B \text{ and } m \in \mathbb{N}.$$

Since the convergence of $m^{-1/2}\varepsilon_m^{-1}$ to 0 is slower than the convergence of $m^{-1/2}$, it is plausible that this is incompatible with the assumption that Q_{P_0} is non-degenerate. This can be made precise by the following lemma (see Pfanzagl 2000b, p. 39, Lemma 4.1):

For nondecreasing functions $F : \mathbb{R} \to [0, 1]$, and $t, s > 0$, the relation $F(u+t) - F(u) \leq s$ for all u in a dense subset implies

$$F(u'') - F(u') \leq 2s(1 + (u'' - u')t^{-1}) \quad \text{for arbitrary} \quad u' < u''.$$

Applied with $F(u) = Q_{P_0}(-\infty, u]$, this lemma implies

$$\begin{aligned} Q_{P_0}(u', u''] &\leq 2m^{-1/2}\sqrt{2\alpha/(1-\alpha)}(1 + (u'' - u')2\lambda^{-1}m^{1/2}\varepsilon_m) \\ &= 2m^{-1/2}\sqrt{2\alpha/(1-\alpha)} + (u'' - u')4\lambda^{-1}\sqrt{2\alpha/(1-\alpha)}\varepsilon_m. \end{aligned}$$

Since $\varepsilon_n \to 0$, this implies the assertion. □

Lemma 5.17.7 *Let* $\alpha \in (0, 1)$. *Then*

$$(1-u)^{1/m} \geq 1 - \frac{u}{1-\alpha} \cdot \frac{1}{m} \quad \textit{for every} \quad u \in [0, \alpha].$$

Proof Let

$$f(u) := (1-u)\exp[u/(1-\alpha)].$$

Then $f(0) = 1$ and $f'(u) \geq 0$ for $u \in [0, \alpha]$ imply $f(u) \geq 1$ for $u \in [0, \alpha]$, hence

$$(1-u) \geq \exp[-u/(1-\alpha)] \geq \left(1 - \frac{u}{1-\alpha} \cdot \frac{1}{m}\right)^m.$$

□

Lemma 5.17.8 *(i)* $H(P^m, Q^m)^2 \leq \alpha < 1$ *implies*

$$H(P, Q)^2 \leq \frac{m^{-1}}{1-\alpha}H(P^m, Q^m)^2.$$

(ii) $d(P^m, Q^m) \leq \alpha < 1$ *implies*

$$d(P, Q) \leq m^{-1/2}\sqrt{2/(1-\alpha)}d(P^n, Q^m)^{1/2}. \tag{5.17.24}$$

Proof (i) Lemma 5.17.7 implies

$$(1 - H(P^m, Q^m)^2)^{1/m} \geq 1 - \frac{m^{-1}}{1-\alpha}H(P^m, Q^m)^2.$$

Relation (i) now follows from

$$1 - H(P, Q)^2 = (1 - H(P^m, Q^m)^2)^{1/m}.$$

(ii) Relation (ii) follows from (i), since

$$H(P, Q)^2 \leq d(P, Q) \leq \sqrt{2}H(P, Q).$$

(See e.g. Strasser 1985, p. 11, Lemma 2.15.) □

5.18 Second Order Optimality

Asymptotic assertions on the performance of estimator sequences more informative than the approximation by limit distributions, require conditions on the family of probability measures going beyond "LAN" as well as regularity conditions on the estimator sequences going beyond "locally uniform convergence to a limit distribution", or "locally uniform median unbiasedness". Though our main interest is in estimation theory, auxiliary results on test sequences will be essential.

Following the historical development, we present the results for parametric families $\mathfrak{P} = \{P_\vartheta^n : \vartheta \in \Theta\}$, $\Theta \subset \mathbb{R}^k$. Corresponding results for general families require more technical preparations. Readers interested in these extensions are referred to Pfanzagl (1985). In this Section we will write $o_p(\varepsilon_n)$ for $o(\varepsilon_n, P_\vartheta^n)$ and use the Einstein convention which omits summation signs over pairs of indices. Besides $L_{ij}(\vartheta) = \int \ell^{(ij)}(\cdot, \vartheta)dP_\vartheta$ and $L_{i,j}(\vartheta) = \int \ell^{(i)}(\cdot, \vartheta)\ell^{(j)}(\cdot, \vartheta)dP_\vartheta$ we introduce

$$L_{ijk}(\vartheta) = \int \ell^{(ijk)}(\cdot, \vartheta)dP_\vartheta, \quad L_{ij,k}(\vartheta) = \int \ell^{(ij)}(\cdot, \vartheta)\ell^{(k)}(\cdot, \vartheta)dP_\vartheta,$$

etc. As before, $L(\vartheta) = (L_{i,j}(\vartheta))_{i,j=1,\ldots,k}$ and $\Lambda(\vartheta) = L(\vartheta)^{-1}$.

As a first step we present asymptotic results of order $o(n^{-1/2})$. Many results familiar from the asymptotic theory of order $o(n^0)$ carry over to $o(n^{-1/2})$. The interpretation of results of order $o(n^{-1})$ is much more complex from the conceptual point of view; the regularity conditions to be imposed on the family $\mathfrak{P}$ are more restrictive.

This is not just a matter of higher order smoothness of the densities, say. Properties which were irrelevant for approximations of order $o(n^0)$ now enter the stage: Assertions of order $o(n^{-1/2})$ about the concentration on even rather simple sets are impossible for families of discrete distributions. If we consider asymptotic assertions as approximations to the reality, we must not ignore that families of probability measures reflect reality to a degree which hardly ever justifies to take serious differences between statistical procedures which are of the order n^{-1} only.

In contrast to an asymptotic theory of order $o(n^{-1})$, some basic results following from an asymptotic theory of order $o(n^{-1/2})$ seem to be of practical relevance. One would expect that an analysis of order $o(n^{-1/2})$ reveals differences of order $n^{-1/2}$ between asymptotically efficient estimator sequences which are indistinguishable by an analysis of order $o(n^0)$. This is, however, not the case. For a large class of estimator sequences, asymptotic efficiency of order $o(n^0)$ implies efficiency of order $o(n^{-1/2})$. It contributes to the relevance of this result that "efficiency of order $o(n^{-1/2})$" holds simultaneously for all loss functions in the case of one-parameter families, and for all symmetric loss functions in the general case.

In this section we write $o_p(a_n)$ for $o(a_n, P_\vartheta^n)$. We also use the Einstein convention and sum over pairs of indices.

Asymptotic theory going beyond the approximation by limit distributions started with papers on the distributions of ML-sequences, say $(\vartheta^{(n)})_{n\in\mathbb{N}}$. For highly regular parametric families $\{P_\vartheta^n : \vartheta \in \Theta\}$, $\Theta \subset \mathbb{R}^k$, the ML estimator sequence has a stochastic expansion (*S-expansion*) of order $o(n^{-1/2})$,

$$n^{1/2}(\vartheta^{(n)} - \vartheta) = S^{(n)}(\cdot, \vartheta) + o_p(n^{-1/2}) \tag{5.18.1}$$

with

$$S^{(n)}(\cdot, \vartheta) = \tilde{\lambda}(\cdot, \vartheta) + n^{-1/2}\Big(\frac{1}{2}\Lambda_{\bullet\ell}(\vartheta)L_{\ell ij}(\vartheta)\tilde{\lambda}_i(\cdot, \vartheta)\tilde{\lambda}_j(\cdot, \vartheta) + \Lambda_{\bullet i}(\vartheta)\tilde{\ell}^{(ij)}(\cdot, \vartheta)\tilde{\lambda}_j(\cdot, \vartheta)\Big). \tag{5.18.2}$$

and $\lambda(\cdot, \vartheta) = \Lambda(\vartheta)\ell^\bullet(\cdot, \vartheta) \in L_*(P_\vartheta)$. Straightforward computation shows that

$$S^{(n)}(\cdot, \vartheta + n^{-1/2}a) = S^{(n)}(\cdot, \vartheta) - a + o_p(n^{-1/2}). \tag{5.18.3}$$

Hence relation (5.18.1) holds with ϑ replaced by $\vartheta + n^{-1/2}a$.

Similar stochastic expansions exist for various estimator sequences: for sequences of ML estimators see Linnik and Mitrofanova (1963, 1965) and Mitrofanova (1967), with imperfect proofs. The first correct proof for the validity of an expansion of the distribution function, i.e., an Edgeworth expansion (*E-expansion*) for one-dimensional minimum contrast estimators was given by Chibisov (1973b) [pp. 651–652, Theorem 2]. A similar result was obtained independently by a different method (which does not extend, however, to the case of vector parameters) in Pfanzagl (1973) [p. 997, Theorem 1], where the polynomials up to the order $o(n^{-1/2})$ are given explicitly. For the ML estimator of a vector parameter see Michel (1975) [p. 70, Theorem 1]. Stochastic expansions for Bayes estimators and for estimators obtained by maximizing the posterior density can be found in Gusev (1975) [p. 476, Theorem 1, p. 489, Theorem 5] and Strasser (1977) [p. 32, Theorem 4]; for estimators obtained as the median of the posterior distribution see Strasser (1978b) [pp. 872–873, Lemma 2].

The Convolution Theorem was the outcome of long-lasting endeavors to prove the asymptotic optimality of ML-sequences. The result was twofold. (i) Under conditions on the basic family, there is a "bound" for regularly attainable limit distribution, and (ii) under additional conditions on the family, estimator sequences attaining this bound do exist.

From the beginning, E-expansions for the distribution of certain sequences of estimators and tests had a less ambitious goal: To obtain more accurate approximations to $P_\vartheta^n \circ n^{1/2}(\vartheta^{(n)} - \vartheta)$, in particular: to distinguish between estimator sequences with identical limit distributions.

One would expect the class of all asymptotically efficient estimator sequences to split up under a closer investigation. If we restrict attention to estimator sequences with distributions approximable by an E-expansion $E_\vartheta^{(n)}$ with λ^k-density $\varphi_{\Lambda(\vartheta)}(u)$ $(1 + n^{-1/2}G(u, \vartheta))$, i.e.,

$$\sup_{C \in \mathscr{C}} \left| P_\vartheta^n\{n^{1/2}(\vartheta^{(n)} - \vartheta) \in C\} - \int_C \phi_{\Lambda(\vartheta)}(u)(1 + n^{-1/2}G(u, \vartheta)) \right| = o(n^{-1/2}),$$

this would mean a variety of $n^{-1/2}$-terms G, and the best one could hope for is the existence of an E-expansion $\overline{E}^{(n)}$ such that, for every $n \in \mathbb{N}$, $\overline{E}^{(n)}(B) \geq E^{(n)}(B)$ for a large class of sets B, say all convex and symmetric sets. It came as a surprise to all of us that this disaster did not occur. It turned out that asymptotically efficient estimator sequences have the same E-expansion of order $o(n^{-1/2})$, except for a difference of order $n^{-1/2}$ in their location.

Remark That different asymptotically efficient statistical procedures agree in their efficiency to a higher order, was already observed by Welch (1965) [p. 6]: confidence procedures based on different statistics like ML estimators or $\tilde{\ell}^\bullet$ have up to $o(n^{-1/2})$ the same covering probability for $\vartheta - n^{-1/2}t$ if the covering probabilities for ϑ agree up to $o(n^{-1/2})$. Hartigan (1966) [p. 56] observes that central Bayes intervals agree in their size for various prior distributions up to the order n^{-1}. For more results of this kind see Strasser (1978b) [p. 874, Theorem 3.] Sharma (1973) [p. 974] observes for certain exponential families: If the minimum variance unbiased estimator sequence has the same asymptotic distribution as the ML-sequence, then the difference between these distributions is of an order—almost—smaller than $n^{-1/2}$. (Readers interested in questions of priority should be aware of J.K. Ghosh, 1991, p. 506, who claims to have been the first to observe the phenomenon of 2nd order efficiency of certain 1st order efficient tests, but gives no information about the date of this happening.)

When Hodges and Hodges and Lehmann (1970) introduced the concept of "deficiency" to characterize the differences between asymptotically efficient statistical procedures, they accepted off-hand that this required an analysis of order n^{-1}, although one would expect that an analysis of order $n^{-1/2}$ should suffice. The same holds true for authors like Albers (1974) and Does (1982) who computed deficiencies for various types of nonparametric statistics. All these statistics are efficient of order $o(n^0)$, and their efficiency is the same up to terms of order n^{-1}, a fact which none of these authors found worth mentioning. An exception is Bickel and van Zwet (1978, p. 988).

That various asymptotically efficient procedures have the same $n^{-1/2}$-term does not necessarily imply that this term is the optimal one. The situation is different for certain tests. Given a univariate family of probability measures, a bound of order $o(n^{-1/2})$ for the power of tests for the hypothesis $\vartheta = \vartheta_0$ against alternatives $\vartheta_0 + n^{-1/2}t$ may be obtained from an E-expansion based on the Neyman–Pearson Lemma (see e.g. Pfanzagl 1973, p. 1000, Theorem 3(iii) and the references cited there). A corresponding result was obtained for tests of composite hypotheses which are similar of order $o(n^{-1/2})$ (see e.g. Chibisov 1973a, p. 40, Theorem 9.1, or Pfanzagl 1974, p. 208, without proof, 1975, p. 14, and 1979, p. 180, Theorem 6.5). Such results imply that the $n^{-1/2}$-terms, common to all tests which are asymptotically efficient of order $o(n^0)$ and similar of order $o(n^{-1/2})$ are, in fact, optimal of order $o(n^{-1/2})$. In view of the variety of such tests, this was surprising (see the remarks in Chibisov 1973a, p. 16 on $C(\alpha)$-tests or Pfanzagl 1974, p. 224). It was mentioned by Takeuchi and Akahira (1976) [p. 629] for the special case of asymptotically efficient tests based on linear combination of order statistics.

These results on tests carry over to estimator sequences which are median unbiased of order $o(n^{-1/2})$. This leads to a bound of order $o(n^{-1/2})$ for the concentration of estimator sequences which are median unbiased of order $o(n^{-1/2})$ in the sense that, for $a \in \mathbb{R}$,

$$P^{(n)}_{\vartheta+n^{-1/2}a}\{\vartheta^{(n)} \le \vartheta + n^{-1/2}a\} = \frac{1}{2} + o(n^{-1/2})$$

and

$$P^{(n)}_{\vartheta+n^{-1/2}a}\{\vartheta^{(n)} \ge \vartheta + n^{-1/2}a\} = \frac{1}{2} + o(n^{-1/2}).$$

There are no further conditions on the structure of the sequence $(\vartheta^{(n)})_{n\in\mathbb{N}}$.

In contrast to the situation for univariate families, there is no "absolute" bound of order $o(n^{-1/2})$ for the concentration of multivariate estimator sequences. This motivates the study of estimator sequences with S-expansion.

Estimator Sequences With S-Expansions

The next step was to consider estimator sequences $\vartheta^{(n)}$, $n \in \mathbb{N}$, with general S-expansion

$$n^{1/2}(\vartheta^{(n)} - \vartheta) = \tilde{\lambda}(\cdot, \vartheta) + n^{-1/2}Q(\tilde{\lambda}(\cdot, \vartheta), \tilde{g}(\cdot, \vartheta), \vartheta) + o_p(n^{-1/2}), \quad (5.18.4)$$

with functions $g(\cdot,\vartheta) \in L_*(\vartheta)^m$ uncorrelated to $\lambda(\cdot,\vartheta)$. In the usual cases, the components $Q_1(\cdot,\vartheta), \dots Q_k(\cdot,\vartheta)$ of $Q(\cdot,\vartheta)$ are polynomials with coefficients depending on ϑ.

Considering the general S-expansion (5.18.4), one would expect a great variety of possible E-expansions. This is, however, not the case. First of all, the generality of the functions $Q_1, \dots, Q_k$ is spurious. If the S-expansion (5.18.4) holds locally uniformly, more precisely: with ϑ replaced by $\vartheta + n^{-1/2}a$, this restricts the possible forms of the functions Q_r. Replacing ϑ by $\vartheta + n^{-1/2}a$ and considering the resulting relation as an identity in a leads to the following *canonical representation* (see Pfanzagl and Wefelmeyer 1978, p. 18, Lemma 5.12 for more details):

$$n^{1/2}(\vartheta^{(n)} - \vartheta) = S^{(n)}(\cdot,\vartheta) + n^{-1/2}R^{(n)}(\cdot,\vartheta) + o_p(n^{-1/2}), \tag{5.18.5}$$

where $S^{(n)}$ is given by (5.18.2) and the remainder term $R^{(n)}$ is of the type

$$R^{(n)}(\cdot,\vartheta) = R(\tilde{f}(\cdot,\vartheta);\vartheta) \tag{5.18.6}$$

with functions $f(\cdot,\vartheta) \in L_*(\vartheta)^m$ that are uncorrelated to $\lambda(\cdot,\vartheta)$. With the seemingly more general S-expansion (5.18.4) we are in the end back to the S-expansion of the ML-sequence except for the remainder term $R^{(n)}$.

According to Pfanzagl (1979) [p. 182, Theorem 7.4], $P^n_\vartheta \circ S^{(n)}(\cdot,\vartheta)$ has an E-expansion with λ^k-density

$$\varphi_{\Lambda(\vartheta)}(u)(1 + n^{-1/2}G(u,\vartheta)), \tag{5.18.7}$$

where

$$G(u,\cdot) = a_i u_i + c_{ijk}u_i u_j u_k \tag{5.18.8}$$

with

$$a_i = -(\frac{1}{2}L_{i,j,k} + L_{ij,k})\Lambda_{jk} \tag{5.18.9}$$

and

$$c_{ijk} = -(\frac{1}{3}L_{i,j,k} + \frac{1}{2}L_{i,jk}). \tag{5.18.10}$$

Let $\mathscr{E}_0$ denote the class of all estimator sequences for ϑ admitting an E-expansion with λ^k-density (5.18.7), (5.18.8) with c_{ijk} given by (5.18.10) and with generic constants $a_1, \dots, a_k$. These constants become unique under additional conditions on the estimator sequence (say ML or componentwise median unbiased.)

By introducing the class $\mathscr{E}_0$, we avoid building an optimality concept which is confined to "standardized" estimator sequences. After all, there is no convincing standardization of order $o(n^{-1/2})$ for multidimensional estimators (except for component-wise median unbiasedness). Moreover, a standardization of $\vartheta^{(n)}$ as an estimator for ϑ does not carry over to a corresponding standardization for $\kappa \circ \vartheta^{(n)}$ as an estimator for $\kappa(\vartheta)$.

$\mathscr{E}_0$ contains all $o(n^{-1/2})$-optimal estimator sequences with S-expansion. Under suitable regularity conditions, any estimator sequence in $\mathscr{E}_0$ can be transformed in any other estimator sequence in this class.

Remark The constants $a_1, \ldots, a_k$ given by (5.18.9) are specific for ML-sequences. They will be different for other asymptotically efficient estimator sequences.

For the following, it will be of interest that an estimator sequence $\vartheta^{(n)}$, $n \in \mathbb{N}$, with an $n^{-1/2}$-term $G(u, \cdot) = a_i u_i + c_{ijk} u_i u_j u_k$ in the density, can be transformed into an estimator sequence $\hat{\vartheta}^{(n)}$, $n \in \mathbb{N}$, with an $n^{-1/2}$-term $\hat{a}_i u_i + c_{ijk} u_i u_j u_k$ by

$$\hat{\vartheta}^{(n)} := \vartheta^{(n)} + n^{-1} \Lambda_{\bullet j}(\vartheta^{(n)})(a_j(\vartheta^{(n)}) - \hat{a}_j(\vartheta^{(n)}))$$

provided Λ_{ij} are continuous and a_j, $\hat{a}_i$ are Lipschitz.

For later use we remark that estimator sequences with E-expansion and $o(n^{-1/2})$-median unbiased marginals have (see Pfanzagl 1979, p. 183, relation (7.9)) the $n^{-1/2}$-term

$$\begin{aligned} G(u, \cdot) = \Big((L_{i,j,k} + L_{ij,k} + \frac{1}{2} L_{i,jk})\Lambda_{jk} + c_{ijk} L_{r,r} \Lambda_{ir} \Lambda_{jr} \Lambda_{kr} L_{r,s}\Big) u_s \\ + c_{ijk} u_i u_j u_k. \end{aligned} \tag{5.18.11}$$

If the concentration of such estimator sequences in $\mathscr{E}_0$ are compared on sets $C \in \mathscr{C}$ symmetric about 0, the difference of order $n^{-1/2}$ in location cancels out, and

$$P_\vartheta^n\{n^{1/2}(\vartheta^{(n)} - \vartheta) \in C\} = N(0, \Lambda(\vartheta))(C) + o(n^{-1/2}).$$

Since estimator sequences with a stochastic expansion (5.18.4) differ, in the end, by a term $n^{-1/2} R^{(n)}$ only—what is the influence of this term on the E-expansion? Theorem 7.4 in Pfanzagl (1979) [p. 182] asserts that such estimator sequences have, except for a shift of order $n^{-1/2}$, the same E-expansion, hence the same efficiency of order $o(n^{-1/2})$. This was the result which justified the phrase "first order efficiency implies second order efficiency". (The paper Pfanzagl 1979, was already presented at the Meeting of Dutch Statisticians in Lunteren in 1975, but was not been published until 1979 because of a delay in the publication of the Hájek Memorial Volume.)

In the following, this result (i.e., the $o(n^{-1/2})$-equivalence of all $o(n^0)$-efficient estimator sequences) will be established under more general conditions.

We consider estimator sequences with the S-expansion

$$n^{1/2}(\vartheta^{(n)} - \vartheta) = S^{(n)}(\cdot, \vartheta) + n^{-1/2} R^{(n)}(\cdot, \vartheta) + o_p(n^{-1/2}) \tag{5.18.12}$$

with $S^{(n)}$ given by (5.18.2). If this representation holds with ϑ replaced by $\vartheta + n^{-1/2} a$, relation (5.18.3) implies that

$$R^{(n)}(\cdot, \vartheta + n^{-1/2} a) = R^{(n)}(\cdot, \vartheta) + o_p(n^0). \tag{5.18.13}$$

Observe that relation (5.18.13) is, in particular, true for remainder terms of the special type (5.18.6). Since $f_i(\cdot,\vartheta)$ are uncorrelated to $\lambda_1(\cdot,\vartheta),\dots,\lambda_k(\cdot,\vartheta)$, this implies $\tilde f_i(\cdot,\vartheta+n^{-1/2}a)=\tilde f_i(\cdot,\vartheta)+o_p(n^0)$. If $R^{(n)}$ is a continuous function of all its arguments, relation (5.18.13) follows.

We shall show that all estimator sequences with a representation (5.18.12) have the same E-expansion, except for a shift of order $n^{-1/2}$. What makes the whole thing tick is the asymptotic independence between $S^{(n)}$ and $R^{(n)}$. This independence follows since $S^{(n)}$ is the contradiction of an asymptotically sufficient and complete statistic, and $R^{(n)}$ is asymptotically invariant (according to (5.18.13) (compare Basu's Theorem).

For the proof of the following Theorem 5.18.1 we need in addition to (5.18.13) that the sequence $R^{(n)}(\cdot,\vartheta+n^{-1/2}a)$, $n\in\mathbb{N}$, is uniformly $P^n_{\vartheta+n^{-1/2}a}$-integrable, a property which is not inherent in the role of $R^{(n)}$ in the representation (5.18.12). A proof under weaker conditions on $R^{(n)}$ is wanted.

Theorem 5.18.1 *If $R^{(n)}(\cdot,\vartheta+n^{-1/2}a)$, $n\in\mathbb{N}$, is uniformly $P^n_{\vartheta+n^{-1/2}a}$-integrable for each a and $P^n_\vartheta\circ(S^{(n)}(\cdot,\vartheta)+n^{-1/2}R^{(n)}(\cdot,\vartheta))$, $n\in\mathbb{N}$, admits an E-expansion, then this E-expansion differs from the E-expansion of $P^n_\vartheta\circ S^{(n)}(\cdot,\vartheta)$, $n\in\mathbb{N}$, by a shift of order $n^{-1/2}$ only.*

More precisely: Under the conditions indicated above, there exist $\rho(\vartheta)\in\mathbb{R}^k$ such that $P^n_\vartheta\circ(S^{(n)}(\cdot,\vartheta)+n^{-1/2}R^{(n)}(\cdot,\vartheta))$ and $P^n_\vartheta\circ(S^{(n)}(\cdot,\vartheta)+n^{-1/2}\rho(\vartheta))$, $n\in\mathbb{N}$, have the same E-expansion.

Remark If ρ is a continuous function of ϑ, then the estimator sequence

$$\hat\vartheta^{(n)}(\mathbf{x}_n)=\vartheta^{(n)}(\mathbf{x}_n)+n^{-1}\rho(\vartheta^{(n)}(\mathbf{x}_n))$$

has the same E-expansion as the estimator sequence $\vartheta^{(n)}$ with the expansion (5.18.12). This shows, in particular, that the class of all estimator sequences which are modifications of the ML-sequence is complete of order $o(n^{-1/2})$ in the class of all estimator sequences with S-expansions. (Earlier completeness theorems, some of these extending to an order going beyond $o(n^{-1/2})$, can be found in Pfanzagl 1973, Sect. 6, Pfanzagl 1974, Sect. 6 Pfanzagl 1975, p. 34, Theorem 6, and Pfanzagl and Wefelmeyer 1978, p. 7, Theorem 1 (iii).)

Example For $\mathfrak{P}=\{N(0,\sigma^2):\sigma^2>0\}$, the estimator sequences

$$s_n^2(\mathbf{x}_n):=n^{-1}\sum_{\nu=1}^n(x_\nu-\overline{x}_n)^2\quad\text{and}\quad\hat s_n^2(\mathbf{x}_n):=n^{-1}\sum_{\nu=1}^n x_\nu^2$$

are both efficient of order $o(n^0)$, and

$$n^{1/2}(s_n^2(\mathbf{x}_n)-\sigma^2)=n^{1/2}(\hat s_n^2(\mathbf{x}_n)-\sigma^2)+n^{-1/2}R^{(n)}(\mathbf{x}_n),$$

with $R^{(n)}(\mathbf{x}_n) = n\overline{x}_n^2$. This is an S-expansion of $n^{1/2}(s_n^2 - \sigma^2)$ with $S^{(n)}(\mathbf{x}_n, \sigma^2) = n^{1/2}(\hat{s}_n^2(\mathbf{x}_n) - \sigma^2)$ and $R^{(n)}(\mathbf{x}_n) = (\tilde{f}(\mathbf{x}_n))^2$ with $f(x) = x$; observe that the correlation between $S^{(n)}(\cdot, \sigma^2)$ and $R^{(n)}$ is of order $o(n^0)$, which follows in this case immediately from the stochastic independence between $n^{1/2}(s_n^2 - \sigma^2)$ and $R^{(n)}$. The distribution of $n^{1/2}(\hat{s}_n^2 - \sigma^2)$ differs from the distribution of $n^{-1/2}(s_n^2 - \sigma^2)$ by a stochastic term of order $o_p(n^{-1/2})$. $N(0, \sigma^2)^n \circ R^{(n)} = \sigma^2 \chi_1^2$ is non-degenerate, with $\int R^{(n)} dN(0, \sigma^2)^n = \sigma^2$. □

Proof of Theorem 5.18.1. Since $R^{(n)}(\cdot, \vartheta)$, $n \in \mathbb{N}$, is uniformly P_ϑ^n-integrable, the sequence $\int |R^{(n)}(\cdot, \vartheta)| dP_\vartheta^n$, $n \in \mathbb{N}$, is bounded. Hence $\rho_n(\vartheta) := \int R^{(n)}(\cdot, \vartheta) dP_\vartheta^n$ is bounded, too, and converges to some $\rho(\vartheta) \in \mathbb{R}^k$ for some subsequence $\mathbb{N}_0$.

Let $g : \mathbb{R}^k \to \mathbb{R}$ be a bounded function with bounded and continuous derivatives $g^{(i)}$, $i = 1, \ldots, k$, so that

$$\begin{aligned} & g(S^{(n)}(\cdot, \vartheta) + n^{-1/2}(R^{(n)}(\cdot, \vartheta) - \rho(\vartheta))) \\ &= g(S^{(n)}(\cdot, \vartheta)) + n^{-1/2} g_i(S^{(n)}(\cdot, \vartheta))(R_i^{(n)}(\cdot, \vartheta) - \rho_i(\vartheta)) + o_p(n^{-1/2}). \end{aligned}$$

Lemma 5.18.9, applied with $h = g^{(i)}$ and $f_n = R^{(n)} - \rho$, implies that

$$\int g_i(S^{(n)}(\cdot, \vartheta))(R_i^{(n)}(\cdot, \vartheta) - \rho_n(\vartheta)) dP_\vartheta^n = o(n^0).$$

(Since $R^{(n)}$ is uniformly integrable, $R^{(n)} - \rho_n$ is uniformly integrable, too.) Hence

$$\int g(S^{(n)}(\cdot, \vartheta) + n^{-1/2}(R^{(n)}(\cdot, \vartheta) - \rho_n(\vartheta))) dP_\vartheta^n = \int g(S^{(n)}(\cdot, \vartheta)) dP_\vartheta^n + o(n^{-1/2}).$$

Applied with $g(u + n^{-1/2}\rho)$ in place of $g(u)$ this implies for $n \in \mathbb{N}_0$,

$$\begin{aligned} & \int g(S^{(n)}(\cdot, \vartheta) + n^{-1/2} R^{(n)}(\cdot, \vartheta)) dP_\vartheta^n \\ &= \int g(S^{(n)}(\cdot, \vartheta) + n^{-1/2}\rho(\vartheta)) dP_\vartheta^n + o(n^{-1/2}). \end{aligned} \quad (5.18.14)$$

Since $P_\vartheta^n \circ g(S^{(n)} + n^{-1/2} R^{(n)})$ admits an approximation by an E-sequence, the convergence of $(\rho_n)_{n \in \mathbb{N}_0}$ to ρ holds, in fact, for the whole sequence.

So far, the equality (5.18.14) holds for certain functions g only. Yet, the class of these functions is large enough to imply that the E-expansions of $P_\vartheta^n(S^{(n)} + n^{-1/2} R^{(n)})$ and $P_\vartheta^{(n)} \circ (S^{(n)} + n^{-1/2}\rho)$ are identical. □

The results indicated so far hold for estimator sequences with an S-expansion. The approximation of $n^{1/2}(\vartheta^{(n)} - \vartheta)$ by an S-expansion is an important device to obtain an approximation to $P_\vartheta^n \circ n^{1/2}(\vartheta^{(n)} - \vartheta)$ by an E-expansion. The S-expansion itself is of no operational significance. Hence it suggests itself to look for estimator sequences with E-expansion which are better than the best estimator sequences with

S-expansion. The purpose of the following section is to show that there are no E-expansions with a better $n^{-1/2}$-term. This assertion will be supported by two different results.

(i) All sufficiently regular E-expansions have an $n^{-1/2}$ term of the particular type (5.18.7), and the one with $b_{ij} \equiv 0$ is optimal in this class.

(ii) Another result is based on the "absolute" concept of $o(n^{-1/2})$-optimality for univariate functionals. It will be shown that

(i) Expansion (5.18.11) is the only E-expansion for estimator sequences $(\vartheta^{(n)})_{n\in\mathbb{N}}$ with $o(n^{-1/2})$-median unbiased marginals.

(ii) For sufficiently regular functionals $\kappa : \mathbb{R}^k \to \mathbb{R}$, the estimator sequence $\kappa \circ \vartheta^{(n)}$, $n \in \mathbb{N}$, is $o(n^{-1/2})$-optimal if $(\vartheta^{(n)})_{n\in\mathbb{N}}$ has an E-expansion (5.18.8).

E-Expansions Without S-Expansions

It is an operationally significant property of the estimator sequences $\vartheta^{(n)}$, $n \in \mathbb{N}$, that $P_\vartheta^n \circ n^{1/2}(\vartheta^{(n)} - \vartheta)$, $n \in \mathbb{N}$, is approximable by a limit distribution. The corresponding refinement is approximability up to $o(n^{-1/2})$ of $P_\vartheta^n \circ n^{1/2}(\vartheta^{(n)} - \vartheta)$, $n \in \mathbb{N}$, by an E-sequence. As for the approximations by limit distributions, it is some kind of local uniformity which makes this idea operationally meaningful and which restricts the possible forms of E-expansions. The idea of local uniformity was applied for S-expansions, and it can be applied as well to the approximation of $P_\vartheta^n \circ n^{1/2}(\vartheta^{(n)} - \vartheta)$ by a sequence $E_\vartheta^{(n)}$ having a λ^k-density of the type (5.18.7), i.e.,

$$\varphi_{\Lambda(\vartheta)}(u)(1 + n^{-1/2}G(u, \vartheta))$$

with some function $G(\cdot, \vartheta)$. The application of this idea leads to the following result:

Assume that for every loss function $\ell \in L_*$ and some $\delta > \frac{1}{2}$,

$$n^\delta \Big| \int \ell(n^{1/2}(\vartheta^{(n)} - \vartheta)) dP_\vartheta^n - \int \ell(u)(1 + n^{-1/2}G(u, \vartheta))\varphi_{\Lambda(\vartheta)}(u)\lambda^k(du)\Big| = o(n^0) \quad (5.18.15)$$

holds uniformly with ϑ replaced by $\vartheta + n^{-1/2}a$. (Condition (5.18.15) with $\delta > 1/2$ is needed in the proof. What corresponds to "approximation of order $o(n^{-1/2})$" is $\delta = 1/2$.)

The use of loss functions ℓ with bounded and continuous first and second order derivatives is needed since approximations of this order on (say convex) sets would be possible only under additional smoothness conditions on the family (which exclude lattice distributions).

If such an estimator sequence admits an E-expansion (5.18.7), then the function G is necessarily of the following type.

$$G(u, \vartheta) = a_i(\vartheta)u_i + b_{ij}(\vartheta)(u_i u_j - \Lambda_{ij}(\vartheta)) + c_{ijk}(\vartheta)u_i u_j u_k, \quad (5.18.16)$$

where the matrix $(b_{ij})_{i,j=1,\ldots,k}$ is positive semidefinite. Recall that the factor c_{ijk} given by (5.18.10) does not depend on the estimator sequence.

This is the parametric version of a more general theorem proved in Pfanzagl (1985, Sect. 9, pp. 288–332) under the assumption that $G(\cdot,\vartheta)$ is bounded by some polynomial and that it fulfills certain smoothness conditions, such as

$$\int \left|G(u,\vartheta_n)\varphi_{\Lambda(\vartheta_n)}(u) - G(u,\vartheta)\varphi_{\Lambda(\vartheta)}(u)\right|\lambda^k(du) = o(n^0) \quad \text{for} \vartheta_n = \vartheta + n^{-1/2}a$$

and

$$\int \left|G(u+y,\vartheta) - G(u,\vartheta)\right|\varphi_{\Lambda(\vartheta)}(u)\lambda^k(du) = O(|y|) \quad \text{if } y \to 0.$$

If we accept the regularity conditions under which this result was proved, then one might say that for the class of all estimator sequences admitting an E-expansion (5.18.7), $G(\cdot,\vartheta)$ is given by (5.18.16). This class of estimator sequences is, therefore, (i) more general than the class of all estimator sequences with S-expansion, and (ii) the optimal estimator sequences fulfill (5.18.16) with $b_{ij} = 0$ for $i, j = 1, \ldots, k$. Hence, any estimator sequence with S-expansion is optimal in the larger class of estimator sequences the distributions of which is approximable up to the order $o(n^{-1/2})$ in the sense described by (5.18.15).

The optimality of E-distributions with $b_{ij} = 0$ for $i, j = 1, \ldots, k$ is based on Lemma 5.18.2. Since the matrix $(b_{ij}), i, j = 1, \ldots, k$ is positive semidefinite, Lemma 5.18.2 implies that $b_{ij}\int \ell(u)\big(u_iu_j - \Lambda_{ij}(\vartheta)\big)\varphi_{\Lambda(\vartheta)}(u)\lambda^k(du) \geq 0$ for any symmetric and subconvex loss function ℓ.

Lemma 5.18.2 *(i) For any symmetric and subconvex loss function $\ell : \mathbb{R}^k \to [0,\infty)$, the matrix*

$$\int \ell(u)\big(u_iu_j - \Sigma_{ij}\big)\varphi_\Sigma(u)\lambda^k(du), \quad i, j = 1, \ldots, k,$$

is positive semidefinite (see Pfanzagl and Wefelmeyer 1978, p. 16, Lemma 5.8 or Pfanzagl 1985, p. 454, Lemma 13.2.4).

(ii) If $k = 1$,

$$\int \ell(u)(u^2 - \sigma^2)\varphi_{\sigma^2}(u)\lambda(du) \geq 0$$

holds for all loss functions ℓ which are subconvex about 0 (Pfanzagl 1985, Sect. 7.11, p. 231.)

The following examples present estimator sequences with an E-expansion with $b_{ij} \neq 0$. More examples can be found in Pfanzagl (1985, Sect. 9.8, pp. 324–330).

Example Let $\mathfrak{P} = \{N(\mu,\sigma^2) : \mu \in \mathbb{R}, \sigma^2 > 0\}$ be the family of all one-dimensional normal distributions. The problem is to estimate $\kappa(N(\mu,\sigma)) = \sigma^2$.

We define the following sequence of estimators. Given a rational number $a>0$, we divide samples of size n into $k_n = an^{1/2}$ subsamples of size $m_n = a^{-1}n^{1/2}$. (To avoid technicalities, we consider only sample sizes n for which k_n and m_n are integers.) With

$$\overline{x}_i := m_n^{-1} \sum_{\nu=1}^{m_n} x_{(i-1)m_n+\nu}$$

we define

$$\kappa^{(n)}(\mathbf{x}_n) := k_n^{-1}(m_n - 1)^{-1} \sum_{i=1}^{k_n} \sum_{\nu=1}^{m_n} (x_{(i-1)m_n+\nu} - \overline{x}_i)^2.$$

(Intuitively speaking, we compute an estimator for each of the k_n subsamples, and take the arithmetic mean of these estimators.)

A straightforward computation shows that the distribution of $n^{1/2}(\kappa^{(n)} - \sigma^2)$ under $N(\mu, \sigma^2)^n$ admits an E-expansion of order $o(n^{-1/2})$ with Lebesgue density

$$\varphi_{2\sigma^4}(u)\Big(1 + n^{-1/2}\Big(- \sigma^{-2}u + \frac{a}{4}\sigma^{-4}(u^2 - 2\sigma^4) + \frac{1}{6}\sigma^{-6}u^3\Big)\Big).$$

It is easy to check that the coefficient of u^3 agrees with that given in (5.18.10).

The important aspect of this example is the occurrence of a quadratic term with *positive* coefficient, $\frac{a}{4}\sigma^{-4}$. Hence the estimator sequence is first-order efficient but not second-order efficient. The following simpler, if somewhat artificial, example shows the same effect.

Example For $\mathfrak{P} = \{N(\vartheta, 1) : \vartheta \in \mathbb{R}\}$, the sample mean $\overline{x}_n$ is the asymptotically optimal regular estimator sequence for ϑ. The S-expansion of $n^{1/2}(\overline{x}_n - \vartheta)$ is $n^{-1/2} \sum_{\nu=1}^{n} (x_\nu - \vartheta)$; the pertaining E-expansion of order $o(n^{-1/2})$ has λ-density φ. The estimator sequence $\vartheta^{(n)}(\mathbf{x}_n) := \overline{x}_n + n^{-3/4}\delta_n(\mathbf{x}_n)$ is regularly approximated by the E-expansion with density

$$u \to \varphi(u)(1 + n^{-1/2}(u^2 - 1)/2)$$

if $\delta_n = \pm 1$ with probability $1/2$, independently of $\overline{x}_n$.

The results indicated above refer to families fulfilling an LAN-condition (resp. Cramér-type regularity conditions). For "irregular" families, E-expansions of a different type may occur.

Example The location parameter family of Laplace distributions, given by the Lebesgue density

$$x \to \frac{1}{2} \exp[-|x - \theta|], \quad \theta \in \mathbb{R},$$

provides an example in which the ML estimator, i.e., the median, admits a stochastic expansion and is first- but not second-order efficient due to the lack of regularity (the density is not differentiable in θ for $\theta = x$).

It was already observed by Fisher (1925) [pp. 716–717] that in this case the distribution of the ML estimator approaches its limiting distribution particularly slowly. Comparing the variance of the ML estimator with the variance of the limiting distribution, he found that the deficiency is of order $n^{1/2}$ (whereas it is of order n^0 in the regular cases).

This example was taken up by Takeuchi (1974, pp. 188–193; see also Akahira, 1976a, pp. 621–622). For the ML-sequence they obtain the E-expansion with density

$$\varphi(u)\Big(1 + n^{-1/2}|u|\Big(\frac{u^2}{2} - 1\Big)\Big)$$

and the distribution function

$$t \to \phi(t) - n^{-1/2}\frac{1}{2}t^2 \operatorname{sign} t\ \varphi(t) + o(n^{-1/2}).$$

Since the ML estimators are median unbiased, it suggests itself to compare this distribution function with the bound of order $o(n^{-1/2})$ for median unbiased estimator sequences, given as

$$t \to \phi(t) - n^{-1/2}\frac{1}{6}t^2 \operatorname{sign} t\ \varphi(t) + o(n^{-1/2}).$$

This bound is attainable in the following sense: For each $s \in \mathbb{R}$ there exists an estimator sequence, median unbiased of order $o(n^{-1/2})$, such that its distribution function coincides with the bound up to $o(n^{-1/2})$ for $t = s$. Hence we have a whole family of first-order efficient estimator sequences with distributions differing by amounts of order $n^{-1/2}$. (For regular families, this phenomenon does not occur until the order n^{-1}.)

"Absolute" Optimality Based on Tests

The purpose of this section is to establish the optimality of E-expansions in $\mathscr{E}_0$. Our starting point is the "absolute" optimality concept for univariate estimator sequences based on the Neyman–Pearson Lemma. Starting from this "absolute" optimality-concept for univariate estimator sequences, a concept of multivariate $o(n^{-1/2})$-optimality will be developed.

The following theorem is a first step towards this "absolute" optimality concept. It provides an "absolute" bound of order $o(n^{-1/2})$ for the concentration of estimator sequences $\kappa^{(n)}$, $n \in \mathbb{N}$, for $\kappa(\vartheta)$ which are median unbiased of order $o(n^{-1/2})$.

Theorem 5.18.3 *Let $\{P_\vartheta : \vartheta \in \Theta\}$, $\Theta \subset \mathbb{R}^k$, be a parametric family fulfilling the conditions indicated above. Let $\kappa : \Theta \to \mathbb{R}$ be a twice continuously differentiable functional. Let $\kappa^{(n)} : X^n \to \mathbb{R}$, $n \in \mathbb{N}$, be an estimator sequence for $\kappa(\vartheta)$ such that*

$$P^m_{\vartheta+n^{-1/2}a}\{\kappa^{(n)} \le \kappa(\vartheta + n^{-1/2}a)\} = \beta + o(n^{-1/2}) \quad \textit{for} \quad a \in \mathbb{R}^k. \qquad (5.18.17)$$

Then the following is true:

$$P_\vartheta^n\{n^{1/2}(\kappa^{(n)} - \kappa(\vartheta)) \le t\}$$
$$\begin{matrix}\le\\ \ge\end{matrix} \Phi\Big(\Phi^{-1}(\beta) + \frac{t}{\sigma} + n^{-1/2}\Big(\sigma^{-3}\frac{1}{6}L_{i,j,k}K_iK_jK_k\Phi^{-1}(\beta)\frac{t}{\sigma} - \bar{c}\frac{t^2}{\sigma^2}\Big)\Big) + o(n^{-1/2})$$
$$\text{for } t \begin{matrix}>\\ <\end{matrix} 0, \tag{5.18.18}$$

with

$$\sigma^2 := \Lambda_{ij}\kappa^{(i)}\kappa^{(j)} = L_{i,j}K_iK_j, \quad K_i := \Lambda_{ij}\kappa^{(j)},$$
$$\bar{c} := \sigma^{-3}\Big(c_{ijk}K_iK_jK_k + \frac{1}{2}\kappa^{(ij)}K_iK_j\Big).$$

This theorem provides a bound for the concentration of confidence bounds with covering probability β. Of interest for our problem is the case $\beta = 1/2$: For estimator sequences $\kappa^{(n)}$, $n \in \mathbb{N}$, which are median unbiased of order $o(n^{-1/2})$, relation (5.18.18) implies that $o(n^{-1/2})$-optimal estimator sequences have an E-expansion with distribution function

$$\Phi\Big(\frac{t}{\sigma} - n^{-1/2}\bar{c}\frac{t^2}{\sigma^2}\Big). \tag{5.18.19}$$

For later use we note the pertaining density

$$\varphi_{\sigma^2}(t)\Big(1 + n^{-1/2}\bar{c}\Big(-2\frac{t}{\sigma} + \frac{t^3}{\sigma^3}\Big)\Big). \tag{5.18.20}$$

For $o(n^{-1/2})$-median unbiased estimator sequences $\kappa^{(n)}$, $n \in \mathbb{N}$, relation (5.18.19) implies

$$P_\vartheta^n\{-t' \le n^{1/2}(\kappa^{(n)} - \kappa(\vartheta)) \le t''\} \tag{5.18.21}$$
$$\le N(0,\sigma^2)\Big(-t' - n^{-1/2}\bar{c}\frac{t'^2}{\sigma}, t'' - n^{-1/2}\bar{c}\frac{t''^2}{\sigma}\Big) + o(n^{-1/2}) \quad \text{for } t', t'' \ge 0.$$

Observe that this bound is absolute in the sense that it holds for every estimator sequence which is median unbiased of order $o(n^{-1/2})$ (without any condition like S-expansion).

Specializing for $\kappa(\vartheta_1, \dots, \vartheta_k) = \vartheta_r$ we obtain the following bound for estimator sequences $\vartheta_r^{(n)}$, $n \in \mathbb{N}$, which are median unbiased of order $o(n^{-1/2})$ for ϑ_r.

$$P_\vartheta^n\{-t' \le n^{-1/2}(\vartheta_r^{(n)} - \vartheta_r) \le t''\} \tag{5.18.22}$$
$$\le N(0,\sigma_r^2)\Big(-t' - n^{-1/2}\bar{c}_r\frac{t'^2}{\sigma}, t'' - n^{-1/2}\bar{c}_r\frac{t''^2}{\sigma}\Big) + o(n^{-1/2}) \quad \text{for } t', t'' \ge 0$$

with $\sigma_r^2 = \Lambda_{rr}$ and $\bar{c}_r := c_{ijk}\Lambda_{ir}\Lambda_{jr}\Lambda_{kr}\Lambda_{rr}^{-2}$. (No summation over r.)

For $k = 1$, relation (5.18.22) occurs in Pfanzagl (1973) [p. 1005, Theorem 6.1, relation (6.4)]. A relation more general than (5.18.22) is given in Pfanzagl (1985, pp. 250/1, Theorem 8.2.3. Observe a misprint in 8.2.4: replace u by $n^{-1/2}u$).

Proof of Theorem 5.18.3. By condition (5.18.17), assumption (5.18.26) of Theorem 5.18.6 below is fulfilled for

$$\varphi_n(\mathbf{x}_n) = 1_{(-\infty,\kappa(\vartheta+n^{-1/2}a)]}(\kappa^{(n)}(\mathbf{x}_n)).$$

The inequality (5.18.18) with $\leq$ follows if we apply assertion (5.18.27) with

$$a_r = t\frac{K_r}{\sigma^2}\Big(1 - n^{-1/2}\frac{t}{2}\frac{\kappa^{(ij)}K_iK_j}{\sigma^4}\Big).$$

This choice of a_r, $r = 1, \dots, k$, minimizes $L_{i,j}a_ia_j$ under the condition

$$n^{1/2}(\kappa(\vartheta + n^{-1/2}a) - \kappa(\vartheta)) = t + o(n^{-1/2}).$$

□

The bound in (5.18.21) for the concentration of $o(n^{-1/2})$-optimal estimator sequences presumes a certain smoothness of $\kappa(\vartheta)$. For such functionals, $o(n^{-1/2})$-optimal estimator sequences $\kappa^{(n)}$ can easily be obtained as $\kappa^{(n)} = \kappa \circ \vartheta^{(n)}$, if $\vartheta^{(n)}$, $n \in \mathbb{N}$, is in $\mathscr{E}_0$. The bound refers to estimator sequences $\kappa^{(n)}$ which are median unbiased of order $o(n^{-1/2})$. It may be achieved by $\kappa \circ \vartheta^{(n)}$ if $\vartheta^{(n)} \in \mathscr{E}_0$ is appropriately chosen. The important point: If $\vartheta^{(n)}$ is in $\mathscr{E}_0$, then $\kappa \circ \vartheta^{(n)}$ will be $o(n^{-1/2})$-optimal.

Observe the following detail: The bound given by (5.18.21) is "absolute", i.e., it holds for any estimator sequence which is median unbiased of order $o(n^{-1/2})$. To show that $\kappa \circ \vartheta^{(n)}$ is $o(n^{-1/2})$ optimal we make use of the fact that, for sufficiently regular families, the estimator sequences $\vartheta^{(n)}$ admit an S-expansion.

Theorem 5.18.4 *For any functional $\kappa : \Theta \to \mathbb{R}$ with a continuous second derivative the following is true.*

(i) There exists $(\vartheta^{(n)})_{n\in\mathbb{N}} \in \mathscr{E}_0$ such that $\kappa \circ \vartheta^{(n)}$, $n \in \mathbb{N}$, is median unbiased of order $o(n^{-1/2})$.
(ii) Any of these estimator sequences is $o(n^{-1/2})$-optimal in the absolute sense.

Proof Let $\vartheta^{(n)}$, $n \in \mathbb{N}$, be an estimator sequence in $\mathscr{E}_0$ with

$$G(u) = a_iu_i + c_{ijk}u_iu_ju_k.$$

From the approximation of

$$n^{1/2}(\kappa \circ \vartheta^{(n)} - \kappa(\vartheta)) = \kappa^{(i)}(\vartheta)n^{1/2}(\vartheta_i^{(n)} - \vartheta_i) + n^{-1/2}\frac{1}{2}\kappa^{(ij)}(\vartheta)n^{1/2}(\vartheta_i^{(n)} - \vartheta_i)(\vartheta_j^{(n)} - \vartheta_j) + o_p(n^{-1/2})$$

we obtain for $n^{1/2}(\kappa \circ \vartheta^{(n)} - \kappa(\vartheta))$ the E-expansion

$$E^{(n)} := \overline{E}^{(n)} \circ (u \to \kappa^{(i)} u_i + n^{-1/2} \frac{1}{2} \kappa^{(ij)} u_i u_j)$$

if $\overline{E}^{(n)}$ is the E-expansion of $n^{1/2}(\vartheta^{(n)} - \vartheta)$. By Lemma 5.18.7, the E-expansion $E^{(n)}$ has the λ-density

$$\varphi_{\sigma^2}(t)\Big(1 + n^{-1/2}\Big(\bar{a}\frac{t}{\sigma} + \bar{c}\frac{t^3}{\sigma^3}\Big)\Big) \tag{5.18.23}$$

with

$$\begin{aligned}
\sigma^2 &= L_{i,j} K_i K_j, \\
\bar{a} &= \sigma^{-3}(a_i K_i + 3c_{ijk}(\Lambda_{ij} K_k - K_i K_j K_k), \\
\bar{c} &= \sigma^{-3}(c_{ijk} K_i K_j K_k + \frac{1}{2}\kappa^{(ij)} K_i K_j).
\end{aligned}$$

The estimator sequence $\kappa \circ \vartheta^{(n)}$ is median unbiased of order $o(n^{-1/2})$ if (5.18.23) holds with $\bar{a} = -2\bar{c}$. This can be achieved by the choice of $(a_1, \ldots, a_k)$. That means: By choosing $(\vartheta^{(n)})_{n\in\mathbb{N}} \in \mathscr{E}_0$ appropriately, one obtains an estimator sequence $\kappa \circ \vartheta^{(n)}$ which is median unbiased of order $o(n^{-1/2})$. Any of these estimator sequences has an E-expansion with λ-density (5.18.20) and is, therefore, $o(n^{-1/2})$-optimal. □

Recall that our purpose is to find an operational concept of $o(n^{-1/2})$-optimality for estimators of multivariate functionals. What is it that qualifies the estimator sequences in $\mathscr{E}_0$ as $o(n^{-1/2})$-optimal? According to Theorem 5.18.4, the estimator sequences in $\mathscr{E}_0$ are a reservoir for generating univariate estimator sequences which are $o(n^{-1/2})$-optimal in an absolute sense. Conversely, any estimator sequence $\vartheta^{(n)}$, $n \in \mathbb{N}$, with this property, belongs to $\mathscr{E}_0$.

It would be nice to have a result of the following kind. If, for every $\alpha \in \mathbb{R}^k$, the estimator sequence $\alpha_i \vartheta_i^{(n)}$ is $o(n^{-1/2})$-optimal for $\alpha_i \vartheta_i$, then $\vartheta^{(n)}$ is in $\mathscr{E}_0$. Yet, this does not work with an optimality concept based on median unbiasedness of order $o(n^{-1/2})$: In general, there is no sequence $\vartheta^{(n)}$ such that

$$P_\vartheta^n\{\alpha_i \vartheta_i^{(n)} \le \alpha_i \vartheta_i\} = \frac{1}{2} + o(n^{-1/2}) \quad \text{for every} \quad \alpha \in \mathbb{R}^k.$$

(See the example in Pfanzagl 1979, pp. 183/4, or Pfanzagl 1985, p. 218, Example 7.8.2.)

Even though there is no estimator sequence $\vartheta^{(n)}$ such that $\alpha_i \vartheta_i^{(n)}$ is $o(n^{-1/2})$-median unbiased for $\alpha_i \vartheta_i$ for every $\alpha \in \mathbb{R}^k$, there are estimator sequences fulfilling this condition for $\alpha_r = \delta_r$, $r = 1, \ldots, k$, i.e., estimator sequences with $o(n^{-1/2})$-median unbiased components.

An interesting converse supporting the interpretation of $\mathscr{E}_0$ as the class of all $o(n^{-1/2})$-optimal estimator sequences reads as follows: Assume that the estimator sequence $\vartheta^{(n)} = (\vartheta_1^{(n)}, \ldots, \vartheta_k^{(n)})$, $n \in \mathbb{N}$, admits an E-expansion with a regular but otherwise unspecified term G. If all marginals of $\vartheta^{(n)}$ are median unbiased of order $o(n^{-1/2})$ and $o(n^{-1/2})$-optimal, then G is uniquely determined (and $(\vartheta^{(n)})_{n\in\mathbb{N}}$ an element of $\mathscr{E}_0$). (Hint: Under suitable regularity conditions, $G(u) = a_i u_i + b_{ij}(u_i u_j - \Lambda_{ij}) + c_{ijk} u_i u_j u_k$. Now $o(n^{-1/2})$-optimality implies $b_{ij} = 0$, and median unbiasedness of order $o(n^{-1/2})$ determines the constants $a_1, \ldots, a_k$.)

Auxiliary Results

The following results hold for highly regular parametric families $\{P_\vartheta^n : \vartheta \in \Theta\}$, $\Theta \subset \mathbb{R}^k$, for which all integrals occurring below are defined. The dependence of $L_{i,j,k}$ etc. on ϑ will be omitted in our notations.

Lemma 5.18.5

$$P_\vartheta^n\Big\{\sum_{\nu=1}^n \log(p(x_\nu, \vartheta + n^{-1/2}a)/p(x_\nu\vartheta)) \le t\Big\} \tag{5.18.24}$$
$$= \Phi\Big(\frac{t}{\tau} + \frac{\tau}{2} + n^{-1/2}(A_{ijk} + tB_{ijk} + t^2 C_{ijk})\frac{a_i a_j a_k}{\tau^3}\Big) + o(n^{-1/2}).$$

$$P_{\vartheta+n^{-1/2}a}^n\Big\{\sum_{\nu=1}^n (\log p(x_\nu, \vartheta + n^{-1/2}a)/p(x_\nu, \vartheta)) \le t\Big\} \tag{5.18.25}$$
$$= \Phi\Big(\frac{t}{\tau} - \frac{\tau}{2} + n^{-1/2}(A'_{ijk} + tB'_{ijk} + t^2 C_{ijk})\frac{a_i a_j a_k}{\tau^3}\Big) + o(n^{-1/2}).$$

In these relations, the following notations are used.

$$\tau^2 = L_{i,j} a_i a_j, \quad C_{ijk} = -\frac{1}{6} L_{i,j,k}/\tau^2,$$
$$A_{ijk} = \frac{1}{6} L_{i,j,k} + \frac{1}{12}\Big(\frac{1}{2} L_{i,j,k} - L_{ijk}\Big)\tau^2, \quad A'_{ijk} = \frac{1}{6} L_{i,j,k} - \frac{1}{12}\Big(\frac{1}{2} L_{i,j,k} - L_{ijk}\Big)\tau^2,$$
$$B_{ijk} = \frac{1}{6} L_{ijk}, \quad B'_{ijk} = -\frac{1}{6}(L_{i,j,k} - L_{ijk}).$$

Theorem 5.18.6 *Let $\varphi_n : X^n \to [0, 1]$ be a critical function. Assume that*

$$\beta_n := \int \varphi_n dP_{\vartheta+n^{-1/2}a}^n, \quad n \in \mathbb{N}, \quad \textit{is bounded away from 0 and 1.} \tag{5.18.26}$$

Then the following is true.

$$\int \varphi_n dP_\vartheta^n \le \Phi\Big(\Phi^{-1}(\beta_n) + \tau$$
$$+ n^{-1/2}\Big(\frac{1}{6} L_{i,j,k}\Phi^{-1}(\beta_n)\tau + \frac{1}{6}(L_{i,j,k} - L_{ijk})\tau^2\Big)\frac{a_i a_j a_k}{\tau^3}\Big) + o(n^{-1/2}).$$

For $\Theta \subset \mathbb{R}$, this relation occurs in Pfanzagl (1973, p. 1003, Theorem 5(i)) or Pfanzagl (1974) [pp. 213/5, Theorem (iv) and pp. 260/1, Proposition].

Proof With $q_n \in dP^n_{\vartheta+n^{-1/2}a}/dP^n_\vartheta$, use relation (5.18.25) to determine $r_n > 0$ such that $P^n_{\vartheta+n^{-1/2}a}\{q_n \le r_n\} = \beta_n + o(n^{-1/2})$. Since $(\varphi_n - 1_{\{q_n \le r\}})(r_n - q_n) \le 0$, this choice of r_n implies that $P^n_\vartheta(\varphi_n) \le P^n_\vartheta\{q_n \le r_n\} + o(n^{-1/2})$. The assertion follows if $P^n_\vartheta\{q_n \le r_n\}$ is approximated by means of relation (5.18.24). □

Lemma 5.18.7 *Assume that $Q_n|\mathbb{B}^k$ has the λ^k-density*

$$u \to \varphi_\Lambda(u)(1 + n^{-1/2}G(u))$$

with $G(u) = a_i u_i + c_{ijk}u_i u_j u_k$. Then

$$Q_n \circ (u \to \alpha_i u_i + n^{-1/2}\beta_{ij}u_i u_j)$$

is up to $o(n^{-1/2})$ approximable by $\overline{Q}_n|\mathbb{B}$ with λ-density

$$v \to \varphi_{\sigma^2}(v)(1 + n^{-1/2}\bar{G}(v)), \tag{5.18.27}$$

where

$$\bar{G}(v) = \bar{a}\frac{v}{\sigma} + \bar{c}\frac{v^3}{\sigma^3} \tag{5.18.28}$$

with

$$\sigma^2 = \Lambda_{ij}\alpha_i\alpha_j, \tag{5.18.29}$$
$$\bar{a}\sigma^3 = \Lambda_{ij}a_i\alpha_j + 3c_{ijk}(\Lambda_{ij}\Lambda_{kr}\alpha_r - \Lambda_{ir}\Lambda_{js}\Lambda_{kt}\alpha_r\alpha_s\alpha_t), \tag{5.18.30}$$
$$\bar{c}\sigma^3 = c_{ijk}\Lambda_{ir}\Lambda_{js}\Lambda_{kt}\alpha_r\alpha_s\alpha_t + \beta_{ij}\Lambda_{ir}\Lambda_{js}\alpha_r\alpha_s. \tag{5.18.31}$$

Proof By definition of $\overline{Q}_n$, the relation

$$\int h(\alpha_i u_i + n^{-1/2}\beta_{ij}u_i u_j)\varphi_\Lambda(u)(1 + n^{-1/2}G(u))\lambda^k(du)$$
$$= \int h(v)\varphi_{\sigma^2}(v)(1 + n^{-1/2}\bar{G}(v))\lambda(dv) + o(n^{-1/2})$$

holds for every continuous $N(0, \Lambda)$-integrable function h. The relations (5.18.27)–(5.18.31) follow from an application with $h(v) = \exp[itv]$. (For a similar proof see Pfanzagl 1985, p. 468, Addendum to Lemma 13.5.12.) □

Lemma 5.18.8 *Assume that $P^n \circ S^{(n)}$ and $P^n \circ \hat{S}^{(n)}$, $n \in \mathbb{N}$, admit E-expansions with λ^k-densities $\varphi_\Lambda(u)(1 + n^{-1/2}G(u))$ and $\varphi_\Lambda(u)(1 + n^{-1/2}\hat{G}(u))$, respectively. If*

$$P^n \circ \{S_i^{(n)} \le t\} \Delta \{\hat{S}_i^{(n)} \le t\} = o(n^{-1/2}) \quad \textit{for } t \in \mathbb{R} \textit{ and } i = 1, \ldots, k, \qquad (5.18.32)$$

then the two E-expansions are identical.

Proof Since

$$\{S_i^{(n)} \le t_i \text{ for } i = 1, \ldots, k\} \Delta \{\hat{S}_i^{(n)} \le t_i \text{ for } i = 1, \ldots, k\} \subset \bigcup_{i=1}^{k} \{S_i^{(n)} \le t_i\} \Delta \{\hat{S}_i \le t_i\},$$

relation (5.18.32) implies that

$$P^n(\{S_i^{(n)} \le t_i \text{ for } i = 1, \ldots, k\} \Delta \{\hat{S}_i^{(n)} \le t_i \text{ for } i = 1, \ldots, k\}) = o(n^{-1/2}).$$

Hence the pertaining E-expansions $E^{(n)}$ and $\overline{E}^{(n)}$ agree for every $n \in \mathbb{N}$ on all rectangles. For $B \in \mathbb{B}^k$ let

$$\mu(B) := \int 1_B(u) G(u) N(0, \Lambda)(du),$$

$$\hat{\mu}(B) := \int 1_B(u) \hat{G}(u) N(0, \Lambda)(du).$$

Since the signed measures μ and $\hat{\mu}$ agree on all rectangles, they agree on $\mathbb{B}^k$. □

A Lemma on Completeness and Independence

Lemma 5.18.9 below refers to the following framework.

For $a \in A \subset \mathbb{R}^k$ let P_a be a probability measure on the measurable space $(X, \mathscr{A})$. Let $T_n : X^n \to \mathbb{R}^m$ be $\mathscr{A}^n, \mathbb{B}^m$-measurable. Assume that T_n is asymptotically sufficient in the sense that for every $a \in A$ there is $p(\cdot, a) : \mathbb{R}^m \to [0, \infty)$ such that

$$P_a^n(B) = \int 1_B(\mathbf{x}_n) p(T_n(\mathbf{x}_n), a) P_0^n(d\mathbf{x}_n) + o(n^0) \quad \text{for} \quad B \in \mathscr{A}^n,$$

i.e., $p(T_n(\cdot), a)$ is asymptotically a P_0^n-density of $P_a^n | \mathscr{A}^n$. Moreover, the family with P_0^n-densities $p(T_n(\cdot), a)$, $a \in A$, is assumed to be asymptotically complete in the sense that, for $Q = P_0^n \circ T_n$ and each measurable $g : \mathbb{R}^m \to \mathbb{R}$,

$$\int g(u) p(u, a) Q(du) = 0 \text{ for } a \in A \quad \text{implies} \quad g = 0 \ Q\text{-a.e.}$$

The interesting application is with $p(u, a) = \exp[a^\top u - \frac{1}{2} a^\top \Lambda a]$.

Lemma 5.18.9 *If $f_n : X^n \to \mathbb{R}$ is $\mathscr{A}^n, \mathbb{B}$-measurable and fulfills*

$$\int f_n dP_a^n = o(n^0) \quad \text{for every} \quad a \in A, \tag{5.18.33}$$

then

$$\int f_n h \circ T_n dP_0^n = o(n^0)$$

for every bounded and continuous function $h : \mathbb{R}^m \to \mathbb{R}$. (5.18.34)

For the proof we need in addition to the essential condition (5.18.33) the following technical condition: f_n, $n \in \mathbb{N}$, is uniformly integrable with respect to P_a^n, i.e., for every $\varepsilon > 0$ there exists $t_{a,\varepsilon}$ such that $\int |f_n| 1_{[t_{a,\varepsilon},\infty)}(|f_n|) dP_a^n < \varepsilon$ for $n \in \mathbb{N}$.

Remark Notice that Lemma 5.18.9 is closely related to Basu's Theorem, which asserts that completeness implies stochastic independence of ancillary statistics. Asymptotic ancillarity could be defined as

$$P_a^n \circ f_n = P_0^n \circ f_n + o(n^0) \quad \text{for } a \in A.$$

Under the weaker condition

$$\int f_n dP_a^n = \int f_n dP_0^n + o(n^0) \quad \text{for} \quad a \in A,$$

relation (5.18.34) applied with f_n replaced by $f_n - \int f_n dP_0^n$, yields

$$\int f_n h \circ T_n dP_0^n = \int f_n dP_0^n \int h \circ T_n dP_0^n + o(n^0),$$

an asymptotic version of the stochastic independence between f_n and T_n.

Proof Let $\tilde{f}_n^+ := f_n 1_{(0,\infty)}(f_n)$ and $\tilde{f}_n^- := -f_n 1_{(-\infty,0)}(f_n)$. Let $\mathbb{N}_0$ be a subsequence such that $c_n^+ := \int f_n^- dP_0^n$, $n \in \mathbb{N}$, converges to some $c^+ \geq 0$. If $c^+ > 0$, we define the probability measures $Q_n^+ | \mathbb{B}^m$ and $Q_n^- | \mathbb{B}^m$ by

$$Q_n^+(B) := \int 1_B \tilde{f}_n^+ dP_0^n / c_n^+, \quad B \in \mathbb{B}^m,$$

and

$$Q_n^-(B) := \int 1_B \tilde{f}_n^- dP_0^n / c_n^-, \quad B \in \mathbb{B}^m.$$

Let $f_n^+ := \tilde{f}_n^+/c_n^+$ and $f_n^- := \tilde{f}_n^-/c_n^-$. Then

$$\int f_n^+ dP_0^n = \int f_n^- dP_0^n = 1.$$

Since $c_n^+ - c_n^- = o(n^0)$, assumption (5.18.33) implies

$$\int f_n^+ dP_a^n = \int f_n^- dP_a^n + o(n^0) \quad \text{for} \quad a \in A.$$

Let $G_n^+ \mathbb{B}^k$ be the probability measure defined by

$$G_n^+(B) = \int f_n^+(x) 1_B(T_n(x)) P_0^n(dx), \quad B \in \mathbb{B}^m. \tag{5.18.35}$$

If f_n^+ is uniformly P_a^n-integrable for every $a \in A$, there exists for every $\varepsilon > 0$ a number $s_{a,\varepsilon} > 0$ such that

$$\int f_n^+ 1_{[s_{a,\varepsilon},\infty)}(f_n^+) dP_a^n < \varepsilon \quad \text{for} \quad n \in \mathbb{N}.$$

Since

$$\begin{aligned}
&\int f_n^+ 1_{[t,\infty)}(|T_n|) dP_a^n \\
&= \int f_n^+ 1_{\{f_n^+ > s_{a,\varepsilon}\}} 1_{[t,\infty)}(|T_n|) dP_a^n + \int f_n^+ 1_{\{f_n^+ \leq s_{a,\varepsilon}\}} 1_{[t,\infty)}(|T_n|) dP_a^n \\
&\leq \int f_n^+ 1_{\{f_n^+ > s_{a,\varepsilon}\}} dP_0^n + s_{a,\varepsilon} P_a^n\{|T_n| > t\},
\end{aligned}$$

the convergence $P_a^n \circ T_n \Rightarrow N(a, \Lambda)$ implies for every fixed neighbourhood A_0 of 0: For every $\varepsilon > 0$ there exists t_ε such that

$$\int f_n^+ 1_{[t_\varepsilon,\infty)}(|T_n|) dP_a^n < \varepsilon \quad \text{for} \quad a \in A_0 \text{ and } n \in \mathbb{N}. \tag{5.18.36}$$

Applied with $a = 0$ this implies (see (5.18.35)) that

$$G_n^+\{u \in \mathbb{R}^m : |u| > t_\varepsilon\} < \varepsilon \quad \text{for} \quad n \in \mathbb{N},$$

i.e., the sequence G_n^+, $n \in \mathbb{N}$, is tight. By Prohorov's Theorem there exists a probability measure $G_*^+|\mathbb{B}^m$ and a subsequence $\mathbb{N}_0$ such that

$$G_n^+ \Rightarrow_{n \in \mathbb{N}_0} G_*^+.$$

By relation (5.18.35),

$$\int f_n^+(x)h(T_n(x))P_0^n(dx) \to_{n\in\mathbb{N}_0} \int h(u))G_*^+(du)$$

for every bounded and continuous function $h|\mathbb{R}^m$.

Since $\int f_n^+ P_a^n - \int f_n^- dP_a^{(n)} = o(n^0)$, relation (5.18.36), applied with f_n^+ and f_n^-, implies

$$\left|\int f_n^+ 1_{[-t,t]}(T_n) dP_a^n - \int f_n^- 1_{[-t,t]}(T_n) dP_a^n\right| < 2\varepsilon \tag{5.18.37}$$

for $n \in \mathbb{N}$ if $t > t_\varepsilon$.

Since

$$\int f_n^\pm 1_{[-t,t]}(T_n) dP_a^n = \int p(u,a) 1_{[-t,t]}(u) G_n^\pm(du),$$

relation (5.18.37) may be rewritten as

$$\left|\int p(u,a)1_{[-t,t]}(u)G_n^+(du) - \int p(u,a)1_{[-t,t]}(u)G_n^-(du)\right| < 2\varepsilon f$$

or $n \in \mathbb{N}$ and $t > t_\varepsilon$.

Since $u \to p(u,a)1_{[-t,t]}(u)$ is bounded and continuous λ^m-a.e.,

$$\lim_{n\in\mathbb{N}_0} \int p(u,a)1_{[-t,t]}(u)G_n^\pm(du) = \int p(u,a)1_{[-t,t]}(u)G_*^\pm(du),$$

hence

$$\left|\int p(u,a)1_{[-t,t]}(u)G_*^+(du) - \int p(u,a)1_{[-t,t]}(u)G_*^-(du)\right| < 3\varepsilon$$

for $a \in A_0$ and $t > t_\varepsilon$.

This implies $G_*^+ = G_*^-$. Hence relation (5.18.35) implies

$$\lim_{n\in\mathbb{N}_0} \left(\int f_n^+(x)h(T_n(x))P_0^n(dx) - \int f_n^-(x)h(T_n(x))P_0^n(dx)\right) = 0$$

for any bounded and continuous h, which implies

$$\lim_{n\in\mathbb{N}_0} \int f_n(x)h(T_n(x))P_0^n(dx) = 0. \tag{5.18.38}$$

Since $\mathbb{N}_0$ may be considered as the subsequence of an arbitrary subsequence, (5.18.38) holds with $\mathbb{N}_0$ replaced by $\mathbb{N}$. □

5.19 Asymptotic Confidence Procedures

Limit distributions provide a first information about the quality of estimator sequences for large samples. Let $\mathfrak{P}|\mathscr{A}$ be a family of probability measures, and let $\kappa : \mathfrak{P} \to \mathbb{R}^k$ be a k-dimensional functional. Let $\kappa^{(n)}$ and $\hat{\kappa}^{(n)}$ be estimators for $\kappa(P)$ such that $P^n \circ c_n(\kappa^{(n)} - \kappa(P)) \Rightarrow Q_P$ and $P^n \circ c_n(\hat{\kappa}^{(n)} - \kappa(P)) \Rightarrow \hat{Q}_P$. If the limit distribution $\hat{Q}_P$ is more concentrated about 0 than Q_P, this suggests to use $\hat{\kappa}^{(n)}$ rather than $\kappa^{(n)}$. Since we do not know the true P, the situation is unclear if it happens that $\hat{Q}_P$ is better than Q_P for P in some subset $\mathfrak{P}_0$ of $\mathfrak{P}$ but worse for P outside $\mathfrak{P}_0$.

If an estimate $\kappa^{(n)}(\mathbf{x}_n)$ has been observed—what does this tell us about the "true" value of the functional $\kappa(P)$? Even if the true distribution of $n^{1/2}(\kappa^{(n)} - \kappa(P))$ under P^n is known, there is no satisfactory answer, because a satisfactory answer must be independent of the unknown P. In special cases, e.g. for real-valued functionals $\kappa : \mathfrak{P} \to \mathbb{R}$, a satisfactory answer can be given by a confidence bound $K^{(n)} : X^n \to \mathbb{R}$ such that $\{\kappa(P) \le K^n(\mathbf{x}_n)\}$ holds under P^n with high probability. An asymptotic solution to this problem can be based on the limit distribution $Q_P|\mathbb{B}$.

(i) Given $\beta \in (0, 1)$, determine the quantile $t_\beta(P)$ defined by

$$Q_P(-\infty, t_\beta(P)] = \beta.$$

The quantile $t_\beta(P)$ is uniquely determined if Q_P has a positive Lebesgue density. The map $P \to t_\beta(P)$ is continuous since $P \to Q_P(-\infty, t]$ is continuous for every $t \in \mathbb{R}$.

(ii) Find a consistent estimator sequence $t_\beta^{(n)}$ for $t_\beta(P)$. Then

$$P^n\{\mathbf{x}_n \in X^n : \kappa(P) \le \kappa^{(n)}(\mathbf{x}_n) + nc_n^{-1} t_\beta^{(n)}(\mathbf{x}_n)\} \to \beta. \tag{5.19.1}$$

The convergence in (5.19.1) should hold uniformly for $P \in \mathfrak{P}$. Regular convergence on parametric subfamilies is all one can achieve through the LAN approach. Confidence bounds are impossible for discontinuous functionals. See Pfanzagl (1998). For more historical remarks on the concept of confidence bounds see Pfanzagl (1994), p. 159.

The approach via quantile estimators generalizes to k-dimensional functionals $\kappa : \mathfrak{P} \to \mathbb{R}^k$. Suppose that $P^n \circ c_n(\kappa^{(n)} - \kappa(P)) \Rightarrow Q_P$. For each P choose a set C_P such that $Q_P(C_P) \ge 1 - \alpha$. If the set C_P depends continuously on P in an appropriate sense and $P^{(n)}$ is a consistent estimator for P, say the empirical estimator, then $K^{(n)} = \kappa^{(n)} + c_n^{-1} C_{P^{(n)}}$ is a confidence procedure for κ with asymptotic confidence level $1 - \alpha$.

In particular, if the limit distribution Q_P has a unimodal density, then it suggests itself to choose C_P bounded by a level set of the density. This may not be unique if the density has plateaus, but natural choices suggest themselves. Such sets C_P are star-shaped, and they obviously have minimal volume among sets C with $Q_P(C) \ge 1 - \alpha$. We can then try to choose an estimator $\kappa^{(n)}$ for which the volume of C_P is minimal. Different such sets need of course not be inclusion ordered. If Q_P is convex unimodal, the sets C_P are convex.

For LAN families and *regular* estimators $\kappa^{(n)}$, an *efficient* estimator leads to the usual confidence procedure with convex C_P of minimal asymptotic volume.

References

Akahira, M. (1975). Asymptotic theory of estimation of location in non-regular cases, I: Order of convergence of consistent estimators; II: Bounds of asymptotic distributions of consistent estimators. *Report Statistical Applications Research Union of Japanese Scientists and Engineers*, *22*, 8–26 and 99–115.

Akahira, M., Takeuchi, K. (1995). *Non-regular Statistical Estimation.* Lecture notes in statistics *107*. New York: Springer.

Akahira, M. (1982). Asymptotic optimality of estimators in non-regular cases. *Annals of the Institute of Statistical Mathematics*, *34*, 69–82.

Akahira, M., & Takeuchi, K. (1979). Remarks on the asymptotic efficiency and inefficiency of maximum probability estimators. *Report Statistics Appllied Research Union of Japanese Scientists and Engineers*, *26*, 22–28.

Albers, W. (1974). *Asymptotic expansions and the deficiency concept in statistics* (Vol. 58). Mathematical centre tracts. Amsterdam: Mathematisch Centrum.

Alexandrov, A. D. (1940–1943). Additive set functions in abstract spaces. *Matematicheskii Sbornik*, *8*, 307–348; *9*, 563–628; *13*, 169–238.

Anderson, T. W. (1955). The integral of a symmetric unimodal function over a symmetric convex set and some probability inequalities. *Proceedings of the American Mathematical Society*, *6*, 170–176.

Ash, R. B. (1972). *Real analysis and probability*. New York: Academic Press.

Ash, R. B. (2000). *Probability and measure theory* (2nd ed.). New York: Harcourt/Academic Press. (with contributions from Doléans-Dade, C.A.).

Bahadur, R. R. (1958). Examples of inconsistency of maximum likelihood estimates. *Sankhyā*, *20*, 207–210.

Bahadur, R. R. (1964). On fisher's bound for asymptotic variances. *The Annals of Mathematical Statistics*, *35*, 1545–1552.

Bar-Lev, S. K., & Reiser, B. (1983). A note on maximum likelihood estimation for the gamma-distribution. *Sankhyā Series B*, *45*, 300–302.

Begun, J. M., Hall, W. J., Huang, W. M., & Wellner, J. A. (1983). Information and asymptotic efficiency in parametric-nonparametric models. *Annals of Statistics*, *11*, 432–452.

Bening, V. E. (2000). *Asymptotic theory of testing statistical hypotheses.* A volume in Modern Probability Theory and Statistics, VSP.

Beran, R. (1974). Asymptotically efficient adaptive rank estimates in location models. *Annals of Statistics*, *2*, 63–74.

Bergström, H. (1945). On the central limit theorem in the space R^k, $k \geq 1$. *Skandinavisk Aktuarietidskrift*, *28*, 106–127.

Bergström, H. (1974). On the central limit theorem in R^k. A correction and conjecture progress in statistics I. *Colloquia Mathematica Societatis Janos Bolyai*, 97–99, North-Holland.

Bernoulli, D. (1778). *Dijudicatio maxime probabilis plurium observationum discrepantium atque verisimillima inductio inde formanda, pars prior.* Acta Academiae Scientiarum Imperialis Petropolitanae *1777*, 3–23.

Bhattacharya, R. N., & Ranga Rao, R. (1976). *Normal approximation and asymptotic expansions.* New York: Wiley.

Bickel, P. J. (1974). Edgeworth expansions in non parametric statistics. *Annals of Statistics*, *2*, 1–20.

Bickel, P. J. (1982). On adaptive estimation. *Annals of Statistics*, *10*, 647–671.
Bickel, P. J., & Hodges, J. L, Jr. (1967). The asymptotic theory of Galton's test and a related simple estimate of location. *The Annals of Mathematical Statistics*, *38*, 73–89.
Bickel, P. J., Klaassen, C. A. J., Ritov, Y., & Wellner, J. A. (1993). *Efficient and adaptive estimation for semiparametric models*. Baltimore: Johns Hopkins University Press. (1998 Springer Paperback.).
Billingsley, P. (1968). *Convergence of probability measures*. New York: Wiley.
Boll, C. H. (1955). *Comparison of estimates in the infinite case*. Thesis: Stanford University.
Brown, L. D., Low, M. G., & Zhao, L. H. (1997). Superefficiency in nonparametric function estimation. *Annals of Statistics*, *25*, 2607–2625.
Carroll, R. J., & Hall, P. (1988). Optimal rates of convergence for deconvolving a density. *Journal of the American Statistical Association*, *83*, 1184–1186.
Casella, G., & Hwang, J. T. (1982). Limit expressions for the risk of James-Stein estimators. *Canadian Journal of Statistics*, *10*, 305–309.
Čentsov, N. N. (1962). Evaluation of an unknown distribution from observations. *Soviet Mathematics*, *3*, 1559–1562.
Chan, L. K. (1967). Remark on a theorem by Gurland. *Skandinavisk Aktuarietidskrift*, *50*, 19–22.
Chanda, K. C. (1954). A note on the consistency and maxima of the roots of likelihood equations. *Biometrika*, *41*, 56–61.
Chernoff, H. (1956). Large-sample theory: parametric case. *The Annals of Mathematical Statistics*, *27*, 1–22.
Chibisov, D. M. (1973a). Asymptotic expansions for Neyman's C(α) tests. In *Proceedings of the Second Japan-USSR Symposium on Probability Theory* (Vol. 330, pp. 16–45). Lecture notes in mathematics. Berlin: Springer.
Chibisov, D. M. (1973b). An asymptotic expansion for the distribution of sums of a special form with an application to minimum contrast estimates. *Theory of Probability and Its Applications*, *18*, 649–661.
Cramér, H. (1946a). *Mathematical methods of statistics*. Princeton: Princeton University Press.
Cramér, H. (1946b). A contribution to the theory of statistical estimation. *Skandinavisk Aktuarietidskrift*, *29*, 85–94.
Czuber, E. (1891). *Theorie der Beobachtungsfehler*. Leipzig: Teubner.
Daniels, H. E. (1961). The asymptotic efficiency of a maximum likelihood estimator. *Proceedings of the Fourth Berkeley Symposium on Mathematical Statistics and Probability*, *1*, 151–163.
Does, R. J. M. M. (1982). *Higher Order Asymptotics for Simple Linear Rank Statistics* (Vol. 151). Mathematical centre tracts. Amsterdam: Mathematisch Centrum.
Doob, J. L. (1934). Probability and statistics. *Transactions of the American Mathematical Society*, *36*, 759–775.
Doob, J. L. (1936). Statistical estimation. *Transactions of the American Mathematical Society*, *39*, 410–421.
Droste, W. (1985). *Lokale asymptotische Normalität und asymptotische Likelihood-Schätzer*. Thesis: University of Cologne.
Droste, W., & Wefelmeyer, W. (1984). On Hájek's convolution theorem. *Statistical Decisions*, *2*, 131–144.
Dugué, D. (1936). Sur le maximum de précision des estimations gaussiennes á la limite. *Comptes Rendus de l'Académie des Sciences Paris*, *202*, 193–196.
Dugué, D. (1937). Application des propriétés de la limite au sens du calcul des probabilités à létude de diverses questions d'estimation. *Journal de l'École Polytechnique*, *3*, 305–374.
Dugué, D. (1958). Traité de statistique théorique et appliquée: Analyse aléatoire, algébre aléatoire. *Collection d'ouvrages de mathématiques l'usage des physiciens*. Paris: Masson.
Dunn, O. J. (1958). Estimation of means of dependent variables. *The Annals of Mathematical Statistics*, *29*, 1095–1111.
Edgeworth, F. Y. (1908/1909). On the probable errors of frequency-constants. *Journal of the Royal Statistical Society*, *71*, 381–97, 499–512, 651–78. Addendum *72*, 81–90.

Edwards, A. W. F. (1974). The history of likelihood. *International Statistical Review, 42*, 9–15.
Edwards, A. W. F. (1992). *Likelihood*. Baltimore: Johns Hopkins University.
Eisenhart, Ch. (1938). Abstract 18. *Bulletin of the American Mathematical Society, 44*, 32.
Elstrodt, J. (2000). *Maß- und Integrationstheorie*. Berlin: Springer.
Encke, J.F. (1832–34). *Über die Methode der kleinsten Quadrate*. Berliner Astronomisches Jahrbuch *1834*, 249–312; *1835*, 253–320; *1836*, 253–308.
Fabian, V. (1970). On uniform convergence of measures. *Zeitschrift für Wahrscheinlichkeitstheorie und verwandte Gebiete, 15*, 139–143.
Fabian, V. (1973). Asymptotically efficient stochastic approximation; the RM case. *Annals of Statistics, 1*, 486–495.
Fabian, V., & Hannan, J. (1982). On estimation and adaptive estimation for locally asymptotically normal families. *Zeitschrift für Wahrscheinlichkeitstheorie, 59*, 459–478.
Farrell, R. H. (1967). On the lack of a uniformly consistent sequence of estimators of a density function in certain cases. *The Annals of Mathematical Statistics, 38*, 471–474.
Farrell, R. H. (1972). On the best obtainable asymptotic rates of convergence in estimation of a density function at a point. *The Annals of Mathematical Statistics, 43*, 170–180.
Ferguson, T. S. (1982). An inconsistent maximum likelihood estimate. *Journal of the American Statistical Association, 77*, 831–834.
Filippova, A. A. (1962). Mises' theorem on the asymptotic behavior of functionals of empirical distribution functions and its statistical applications. *Theory of Probability and Its Applications, 7*, 24–57.
Fisher, R. A. (1912). On an absolute criterion for fitting frequency curves. *Messenger of Mathematics, 41*, 155–160.
Fisher, R. A. (1922). On the mathematical foundation of theoretical statistics. *Philosophical transactions of the Royal Society of London Series A, 222*, 309–368.
Fisher, R. A. (1925). Theory of statistical estimation. *Proceedings of the Cambridge Philosophical Society, 22*, 700–725.
Fisher, R. A. (1959). *Statistical methods in scientific inference* (2nd edn.). Oliver and Boyd.
Fix, E., & Hodges, J. L., Jr. (1951). Discriminatory analysis—nonparametric discrimination: Consistency properties. *Report 4*, USAF School of Aviation Medicine. Reprinted in International Statistical Review *57*, 238–247 (1989).
Foutz, R. V. (1977). On the unique consistent solution of the likelihood equations. *Journal of the American Statistical Association, 72*, 147–148.
Gänssler, P., & Stute, W. (1976). On uniform convergence of measures with applications to uniform convergence of empirical distributions. In *Empirical Distributions and Processes* (Vol. 566, pp. 45–56). Lecture notes in mathematics.
Gauss, C. F. (1816). Bestimmung der Genauigkeit der Beobachtungen. Carl Friedrich Gauss Werke, 4, 109–117, Königliche Gesellschaft der Wissenschaften, Göttingen.
Ghosh, J. K. (1985). Efficiency of estimates. *Part I. Sankhyā Series A, 47*, 310–325.
Ghosh, J. K. (1991). Higher order asymptotics for the likelihood ratio, Rao's and Wald's tests. *Statistics and Probability Letters, 12*, 505–509.
Ghosh, M. (1995). Inconsistent maximum likelihood estimators for the Rasch model. *Statistics and Probability Letters, 23*, 165–170.
Ghosh, J. K., & Subramanyam, K. (1974). Second order efficiency of maximum likelihood estimators. *Sankhyā Series A, 36*, 325–358.
Glivenko, V. I. (1939). *Course in probability theory* (Russian). Moscow: ONTI.
Godambe, V. P. (1960). An optimum property of regular maximum likelihood estimation. *The Annals of Mathematical Statistics, 31*, 1208–1211.
Gong, G., & Samaniego, F. J. (1981). Pseudo maximum likelihood estimation: theory and applications. *Annals of Statistics, 9*, 861–869.
Grossmann, W. (1981). *Efficiency of estimates in nonregular cases.* In P. Révész, L. Schmetterer, V. M. Zolotarev (Eds.), *First Pannonian Symposium on Mathematical Statistics*. Lecture notes in statistics (Vol. 8, pp. 94–109). Berlin: Springer.

Grossmann, W. (1979). Einige Bemerkungen zur Theorie der Maximum Probability Schätzer. *Metrika*, *26*, 129–137.
Gurland, J. (1954). On regularity conditions for maximum likelihood estimation. *Skandinavisk Aktuarietidskrift*, *37*, 71–76.
Gusev, S. I. (1975). Asymptotic expansions that are connected with certain statistical estimates in the smooth case. I. Expansions of random variables (Russian). *Teoriya Veroyatnostei i ee Primeneniya*, *20*, 488–514.
Hahn, H. (1932). *Reelle Funktionen*. Leipzig: Akademische Verlagsgesellschaft.
Hájek, J. (1962). Asymptotically most powerful rank-order tests. *The Annals of Mathematical Statistics*, *33*, 1124–1147.
Hájek, J. (1970). A characterization of limiting distributions of regular estimates. *Zeitschrift für Wahrscheinlichkeitsheorie und verwandte Gebiete*, *14*, 323–330.
Hájek, J. (1971). Limiting properties of likelihood and inference. In V. P. Godambe & O. A. Sprott (Eds.), *Foundation of Statistical Inference* (pp. 142–162). Rinehart and Winston: Holt.
Hájek, J. (1972). Local asymptotic minimax and admissibility in estimation. In *Proceedings of the Sixth Berkeley Symposium on Mathematical Statistics and Probability I* (pp. 175–194)
Hájek, J., & Šidák, Z. (1967). *Theory of rank tests* (1st edn.). Prague: Academia Publishing House.
Hald, A. (1998). *A history of mathematical statistics from 1750 to 1930*. New York: Wiley.
Hald, A. (1999). On the history of maximum likelihood in relation to inverse probability and least squares. *Statistical Science*, *14*, 214–222.
Hall, P. (1989). On convergence rates in nonparametric problems. *International Statistical Review*, *57*, 45–58.
Hall, P., & Welsh, A. M. (1984). Best attainable rates of convergence for estimates of parameters of regular variation. *Annals of Statistics*, *12*, 1079–1084.
Hartigan, J. A. (1966). Note on the confidence-prior of Welch and Peers. *Journal of the Royal Statistical Society Series B*, *28*, 55–56.
Has'minskii, R. Z., & Ibragimov, I. A. (1979). On the nonparametric estimation of functionals. In P. Mandl & H. Huškova (Eds.) *Proceedings of the Second Prague Symposium on Asymptotic Statistics* (pp. 41–51), North-Holland.
Helmert, C. F. (1876). Über die Wahrscheinlichkeit der Potenzsummen der Beobachtungsfehler und über einige damit in Zusammenhang stehende Fragen. *Zeitschrift für Mathematik und Physik*, *21*, 192–219.
Hewitt, E., & Stromberg, K. (1965). *Real and abstract analysis*. New York: Springer.
Hodges, J. L, Jr., & Lehmann, E. L. (1970). Deficiency. *The Annals of Mathematical Statistics*, *41*, 783–801.
Hoeffding, W. (1948). A class of statistics with asymptotically normal distribution. *The Annals of Mathematical Statistics*, *19*, 293–325.
Hoeffding, W. (1962). Review of Wilks: Mathematical statistics. *The Annals of Mathematical Statistics*, *33*, 1467–1473.
Hoeffding, W., & Wolfowitz, J. (1958). Distinguishability of sets of distributions. (The case of independent and identically distributed chance variables). *The Annals of Mathematical Statistics*, *29*, 700–718.
Hotelling, H. (1930). The consistency and ultimate distribution of optimum statistics. *Transactions of the American Mathematical Society*, *32*, 847–859.
Huber, P. J. (1966). Strict efficiency excludes superefficiency. Preprint. ETH Zürich. Abstract 25. *The Annals of Mathematical Statistics*, *37*, 1425/6.
Huber, P. J. (1967). The behavior of maximum likelihood estimates under nonstandard conditions. In *Proceedings of the Fifth Berkeley Symposium on Mathematical Statistics and Probability* (Vol. 1). Berkeley & Los Angeles: University of California Press.
Huzurbazar, V. S. (1948). The likelihood equation, consistency and the maximum likelihood function. *Annals of Eugenics*, *14*, 185–200.
Ibragimov, I. A. (1956). On the composition of unimodal distributions. *Theory of Probability and Its Applications*, *1*, 255–260.

Ibragimov, I. A., & Has'minskii, R. Z. (1981). *Statistical estimation: Asymptotic theory*. New York: Springer. Translation of the Russian original (1979).

Ibragimov, I. A., & Has'minskii, R. Z. (1970). On the asymptotic behavior of generalized Bayes estimators. *Doklady Akademii Nauk SSSR*, *194*, 257–260.

Ibragimov, I. A., & Has'minskii, R. Z. (1973). Asymptotic analysis of statistical estimators in the "almost smooth" case. *Theory of Probability and Its Applications*, *18*, 241–252.

Inagaki, N. (1970). On the limiting distribution of a sequence of estimators with uniformity property. *Annals of the Institute of Statistical Mathematics*, *22*, 1–13.

Janssen, A., & Ostrovski, V. (2013). *The convolution theorem of Hájek and Le Cam-revisited*. Düsseldorf: Mathematical Institute.

Jogdeo, K. (1970). An inequality for a strict unimodal function with an application to a characterization of independence. *Sankhyā Series A*, *32*, 405–410.

Kallianpur, G., & Rao, C. R. (1955). On Fisher's lower bound to asymptotic variance of a consistent estimate. *Sankhyā Series A*, *15*, 331–342.

Kati, S. (1983). Comment on Bar-Lev and Reiser (1983). *Sankhyā Series B*, *45*, 302.

Kaufman, S. (1966). Asymptotic efficiency of the maximum likelihood estimator. *Annals of the Institute of Statistical Mathematics*, *18*, 155–178. See also Abstract. *The Annals of Mathematical Statistics36*, 1084 (1965)

Kelley, J. L. (1955). *General topology*. Toronto: Van Nostrand.

Kendall, M. G., & Stuart, A. (1961). *The advanced theory of statistics* (Vol. 1). London: Griffin.

Kersting, G. D. (1978). Die Geschwindigkeit der Glivenko-Cantelli-Konvergenz gemessen in der Prohorov-Metrik. *Mathematische Zeitschrift*, *163*, 65–102.

Kiefer, J. (1982). Optimum rates for nonparametric density and regression estimates under order restrictions. In C. R. Rao, G. Kallianpur, P. R. Krishnaiah, & J. K. Ghosh (Eds.), *Statistics and Probability: Essays in Honor* (pp. 419–429). Amsterdam: North-Holland.

Klaassen, C. A. J. (1979). Nonuniformity of the convergence of location estimators. In P. Mandl, & M. Hušková (Eds.), *Proceedings of the Second Prague Symposium on Asymptotic Statistics* (pp. 251–258).

Klaassen, C. A. J. (1981). Statistical performance of location estimators. In *Mathematical Centre Tracts* (Vol. 133). Amsterdam: Mathematisch Centrum.

Klaassen, C. A. J. (1987). Consistent estimation of the influence function of locally asymptotically linear estimators. *Annals of Statistics*, *15*, 1548–1562.

Kolmogorov, A. N. (1933). *Grundbegriffe der Wahrscheinlichkeitstheorie*. Berlin: Springer.

Koshevnik, Yu A, & Levit, B. Ya. (1976). On a nonparametric analogue of the information matrix. *Theory of Probability and Its Applications*, *21*, 738–753.

Kraft, C., & Le Cam, L. (1956). A remark on the roots of the maximum likelihood equation. *The Annals of Mathematical Statistics*, *27*, 1174–1177.

Kronmal, R., & Tarter, M. (1968). The estimation of probability densities and cumulatives by Fourier series methods. *Journal of the American Statistical Association*, *63*, 925–952.

Kulldorff, G. (1956). On the condition for consistency and asymptotic efficiency of maximum likelihood estimates. *Skandinavisk Aktuarietidskrift*, *40*, 129–144.

Lambert, J. H. (1760). Photometria, sive de mensura et gradilus luminis colorum et umbrae. Augustae Vindelicorum.

Landers, D. (1972). Existence and consistency of modified minimum contrast estimates. *The Annals of Mathematical Statistics*, *43*, 74–83.

Laplace, P. S. (1812). *Théorie analytique des probabilités*. Paris: Courcier.

Le Cam, L. (1953). On some asymptotic properties of maximum likelihood estimates and related Bayes' estimates. *University of California Publications in Statistics*, *1*, 277–330.

Le Cam, L. (1956). On the asymptotic theory of estimation and testing hypotheses. In *Proceedings of the Third Berkeley Symposium on Mathematical Statistics and Probability* (pp. 129–156). Berkeley & Los Angeles: University of California Press.

Le Cam, L. (1958). Les propriétés asymptotiques des solutions de Bayes. *Publication Institute Statistics University of Paris VII*, 17–35.

Le Cam, L. (1960). Locally asymptotically normal families of distributions. *University of California Publications in Statistics, 3*, 37–98.
Le Cam, L. (1964). Sufficiency and approximate sufficiency. *The Annals of Mathematical Statistics, 35*, 1419–1455.
Le Cam, L. (1966). *Likelihood functions for large numbers of independent observations.* Research Papers in Statistics (Festschrift J. Neyman), 167–187. London: Wiley.
Le Cam, L. (1969). *Théorie asymptotique de la décision statistique.* Séminaire de Mathématiques Suprieures 33. Les Presses de l'Université de Montréal.
Le Cam, L. (1972). *Limits of experiments.* In *Proceedings of the Sixth Berkeley Symposium on Mathematical Statistics and Probability I* (pp. 249–261).
Le Cam, L. (1974). *Notes on asymptotic methods in statistical decision theory.* Centre de Recherches Mathématiques: Université de Montréal.
Le Cam, L. (1979). On a theorem of Hájek. In J. Jurečkova (Ed.), *Contributions to Statistics: Journal of the Hájek Memorial Volume.* Akademia: Prague.
Le Cam, L. (1984). *Review of I. A. Ibragimov and R. Z. Has'minskii, statistical estimation, asymptotic theory, and J. Pfanzagl, Contributions to a general asymptotic statistical theory.* Bull. Amer. Math. Soc. (N.S.) *11*, 392–400.
Le Cam, L. (1985). Sur l'approximation de familles de mesures par des familles gaussiennes. *Annales de l'Institut Henri Poincare The Probability and Statistics, 21*, 225–287.
Le Cam, L. (1986). *Asymptotic methods in statistical decision theory.* New York: Springer.
Le Cam, L. (1994). An infinite dimensional convolution theorem. In S. S. Gupta & J. O. Berger (Eds.), *Statistical Decision Theory and Related Topics V* (pp. 401–411). Berlin: Springer.
Le Cam, L., & Yang, G. L. (1990). *Asymptotics in statistics: Some basic concepts.*, Springer series in statistics New York: Springer.
Lehmann, E. L. (1959). *Testing statistical hypotheses.* New York: Wiley.
Lehmann, E. L., & Casella, G. (1998). *Theory of point estimation* (2nd ed.). Berlin: Springer.
Levit, B. Y. (1974). On optimality of some statistical estimates. In J. Hájek (Ed.), *Proceedings of the Prague symposium on asymptotic statistics* (Vol. 2, pp. 215–238) Charles University Press. (Reprinted from *Breakthroughs in Statistics III* (pp. 330–346), by S. Kotz & N. L. Johnson (Eds.), Berlin: Springer.
Levit, B. Ya. (1975). On the efficiency of a class of nonparametric estimates. *Theory of Probability and Its Applications, 20*, 723–740.
Linnik, Yu V, & Mitrofanova, N. M. (1963). On the asymptotics of the distribution of a maximum likelihood statistic (Russian). *Doklady Akademii Nauk SSSR, 149*, 518–520.
Linnik, Yu V, & Mitrofanova, N. M. (1965). Some asymptotic expansions for the distribution of the maximum likelihood estimate. *Sankhyā Series A, 27*, 73–82.
Liu, R. C., & Brown, L. D. (1993). Nonexistence of informative unbiased estimators in singular problems. *Annals of Statistics, 21*, 1–13.
Lukacs, E. (1960). *It characteristic functions.*, Griffin's Statistical Monographs & Courses 5 New York: Hafner.
Michel, R. (1974). Bounds for the efficiency of approximately median-unbiased estimates. *Metrika, 21*, 205–211.
Michel, R. (1975). An asymptotic expansion for the distribution of asymptotic maximum likelihood estimators of vector parameters. *Journal of Multivariate Analysis, 5*, 67–85.
Michel, R., & Pfanzagl, J. (1970). Asymptotic normality. *Metrika, 16*, 188–205.
Milbrodt, H. (1983). *Global asymptotic normality. Statistics and Decisions, 1*, 401–425.
Millar, P. W. (1983). The minimax principle in asymptotic statistical theory. In P. L. Hennequin (Ed.), *Ecole d'Eté de Probabilités de Saint-Flour XI-1981* (Vol. 976, pp. 75–265). Lecture notes in mathematics. New York: Springer.
Mitrofanova, N. M. (1967). An asymptotic expansion for the maximum likelihood estimate of a vector parameter. *Theory of Probability and Its Applications, 12*, 364–372.
Neyman, J., & Pearson, K. (1933). The testing of statistical hypotheses in relation to probabilities a priori. *Philosophical Transactions of the Royal Society of London (A), 231*, 289–337.

Neyman, J., & Pearson, E. S. (1936). Sufficient statistics and uniformly most powerful tests of statistical hypotheses. *Statistical Research Memory, 1*, 113–137.

Neyman, J., & Scott, E. L. (1948). Consistent estimates based on partially consistent observations. *Econometrica, 16*, 1–32.

Osgood, W. F. (1897). Non-uniform convergence on the integration of series term by term. *American Journal of Mathematics, 19*, 155–190.

Parke, W. R. (1986). Pseudo-maximum-likelihood estimation: The asymptotic distribution. *Annals of Statistics, 14*, 355–357.

Parthasarathy, K. R. (1967). *Probability measures on metric spaces*. New York: Academic Press.

Parzen, E. (1962). On estimation of a probability density function and mode. *The Annals of Mathematical Statistics, 33*, 1065–1076.

Pearson, K. (1896). Mathematical contributions to the theory of evolution III. Regression, heredity and panmixia. *Philosophical Transactions of the Royal Society of London (A), 187*, 253–318.

Pearson, K., & Filon, L. N. G. (1898). Mathematical contributions to the theory of evolution IV. On the probable errors of frequency constants and on the influence of random selection on variation and correlation. *Philosophical Transactions of the Royal Society of London (A), 191*, 229–311.

Pfanzagl, J. (1969a). On the measurability and consistency of minimum contrast estimators. *Metrika, 14*, 249–272.

Pfanzagl, J. (1969b). On the existence of product measurable densities. *Sankhyā, 31*, 13–18.

Pfanzagl, J. (1970). On the asymptotic efficiency of median-unbiased estimates. *The Annals of Mathematical Statistics, 41*, 1500–1509.

Pfanzagl, J. (1972a). On median-unbiased estimates. *Metrika, 18*, 154–173.

Pfanzagl, J. (1972b). Further results on asymptotic normality I. *Metrika, 18*, 174–198.

Pfanzagl, J. (1972c). Further results on asymptotic normality II. *Metrika, 18*, 89–97.

Pfanzagl, J. (1973). Asymptotic expansions related to minimum contrast estimators. *Annals of Statistics, 1*, 993–1026.

Pfanzagl, J. (1974). Asymptotically optimum estimation and test procedures. In J. Hájek (Ed.), *Proceedings of the Prague Symposium on Asymptotic Statistics* (Vol. 1, pp. 201–272). Prague: Charles University.

Pfanzagl, J. (1975). *On asymptotically complete classes* (Vol. 2, pp. 1–43). Statistical inference and related topics. New York: Academic Press.

Pfanzagl, J. (1979). First order efficiency implies second order efficiency. In J. Jurečková (Ed.), *Contribution to Statistics* (pp. 167–196). Prague: Academia.

Pfanzagl, J. (1982). (with the assistance of W. Wefelmeyer). *Contributions to a general asymptotic statistical theory.* Lecture Notes in Statistics (Vol. 13). Berlin: Springer.

Pfanzagl, J. (1985) (With the assistance of W. Wefelmeyer). *Asymptotic expansions for general statistical models* (Vol. 31). Lecture notes in statistics. Berlin: Springer.

Pfanzagl, J. (1990). *Estimation in semiparametric models. Some recent developments* (Vol. 63). Lecture notes in statistics. New York: Springer.

Pfanzagl, J. (1994). *Parametric Statistical Theory*. Berlin: De Gruyter.

Pfanzagl, J. (1995). On local and global asymptotic normality. *Mathematical Methods of Statistics, 4*, 115–136.

Pfanzagl, J. (1998). The nonexistence of confidence sets for discontinuous functionals. *Journal of Statistical Planning and Inference, 75*, 9–20.

Pfanzagl, J. (1999a). On the optimality of limit distributions. *Mathematical Methods of Statistics, 8*, 69–83. Erratum *8*, 550.

Pfanzagl, J. (1999b). On rates and limit distributions. *Annals of the Institute of Statistical Mathematics, 51*, 755–778.

Pfanzagl, J. (2000a). Subconvex loss functions, unimodal distributions, and the convolution theorem. *Mathematical Methods of Statistics, 9*, 1–18.

Pfanzagl, J. (2000b). On local uniformity for estimators and confidence limits. *Journal of Statistical Planning and Inference, 84*, 27–53.

Pfanzagl, J. (2001). A nonparametric asymptotic version of the Cramér–Rao bound. In W. R. van Zwet, M. De Gunst, C. Klaassen, A. W. van der Vaart (Eds.), *The State of Art in Probability and Statistics* (Vol. 36, pp. 499–517). Lecture notes and monograph series. Institute of mathematical statistics, Festschrift.
Pfanzagl, J. (2002a). Asymptotic optimality of estimator sequences: "Pointwise" versus "locally uniform". *Mathematical Methods of Statistics*, *11*, 69–97.
Pfanzagl, J. (2002b). On distinguished LAN-representations. Mathematical Methods of Statistics, 11, 477–488. *Correction note*, *14*, 377–378.
Pfanzagl, J. (2003). Asymptotic bounds for estimators without limit distribution. *Annals of the Institute of Statistical Mathematics*, *55*, 95–110.
Pfanzagl, J., & Wefelmeyer, W. (1978). *A third-order optimum property of the maximum likelihood estimator.* Journal of Multivariate Analysis *8*, 1–29. Addendum *9*, 179–182 (1979).
Prakasa Rao, B. L. S. (1968). Estimation of the location of the cusp of a continuous density. *The Annals of Mathematical Statistics*, *39*, 76–87.
Prakasa Rao, B. L. S. (1969). Estimation of a unimodal density. *Sankhyā Series A*, *31*, 23–36.
Prakasa Rao, B. L. S. (1983). *Nonparametric Functional Estimation*. Academic Press.
Pratt, J. W. (1976). F.Y. Edgeworth and R.A. Fisher on the efficiency of the maximum likelihood estimation. *Annals of Statistics*, *4*, 501–514.
Ranga Rao, R. (1962). Relations between weak and uniform convergence of measures with applications. *The Annals of Mathematical Statistics*, *33*, 659–680.
Rao, C. R. (1945). Information and accuracy attainable in the estimation of statistical parameters. *Bulletin of Calcutta Mathematical Society*, *37*, 81–91.
Rao, C. R. (1947). Minimum variance estimation of several parameters. *Proceedings of the Cambridge Philosophical Society*, *43*, 280–283.
Rao, C. R. (1961). Asymptotic efficiency and limiting information. *Proceedings of the Fourth Berkeley Symposium on Mathematical Statistics and Probability*, *1*, 531–545.
Rao, C. R. (1962a). Efficient estimates and optimum inference procedures in large samples. *Journal of the Royal Statistical Society Series B*, *24*, 46–72.
Rao, C. R. (1962b). Apparent anomalities and irregularities in maximum likelihood estimation (with discussion). *Sankhyā Series A*, *24*, 73–102.
Rao, C. R. (1963). Criteria for estimation in large samples. *Sankhyā Series A*, *25*, 189–206.
Rao, P. V., Schuster, E. F., & Littell, R. C. (1975). Estimation of shift and center of symmetry based on Kolmogorov-Smirnov statistics. *Annals of Statistics*, *3*, 862–873.
Reeds, J. A. (1985). Asymptotic number of roots of Cauchy location likelihood equations. *Annals of Statistics*, *13*, 775–784.
Reiss, R. D. (1973). On the measurability and consistency of maximum likelihood estimates for unimodal densities. *Annals of Statistics*, *1*, 888–901.
Riesz, F. (1928/9). Sur la convergence en moyenne. *Acta Scientiarum Mathematicarum*, *4*, I 58–64, II 182–185.
Ritov, Y., & Bickel, P. J. (1990). Achieving information bounds in non and semiparametric models. *Annals of Statistics*, *18*, 925–938.
Rosenblatt, M. R. (1956). Remarks on some nonparametric estimates of a density function. *The Annals of Mathematical Statistics*, *27*, 832–837.
Roussas, G. G. (1965). Asymptotic inference in Markov processes. *The Annals of Mathematical Statistics*, *36*, 978–992.
Roussas, G. G. (1968). Some applications of asymptotic distribution of likelihood functions to the asymptotic efficiency of estimates. *Zeitschrift für Wahrscheinlichkeitsheorie und verwandte Gebiete*, *10*, 252–260.
Roussas, G. G. (1972). *Contiguity of probability measures: Some applications in statistics*. Cambridge: Cambridge University Press.
Roussas, G. G., & Soms, A. (1973). On the exponential approximation of a family of probability measures and a representation theorem of Hájek-Inagaki. *Annals of the Institute of Statistical Mathematics*, *25*, 27–39.

Rüschendorf, L. (1988). *Asymptotische Statistik.* Teubner Skripten zur Mathematischen Stochastik.
Sacks, J. (1975). An asymptotically efficient sequence of estimators of a location parameter. *Annals of Statistics, 3*, 285–298.
Savage, L. J. (1976). On rereading Fisher. *Annals of Statistics, 4*, 441–500.
Scheffé, H. (1947). A useful convergence theorem for probability distributions. *The Annals of Mathematical Statistics, 18*, 434–438.
Schick, A. (1986). On asymptotically efficient estimation in semiparametric models. *Annals of Statistics, 14*, 1139–1151.
Schmetterer, L. (1966). On the asymptotic efficiency of estimates. In *Research Papers in Statistics.* Festschrift for J. Neyman. David, F.N. (ed) (pp. 301–317). New Jersey: Wiley.
Schmetterer, L. (1956). *Einführung in die Mathematische Statistik* (2nd ed.). Wien: Springer.
Schmetterer, L. (1959). Über nichtparametrische Methoden in der Mathematischen Statistik. *Jahresbericht der Deutschen Mathematiker-Vereinigung, 61*, 104–126.
Schmetterer, L. (1974). *Introduction to mathematical statistics.* Berlin: Springer.
Schrödinger, E. (1915). Zur Theorie der Fall- und Steigversuche an Teilchen mit Brownscher Bewegung. *Physikalische Zeitschrift, 16*, 289–295.
Scott, A. (1967). A note on conservative confidence regions for the mean of a multivariate normal. *The Annals of Mathematical Statistics, 38*, 278–280.
Serfling, R. J. (1980). *Approximation theorems of mathematical statistics.* New York: Wiley.
Sharma, D. (1973). Asymptotic equivalence of two estimators for an exponential family. *Annals of Statistics, 1*, 973–980.
Sheynin, O. B. (1966). Origin of the theory of errors. *Nature, 211*, 1003–1004.
Šidak, Z. (1967). Rectangular confidence regions for the means of multivariate normal distributions. *Journal of the American Statistical Association, 62*, 626–633.
Smirnov, N. V. (1950). On the approximation of probability densities of random variables. *Doklady Akademii Nauk SSSR, LXXIV*, 189–191.
Stein, C. (1956). Efficient nonparametric testing and estimation. In *Proceedings of the Third Berkeley Symposium on Mathematical Statistics and Probability I* (pp. 187–195). California: University of California Press.
Stigler, S. M. (2005). Fisher in 1921. *Statistical Science, 20*, 32–49.
Stigler, S. M. (2007). The epic story of maximum likelihood. *Statistical Science, 22*, 598–620.
Stoica, P., & Ottersten, B. (1996). The evil of superefficiency. *Signal Process, 55*, 133–136.
Stone, C. J. (1975). Adaptive maximum likelihood estimation of a location parameter. *Annals of Statistics, 3*, 267–284.
Stone, C. J. (1980). Optimal rates of convergence for nonparametric estimators. *Annals of Statistics, 8*, 1348–1360.
Stone, C. J. (1983). Optimal uniform rate of convergence for nonparametric estimators of a density function or its derivatives. *Recent Advances in Statistics* (pp. 393–406). New York: Academic Press.
Strasser, H. (1977). *Asymptotic expansions for Bayes procedures.* Recent Developments in Statistics, 9–35. North-Holland: Amsterdam.
Strasser, H. (1978b). Global asymptotic properties of risk functions in estimation. *Zeitschrift für Wahrscheinlichkeitsheorie und verwandte Gebiete, 45*, 35–48.
Strasser, H. (1978b). Global asymptotic properties of risk functions in estimation. *Zeitschrift für Wahrscheinlichkeitsheorie und verwandte Gebiete, 45*, 35–48.
Strasser, H. (1985). *Mathematical theory of statistics.* Berlin: De Gruyter.
Strasser, H. (1997). Asymptotic admissibility and uniqueness of efficient estimates in semiparametric models. In D. Pollard, E. Torgersen, & G. L. Yang (Eds.), *Festschrift for Lucien Le Cam, Research papers in probability and statistics.* Berlin: Springer.
Stuart, A., & Ord, K. (1991). *Kendall's advanced theory of statistics. Classical inference and relationship* (5th ed., Vol. 2). London: Edward Arnold.

Takeuchi, K., & Akahira, M. (1976). On the second order asymptotic efficiencies of estimators. In G. Maruyama, & J. V. Prohorov (Eds.), *USSR Symposium on Probability Theory* (Vol. 550, pp. 604–638). Lecture notes in mathematics. Berlin: Springer.
Takeuchi, K. (1971). A uniformly asymptotically efficient estimator of a location parameter. *Journal of the American Statistical Association*, *66*, 292–301.
Takeuchi, K. (1974). *Tokei-teki Suitei no Zenkinriron (Asymptotic theory of statistical estimation) (Japanese)*. Tokyo: Kyoiku-Shuppan.
Tarone, P. E., & Gruenhage, G. (1975). A note on the uniqueness of roots of the likelihood equations for vector-valued parameters. *Journal of the American Statistical Association*, *70*, 903–904.
Tierney, L. (1987). An alternative regularity condition for Hájek's representation theorem. *Annals of Statistics*, *15*, 427–431.
Todhunter, I. (1865). *A history of mathematical theory of probability*. Cambridge: Macmillan.
Tong, Y. L. (1990). *The multivariate normal distribution*. Berlin: Springer.
van der Vaart, A. W. (1988). *Statistical Estimation in Large Parameter Spaces.*, CWI Tract 44 Amsterdam: Centrum foor Wiskunde en Informatica.
van der Vaart, A. W. (1991). On differentiable functionals. *Annals of Statistics*, *19*, 178–204.
van der Vaart, A. W. (1997). Super efficiency. In D. Pollard, E. Torgersen, & G. L. Yang (Eds.), *Festschrift for Lucien Le Cam* (pp. 397–410). Research Papers in Probability and Statistics. Springer.
van der Vaart, A. W. (2002). The statistical work of Lucien LeCam. *Annals of Statistics*, *30*, 631–682.
van Eeden, C. (1968). Nonparametric estimation. *Séminaire de Mathématiques Supérieures* (Vol. 35). Montréal: Les Presses de l'Université de Montréal.
van Eeden, C. (1970). Efficiency-robust estimation of location. *The Annals of Mathematical Statistics*, *41*, 172–181.
von Mises, R. (1947). On the asymptotic distribution of differentiable statistical functions. *The Annals of Mathematical Statistics*, *18*, 309–348.
Wahba, G. (1971). A polynomial algorithm for density estimation. *The Annals of Mathematical Statistics*, *42*, 1870–1886.
Wahba, G. (1975). Optimal convergence properties of variable knot, kernel, and orthogonal series methods for density estimation. *Annals of Statistics*, *3*, 15–29.
Wald, A. (1941a). *On the principles of statistical inference*. Four Lectures Delivered at the University of Notre Dame, February 1941. Published as Notre Dame Mathematical Lectures (Vol. 1), 1952.
Wald, A. (1941b). Asymptotically most powerful tests of statistical hypotheses. *The Annals of Mathematical Statistics*, *12*, 1–19.
Wald, A. (1943). Test of statistical hypotheses concerning several parameters when the number of observations is large. *Transactions of the American Mathematical Society*, *54*, 426–482.
Wald, A. (1949). Note on the consistency of the maximum likelihood estimate. *The Annals of Mathematical Statistics*, *20*, 595–601.
Weiss, L., & Wolfowitz, J. (1974). *Maximum probability estimators and related topics*. Lecture notes in mathematics (Vol. 424). Berlin: Springer.
Weiss, L., & Wolfowitz, J. (1966). Generalized maximum likelihood estimators. *Theory of Probability and Its Applications*, *11*, 58–83.
Weiss, L., & Wolfowitz, J. (1967a). Estimating a density function at a point. *Zeitschrift für Wahrscheinlichkeitsheorie und verwandte Gebiete*, *7*, 327–335.
Weiss, L., & Wolfowitz, J. (1967b). Maximum probability estimators. *Annals of the Institute of Statistical Mathematics*, *19*, 193–206.
Weiss, L., & Wolfowitz, J. (1970). Asymptotically efficient non-parametric estimators of location and scale parameters. *Zeitschrift für Wahrscheinlicheitstheorie Verw Gebiete*, *16*, 134–150.
Weiss, L., & Wolfowitz, J. (1973). Maximum likelihood estimation of a translation parameter of a truncated distribution. *Annals of Statistics*, *1*, 944–947.
Welch, B. L. (1965). On comparisons between confidence point procedures in the case of a single parameter. *Journal of the Royal Statistical Society Series B*, *27*, 1–8.

Wendel, J. G. (1952). Left centralizers and isomorphisms of group algebras. *Pacific Journal of Mathematics*, *2*, 251–261.

Wilks, S. S. (1962). *Mathematical statistics*. New York: Wiley.

Williamson, J. A. (1984). A note on the proof of H.E. Daniels of the asymptotic efficiency of a maximum likelihood estimator. *Biometrika*, *71*, 651–653.

Witting, H. (1985). *Mathematische Statistik I: Parametrische Verfahren bei festem Stichprobenumfang*. Teubner.

Witting, H. (1998). Nichtparametrische Statistik: Aspekte ihrer Entwicklung 1957–1997. *Jahresbericht der Deutschen Mathematiker-Vereinigung*, *100*, 209–237.

Witting, H., & Müller-Funk, U. (1995). *Mathematische Statistik II. Asymptotische Statistik: parametrische Modelle und nichtparametrische Funktionale*. Wiesbaden: Teubner.

Wolfowitz, J. (1963). *Asymptotic efficiency of the maximal likelihood estimator.* Seventh All-Soviet Union Conference on Probability and Mathematical Statistics, Tbilisi.

Wolfowitz, J. (1970). Reflections on the future of mathematical statistics. In R. C. Bose et al. (Eds.), *Essays in Probability and Statistics* (pp. 739–750). University of North Carolina Press.

Wolfowitz, J. (1952). Consistent estimators of the parameters of a linear structural relation. *Skandinavisk Aktuarietidskrift*, *35*, 132–151.

Wolfowitz, J. (1953a). The method of maximum likelihood and the Wald theory of decision functions. *Indagationes Mathematicae*, *15*, 114–119.

Wolfowitz, J. (1953b). Estimation by the minimum distance method. *Annals of the Institute of Statistical Mathematics*, *5*, 9–23.

Wolfowitz, J. (1957). The minimum distance method. *The Annals of Mathematical Statistics*, *28*, 75–88.

Wolfowitz, J. (1965). Asymptotic efficiency of the maximum likelihood estimator. *Theory of Probability and Its Applications*, *10*, 247–260.

Woodroofe, M. (1972). Maximum likelihood estimation of a translation parameter of a truncated distribution. *The Annals of Mathematical Statistics*, *43*, 113–122.

Yang, G. L. (1999). A conversation with Lucien Le Cam. *Statistical Science*, *14*, 223–241.

Zacks, S. (1971). *The theory of statistical inference*. New York: Wiley.

Zermelo, E. (1929). Die Berechnung der Turnier-Ergebnisse als ein Maximumproblem der Wahrscheinlichkeitsrechnung. *Mathematische Zeitschrift*, *29*, 436–460.

Author Index

J. Pfanzagl, *Mathematical Statistics*, Springer Series in Statistics,
DOI 10.1007/978-3-642-31084-3

Subject Index

J. Pfanzagl, *Mathematical Statistics*, Springer Series in Statistics,
DOI 10.1007/978-3-642-31084-3

Symbol Index

J. Pfanzagl, *Mathematical Statistics*, Springer Series in Statistics,
DOI 10.1007/978-3-642-31084-3

Printed by Printforce, the Netherlands